ALGEBRA–
PROGRAMMED
part 2

ALGEBRA-

PRENTICE-HALL, INC.
Englewood Cliffs, New Jersey 07632

PROGRAMMED

Second Edition

part 2

ROBERT H. ALWIN

ROBERT D. HACKWORTH

Division of Science and Mathematics
St. Petersburg Junior College

Library of Congress Cataloging in Publication Data

Alwin, Robert H
Algebra—programmed.

Includes index.

1. Algebra—Programmed instruction. I. Hackworth, Robert D., joint author. II. Title.
QA154.2.A38 1978 512'.0077 77-24274
ISBN 0-13-022020-5 pbk. (v.2)

ALGEBRA—PROGRAMMED
Second Edition
part 2
Robert H. Alwin / Robert D. Hackworth

10 9 8 7

Printed in the United States of America

PRENTICE-HALL INTERNATIONAL, INC., *London*
PRENTICE-HALL OF AUSTRALIA PTY. LIMITED, *Sydney*
PRENTICE-HALL OF CANADA, LTD., *Toronto*
PRENTICE-HALL OF INDIA PRIVATE LIMITED, *New Delhi*
PRENTICE-HALL OF JAPAN, INC., *Tokyo*
PRENTICE-HALL OF SOUTHEAST ASIA PTE. LTD., *Singapore*
WHITEHALL BOOKS LIMITED, *Wellington, New Zealand*

Contents

2 *Solving Pairs of Equations*, 91

3 *An Introduction to Polynomials*, 129

4 *Factoring Polynomials*, 183

5 *Polynomial Fractions,* 229

6 *Solving Linear Equations,* 271

7 *Fractional and Quadratic Equations,* 305

8 *Rational and Irrational Numbers,* 341

9 *Solving Quadratic Equations with Irrational Solutions,* 408

10 *Applications,* 471

Preface

This is a programmed text, designed for effective self-teaching of the student reader. Its use in an individualized study course affords the student a real opportunity to learn at a pace that is best for the individual. Students with good mathematical backgrounds, good study habits and skills, or both may complete this material quickly without being restricted by the speed of a lecture class. Students with poor mathematical backgrounds can make continual and satisfactory progress without being pushed into failure and frustration by the schedule of a lecture class.

The text accepts full responsibility for teaching its contents. There is no need for group instruction, supplementary lectures, or work sessions. This does not eliminate the need for a teacher, because the student reader will benefit if help is available whenever any difficulty is encountered. However, help will be most constructive when it leads the student directly back into the flow of the program without distraction from the sequence of the material.

The first edition of *Algebra—Programmed, Part 2* was published in 1969. It quickly became and has remained the most successful mathematics text of its kind. The extensive and continued use of the text over an eight-year period proves the quality of the first edition. That eight-year experience by students and institutions throughout the country also provided a wealth of classroom testing which has been the basis for creating this second edition.

Because the first edition proved to be successful both in terms of its content and its teachability, this second edition retains the expert programming and good mathematics that made the text the leader in its field. At the same time, every area of the first edition has been carefully restudied; those particular frames and sections which evoked any concern from past users have been rewritten. Teachers who used the first edition can confidently select this second edition with full expectation of a text that has been greatly improved by its revision.

The mathematical content of *Algebra—Programmed, Part 2* is equivalent to the second half of a beginning algebra course. This second edition includes some topics that have been added to those of the first edition. These additional topics include graphing by intercepts, FOIL multiplication and factoring of polynomials, use of the least common multiple (LCM) in solving fractional equations, and word problems.

There are now ten chapters in Part 2 compared with four in the first edition. The shorter chapters encourage the readers to engage in self-evaluation and review more frequently. Chapter 1 covers the graphing of two-variable equations, while Chapter 2 covers solving two equations with two variables. Chapter 3 is an introduction to polynomials, while Chapter 4 is devoted entirely to the factoring of polynomials, and Chapter 5 covers all of the operations of polynomial fractions. Chapter 6 is a review of solving linear equations, while Chapter 7 covers fractional and quadratic equation solutions. Rational and irrational numbers are presented in Chapter 8, while Chapter 9 is devoted to solving quadratic equations with irrational solutions. Chapter 10 is applications or word problems.

We have received far too many constructive comments from our readers over these past eight years to make it possible to name all of the people who have played significant roles in revising and improving the text. We are, however, most appreciative of the contributions of these many readers and wish to express our thanks to each of them.

At least two important research studies have been completed using *Algebra—Programmed, Part 2*. Both of these studies compared programmed instruction with traditional teaching techniques. We are honored by the choice of our text as the programmed material to make such research valid, and we are pleased that both studies confirmed the effectiveness of *Algebra—Programmed, Part 2.* We thank the researchers, Dr. David Conroy of Northern Virginia Community College and Dr. Ernest Ross of St. Petersburg Junior College, for their willingness to share the insights gained in their studies, which have been very helpful to us in preparing this revision.

Much has changed here at St. Petersburg Junior College, Clearwater Campus, since we began writing the material a decade ago. Our mathematics laboratory program has grown from a handful to well over a thousand students each year, and from a one-course offering to a multicourse offering through the use of programmed materials. Much of the credit for our success in the laboratory is due to the leadership and assistance of Dr. Joseph Gould, Director of the Division of Science and Mathematics.

Directions for Using This Text

This will probably be your first experience with a programmed textbook. It is different from the usual mathematics book, and you should read these directions for its proper use carefully.

Each page of the programmed text has two columns. The left column contains a series of questions or problems called *frames*. The right column shows the correct answer for each question or problem. You should cover the right column with a sheet of paper so that you won't see the answers until you have written in your response in the left column.

Most frames include some new information which must be understood to answer the questions correctly. Each question will be relatively easy to answer by the time you encounter it, but your correct response depends upon the knowledge you acquired in previous frames. For example, Frame 200 in Chapter 5 will not be difficult because of the knowledge you acquired in the preceding frames and chapters.

Programmed learning is based on the work of the psychologist B. F. Skinner. Programmed materials are carefully designed to ensure that you are actively involved in the learning process, your acquisition of knowledge is reinforced so that you remember the new skills and concepts, and you enjoy repeated success which encourages you to study further.

To receive the greatest possible benefit from programmed material, you must *always* do three things:

1. Read the question carefully. Most of the wrong answers that students give are caused by misreading the question.
2. Write the answer in the answer blank provided. It is extremely important for you to commit yourself to an answer. Without

writing in the answer, this commitment tends to be incomplete and your learning suffers.

3. Check your written response immediately. Uncover the answer in the right column and compare it with what you have written. This will reinforce your learning or alert you to an incorrect response. If your answer is correct, continue to the next frame. If your answer is incorrect, it is important to find the reason for your error. Do not proceed in the program if your answers continue to be incorrect. Review the preceding section until you overcome the difficulty. If a teacher is available, do not hesitate to ask for help.

Two kinds of evaluating tools are incorporated into the text. Each chapter begins with a Pre-Test and ends with a Post-Test. The Pre-Test tells you if your previous learning includes the material to be covered in that chapter. A score of 90% or more on the Pre-Test indicates that you already have sufficient mastery of the material, and you can immediately take the Post-Test without going through the frames in that chapter. Usually, though, you will go through the entire chapter frame by frame. Then you will take the Post-Test to check your mastery of all the concepts and skills you developed in the chapter. A score of 90% or more on the Post-Test indicates that you have sufficient mastery of the material to proceed to the next chapter.

At frequent intervals throughout each chapter you will find Self-Quizzes. Each Self-Quiz is a short test which reviews your mastery of the material immediately preceding the quiz. These Self-Quizzes also serve as good starting and stopping points in planning your study.

Whenever you miss a question, you should mark that frame clearly in the text margin so that you can easily locate it when you are reviewing for a Self-Quiz. Similarly, if you miss a question on a Self-Quiz you should mark it clearly so that you can easily locate it when you review for a Post-Test. In this way your incorrect responses become very helpful study guides, because they indicate the probable areas of difficulty that you will encounter on a Post-Test.

Answers to all Pre-Tests, Self-Quizzes, and Post-Tests are given at the back of the book.

ALGEBRA– PROGRAMMED

part 2

Graphing Equations with Two Variables

1 2 3 4 5 6 7 8 9 10

CHAPTER PRE-TEST

The following questions indicate the objectives of this chapter. A score of 90% indicates sufficient mastery, and the student may immediately take the Chapter Post-Test.

1. Is $(4, {}^{-}1)$ a solution for $x + y = 3$?
2. Is $(3, {}^{-}11)$ a solution for $2x - y = 17$?
3. Is $({}^{-}2, {}^{-}3)$ a solution for $2x - 3y = 6$?
4. Is $(5, {}^{-}1)$ a solution for $5x - 2y = 23$?
5. (6, ______) is a solution for $3x - y = 9$.
6. (______, 5) is a solution for $2x - 3y = 9$.
7. (7, ______) is a solution for $x + 2y = 8$.
8. (______, $^{-}4$) is a solution for $5x - 3y = 13$.
9. Write ordered pairs (x, y) for each point shown on the graph by a capital letter.

   A ______ .
   B ______ .
   C ______ .
   D ______ .
   E ______ .

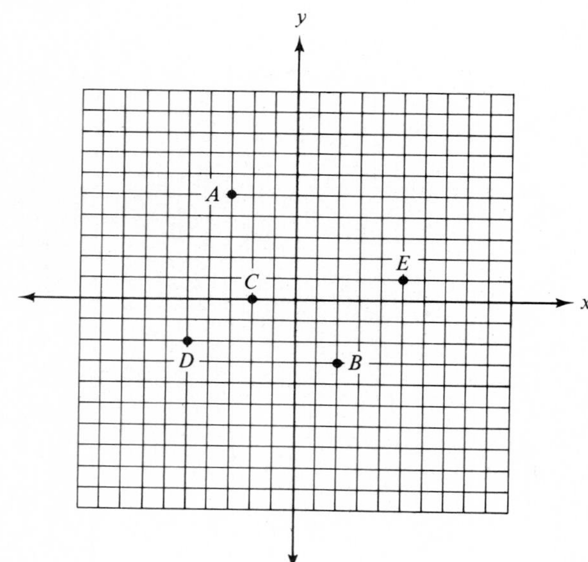

10. Mark the following points (by their capital letters) on the graph:

A (2, 5).
B (4, $^-1$).
C (0, $^-5$).
D ($^-6$, 4).
E ($^-2$, $^-5$).

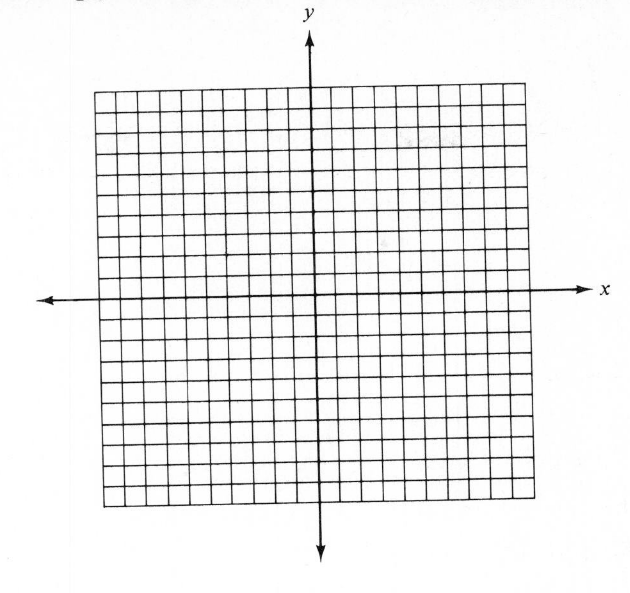

11. Graph $2x + y = 7$.

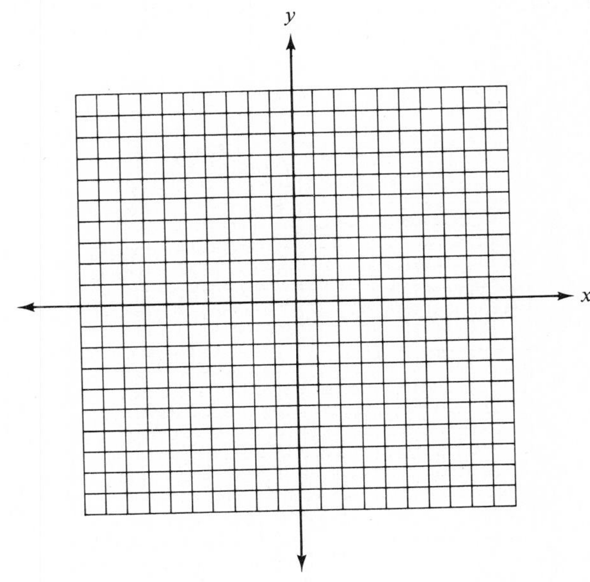

12. Graph $2x - 3y = 12$.

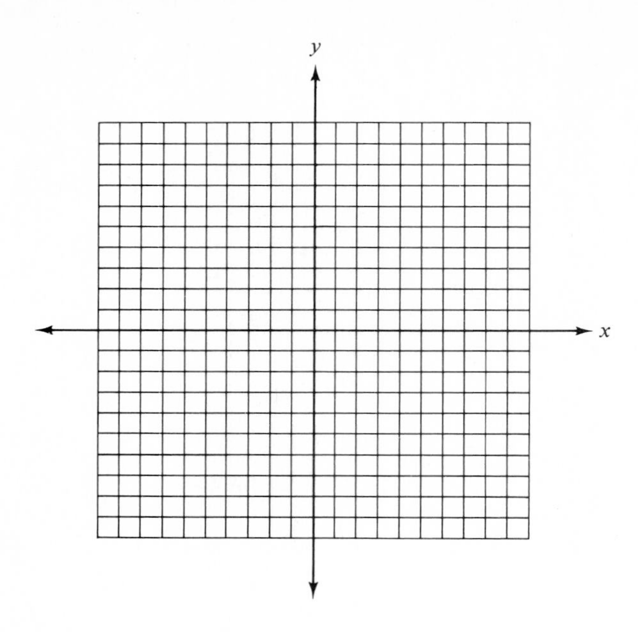

13. Graph both $3x + y = 11$ and $2x + 3y = 5$ and find the point common to both lines.

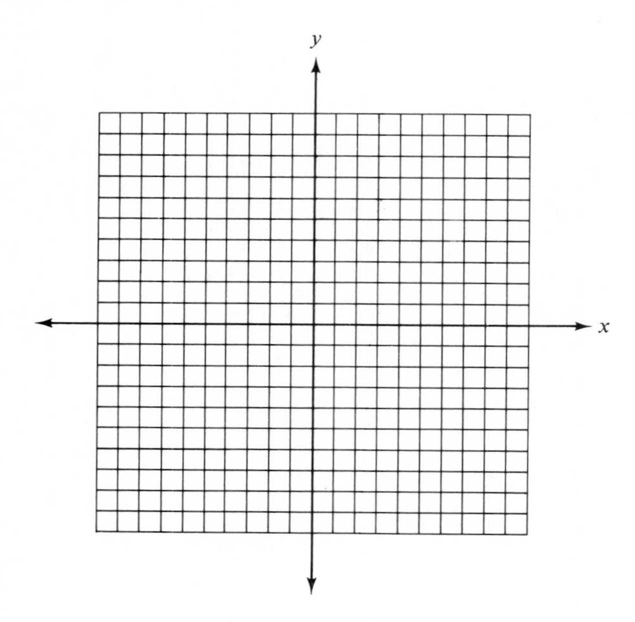

$x + y = 7$ is an open sentence or equation, but it is different from the equations solved in *Algebra Programmed, Part 1*, because there are two letters to be replaced by numbers.

This chapter deals with equations involving two letters, or variables.

---

1 $x + y = 10$ is an equation. *Any* rational number may be used to replace x and *any* rational number may be used to replace y. In the equation $x + y = 15$, *any* rational number may be used to replace x and _______ rational number may be used to replace y.

any

2 $x + y = 7$ is an equation. *Any* rational number may be used to replace x and _______ rational number may be used to replace y.

any

3 For the open sentence $x + y = 7$, if x is replaced by 3, the equation $3 + y = 7$ is obtained. For the open sentence $x + y = 7$, what equation is obtained if x is replaced by $^{-}2$? ______________ .

$^{-}2 + y = 7$

4 For the open sentence $x + y = 7$, if x is replaced by 10, what equation is obtained? ______________ .

$10 + y = 7$

5 For the open sentence $x + 6y = 9$, if x is replaced by 5, what equation is obtained? ______________ .

$5 + 6y = 9$

6 For the open sentence $2x - 5y = 8$, if x is replaced by 6, the equation $2 \cdot 6 - 5y = 8$ or $12 - 5y = 8$ is obtained. For the open sentence $2x - 5y = 8$, if x is replaced by 5, what equation is obtained?

______________ .

$2 \cdot 5 - 5y = 8$ or $10 - 5y = 8$

7 For the open sentence $3x - y = {}^{-}4$, if x is replaced by 9, what equation is obtained? ______________ .

$27 - y = {}^{-}4$

8 For the open sentence $2x + 7y = 6$, if x is replaced by 0, what equation is obtained? ________________ .

$7y = 6$

9 For the open sentence $^{-}2x + 7y = {}^{-}1$, if x is replaced by 4, what equation is obtained?

________________ .

$^{-}8 + 7y = {}^{-}1$

10 For the open sentence $^{-}3x - 2y = 9$, if x is replaced by $^{-}5$, what equation is obtained? ________________ .

$15 - 2y = 9$

11 For the open sentence $^{-}x + 8y = {}^{-}3$, if x is replaced by 7, what equation is obtained?
(^{-}x is equivalent to $^{-}1x$.) ________________ .

$^{-}7 + 8y = {}^{-}3$

12 For the open sentence $^{-}x - 3y = 6$, if x is replaced by $^{-}5$, what equation is obtained?

________________ .

$^{-}1 \cdot {}^{-}5 - 3y = 6$ or
$5 - 3y = 6$

13 For the open sentence $4x - 7y = {}^{-}1$, if x is replaced by 0, what equation is obtained? ________________ .

$^{-}7y = {}^{-}1$

14 To obtain a statement from the equation $x + y = 7$, both x and y must be replaced by numbers. What statement is obtained by replacing x by 4 and y by 6 in $x + y = 7$? ________________ .

$4 + 6 = 7$

15 What statement is obtained from the equation $x + y = 11$ if x is replaced by 5 and y is replaced by 6? ________________ .

$5 + 6 = 11$

16 What statement is obtained from the equation $x + y = 5$ if x is replaced by 4 and y is replaced by 3?

________________ .

$4 + 3 = 5$

17 $2 + 3 = 8$ is a false statement obtained from the equation $x + y = 8$ when $x = 2$ and $y = 3$. What statement is obtained from $x + y = 8$ when $x = 4$ and $y = 9$? ________ .

$4 + 9 = 8$

18 What statement is obtained from the equation $x + y = 13$ when $x = 2$ and $y = 11$?

________ .

$2 + 11 = 13$

19 $2 + 5 = 7$ is a *true* statement obtained from $x + y = 7$ when $x = 2$ and $y = 5$. $4 + 3 = 7$ is a ________ statement obtained from $x + y = 7$ when $x = 4$ and $y = 3$.

true

20 If $x = 2$ and $y = 6$, is the statement obtained from the equation $x + y = 8$ true or false? ________ .

True

21 If $x = 11$ and $y = 4$, is the statement obtained from the equation $x + y = 15$ true or false? ________ .

True

22 $3 + 7 = 5$ is a false statement obtained from $x + y = 5$ when $x = 3$ and $y = 7$. $4 + 2 = 5$ is a ________ statement obtained from $x + y = 5$ when $x = 4$ and $y = 2$.

false

23 If $x = 6$ and $y = 9$, is the statement obtained from the equation $x + y = 11$ true or false? ________ .

False

24 If $x = 0$ and $y = 12$, is the statement obtained from the equation $x + y = 17$ true or false? ________ .

False

25 If $x = 7$ and $y = 10$, is the statement obtained from the equation $x + y = 19$ true or false? ________ .

False

26 If $x = 2$ and $y = 2$, is the statement obtained from the equation $x + y = 4$ true or false? ________ .

True

27 If $x = 10$ and $y = 5$, is the statement obtained from the equation $x + y = 15$ true or false? ________ . True

28 If $x = 14$ and $y = 5$, is the statement obtained from the equation $x + y = 12$ true or false? ________ . False

29 $15 - 16 = {}^{-}1$ is the statement obtained from the equation $5x - 2y = {}^{-}1$ when $x = 3$ and $y = 8$. Is $15 - 16 = {}^{-}1$ a true statement? ________ . Yes

30 $10 + 3 = 8$ is the statement obtained from the equation $2x - 3y = 8$ when $x = 5$ and $y = {}^{-}1$. Is $10 + 3 = 8$ a true statement? ________ . No

31 If $x = 5$ and $y = 0$, is the statement obtained from the equation $x - 2y = 5$ true or false? ________ . True

32 If $x = 1$ and $y = 7$, is the statement obtained from the equation $3x + 2y = 15$ true or false? ________ . False

33 If $x = 1$ and $y = {}^{-}3$, is the statement obtained from the equation $3x - 2y = 6$ true or false? ________ . False

34 If $x = 5$ and $y = 5$, is the statement obtained from $2x - 3y = {}^{-}5$ true or false? ________ . True

35 If $x = {}^{-}2$ and $y = 4$, is the statement obtained from $5x - 2y = 15$ true or false? ________ . False

36 If $x = {}^{-}5$ and $y = {}^{-}4$, is the statement obtained from $2x - 3y = {}^{-}2$ true or false? ________ . False

37 If $x = {}^{-}3$ and $y = 0$, is the statement obtained from $3x - 4y = {}^{-}9$ true or false? ________ . True

38 For the equation $2x - y = 8$, both x and y must be replaced to obtain a statement. The equation $2x - y = 8$ contains two letters and requires ________ (how many) number replacements to obtain a statement.

two

39 Both x and y must be replaced by numbers in $7x - y = 4$ to obtain a statement. How many letters must be replaced in $4x - y = 9$ to obtain a statement? ________ .

Two

40 For the equation $x + y = 7$, how many letters must be replaced by numbers to obtain a mathematical statement? ________ .

Two

41 For the equation $7x - 3y = 6$, how many letters must be replaced by numbers to obtain a statement? ________ .

Two

42 For the equation $3x - 4y = 8$, how many letters must be replaced by numbers to obtain a statement? ________ .

Two

43 Two numbers or a pair of numbers are necessary to obtain a statement from the equation $2x - 3y = 6$. How many numbers are in the parentheses of (7, 1)? ________ .

Two

44 $(6, {}^{-}5)$ shows two numbers or a pair of numbers. (7, 5) shows ________ numbers.

two

45 (5, 3) is called an *ordered pair* of numbers. (7, 2) is another ________ ________ of numbers.

ordered pair

46 $(3, {}^{-}8)$ is an ordered pair of numbers. 3 is the first number in the ordered pair and ________ is the second number in the ordered pair.

${}^{-}8$

47 (5, 0) is an ordered pair of numbers. What number appears first in this ordered pair? ________ . [*Note:* This pair is called *ordered* because the *order* in which the two numbers are written makes a difference.]

5

48 (2, 7) is an ordered pair of numbers. 2 is the first number in the ordered pair and ________ is the second number in the ordered pair.

7

49 An ordered pair in the form (x, y) shows that the first number is to replace x and the second number is to replace y in an equation involving both x and y. (4, 3) shows that x is to be replaced by 4 and y by ________ in an equation involving both x and y.

3

50 The ordered pair (5, 9) is used to show that $x = 5$ and $y = 9$. The ordered pair $(8, {}^{-}1)$ is used to show that $x = 8$ and $y =$ ________ .

${}^{-}1$

51 The ordered pair $({}^{-}2, 4)$ is used to show that $x = {}^{-}2$ and $y = 4$. The ordered pair $({}^{-}5, 7)$ is used to show that $x = {}^{-}5$ and $y =$ ________ .

7

52 The ordered pair (10, 5) is used to show that $x = 10$ and $y =$ ________ .

5

53 The ordered pair (8, 3) is used to show that $x =$ ________ and $y = 3$.

8

54 The ordered pair $({}^{-}9, {}^{-}3)$ is used to show that $x =$ ________ and $y = {}^{-}3$.

${}^{-}9$

55 The ordered pair $(11, {}^{-}3)$ is used to show that x is to be replaced by 11 and y by ${}^{-}3$. What ordered pair shows that x is to be replaced by 6 and y by ${}^{-}8$? ________ .

$(6, {}^{-}8)$

56 The ordered pair $(4, 0)$ is used to show that $x = 4$ and $y = 0$. Write an ordered pair (x, y) to show that $x = {}^{-}4$ and $y = 7$. ________ .

$({}^{-}4, 7)$

57 Write an ordered pair (x, y) to show that y is to be replaced by 9 and x by ${}^{-}14$. ________ . [*Note:* The ordered pair must show x as the first number.]

$({}^{-}14, 9)$

58 To use the ordered pair $(2, 5)$ with the equation $3x + y = 9$, the first number, 2, in the ordered pair is to replace the x and the second number, 5, is to replace the ________ .

y

59 To use the ordered pair $(7, 4)$ with the equation $5x - 2y = 11$, x is to be replaced by 7 and y is to be replaced by ________ .

4

60 To use the ordered pair $(2, {}^{-}5)$ with the equation $x + y = 3$, x is to be replaced by 2 and y is to be replaced by ________ .

${}^{-}5$

61 The first number in the ordered pair $(2, 9)$ is to replace x in the equation $4x + 5y = 0$. x is to be replaced by ________ .

2

62 $(6, 3)$ with the equation $2x - 3y = 8$ means that y is to be replaced by ________ .

3

63 $(8, {}^{-}6)$ with the equation $3x + 4y = 0$ means that x is to be replaced by ________ .

8

64 $({}^{-}1, 7)$ with the equation $2x + 5y = 20$ means that x is to be replaced by ________ .

${}^{-}1$

65 $(4, {}^{-}6)$ with the equation $2x + 3y = 5$ means that y is to be replaced by ________ .

${}^{-}6$

66 $(4, {}^{-}7)$ with the equation $x + y = 6$ gives the statement $4 + {}^{-}7 = 6$. What statement is obtained from $x + y = 6$ using the ordered pair $(5, 3)$?

______________ .

$5 + 3 = 6$

67 What statement is obtained from $x + y = 13$ using the ordered pair $(8, 5)$? ______________ .

$8 + 5 = 13$

68 For $2x + 3y = 8$, the ordered pair $(3, {}^{-}5)$ gives the statement $6 - 15 = 8$. For $2x + 3y = 8$, the ordered pair $(1, 5)$ gives the statement

______________ .

$2 + 15 = 8$

69 Using the ordered pair $(2, 4)$ with the equation $5x + y = 14$ gives the statement ______________ .

$10 + 4 = 14$

70 The ordered pair $({}^{-}2, 5)$ with the equation $3x - y = {}^{-}8$ gives the statement ${}^{-}6 - 5 = {}^{-}8$. The ordered pair $({}^{-}3, 1)$ with the equation $3x - y = {}^{-}8$ gives the statement ______________ .

${}^{-}9 - 1 = {}^{-}8$

71 Using the ordered pair $(2, 8)$ with the equation $6x - 2y = {}^{-}8$ gives the statement $12 - 16 = {}^{-}8$. Using the ordered pair $(3, 5)$ with the equation $6x - 2y = {}^{-}8$ gives the statement ______________.

$18 - 10 = {}^{-}8$

72 $(3, 4)$ is a *solution* for $x + y = 7$ because $3 + 4 = 7$ is a *true statement.* $(5, 2)$ is another *solution* for $x + y = 7$ because $5 + 2 = 7$ is a ________ statement.

true

73 $(1, 4)$ is a solution for $2x + y = 6$ because $2 + 4 = 6$ is a true statement. Is $(2, 2)$ also a solution for $2x + y = 6$? ________ .

Yes

74 To determine whether or not $(5, 1)$ is a solution for $2x - y = 9$, replace x by 5 and y by 1. Is $10 - 1 = 9$ a true statement? ________ .

Yes

75 (5, 1) is a solution for $2x - y = 9$ because $10 - 1 = 9$ is a true statement. Is (6, 3) also a solution for $2x - y = 9$? ________ . — Yes

76 Is (7, 2) a solution for $x + 2y = 11$? ________ . — Yes

77 Is (8, 1) a solution for $x - 3y = 5$? ________ . — Yes

78 Is $(2, {}^{-}1)$ a solution for $3x + y = 5$? ________ . — Yes

79 To determine whether or not (2, 5) is a solution for $4x - y = 1$, replace x by 2 and y by 5. Is $8 - 5 = 1$ a true statement? ________ . — No

80 Since (2, 5) with $4x - y = 1$ gives the false statement $8 - 5 = 1$, (2, 5) is not a solution for $4x - y = 1$. Is (3, 5) a solution for $4x - y = 1$? ________ . — No

81 Is (4, 1) a solution for $2x - y = 11$? ________ . — No

82 Is $({}^{-}1, 3)$ a solution for $5x + y = 7$? ________ . — No

83 Is (4, 2) a solution for $x - 3y = 7$? ________ . — No

84 (7, 3) is a solution for $x - y = 4$ because $7 - 3 = 4$ is a true statement. Is (9, 5) a solution for $x - y = 4$? ________ . — Yes

85 $(3, {}^{-}5)$ is not a solution for $2x + y = 3$ because $6 + {}^{-}5 = 3$ is a false statement. Is (4, 3) a solution for $2x + y = 3$? ________ . — No

86 Is (8, 6) a solution for $x - y = 2$? ________ . — Yes

87 Is (4, 7) a solution for $x - y = 3$? ________________ . — No; $4 - 7 \neq 3$

88 Is (5, 3) a solution for $2x - y = 7$? (Consider the statement $10 - 3 = 7$.) ________ . — Yes

89 Is (5, 4) a solution for $2x - 3y = 0$? ________ . No

90 Is (2, 7) a solution for $3x + y = 13$? ________ . Yes

91 Is $(3, {}^{-}1)$ a solution for $2x - 3y = 9$? (Consider $2 \cdot 3 + {}^{-}3 \cdot {}^{-}1 = 9$.) ________ . Yes

92 Is $(5, {}^{-}1)$ a solution for $x - y = 6$? ________ . Yes

93 Is $(4, {}^{-}3)$ a solution for $2x - y = 11$? ________ . Yes

94 Is (2, 3) a solution for $5x - 2y = 16$? ________ . No

95 Is $(2, {}^{-}3)$ a solution for $5x - 2y = 16$? ________ . Yes

96 Is (8, 3) a solution for $x + y = 11$? ________ . Yes

97 Is (4, 7) a solution for $x + y = 11$? ________ . Yes

98 Is $(12, {}^{-}1)$ a solution for $x + y = 11$? ________ . Yes

99 Is $({}^{-}5, 16)$ a solution for $x + y = 11$? ________ . Yes

100 (8, 3), (4, 7), $(12, {}^{-}1)$, and $({}^{-}5, 16)$ are all solutions for $x + y = 11$. Is $(20, {}^{-}9)$ also a solution for $x + y = 11$? ________ . Yes

Self-Quiz # 1

The following questions test the objectives of the preceding section. 100% mastery is desired.

1. Write an ordered pair (x, y) to show that $x = 3$ and $y = 5$.

______ .

2. Write an ordered pair (x, y) to show that $y = 1$ and $x = 16$.

______ .

3. $(2, 5)$, $(4, 3)$, $(0, 7)$, and $(12, {}^{-}5)$ are all solutions for $x + y = 7$. Is $({}^{-}3, 10)$ also a solution for $x + y = 7$? ______ .

4. $(5, 1)$, $({}^{-}2, {}^{-}13)$, $(0, {}^{-}9)$, and $(6, 3)$ are all solutions for $2x - y = 9$. Is $(3, {}^{-}3)$ also a solution for $2x - y = 9$? ______ .

5. $(5, 0)$, $(2, {}^{-}1)$, $(14, 3)$, and $({}^{-}7, {}^{-}4)$ are all solutions for $x - 3y = 5$. Is $(8, 1)$ also a solution for $x - 3y = 5$? ______ .

101 For the equation $x + y = 11$, any rational number may replace x and any rational number may replace y. If $x = 2$, the equation $2 + y = 11$ is obtained. What replacement for y in $2 + y = 11$ will give a true statement? ______ .

9

102 The ordered pair $(2, 9)$ is a solution for $x + y = 11$ because $2 + 9 = 11$ is a true statement. Complete the ordered pair (3, ______) to obtain another solution for $x + y = 11$. (Solve $3 + y = 11$.)

$(3, \underline{8})$

103 To complete the ordered pair (4, __) as a solution for $2x - y = 7$, the following steps are used:

$$\begin{aligned} 2x - y &= 7 \quad (4, \_\_) \\ 2 \cdot 4 - y &= 7 \\ 8 - y &= 7 \\ ^{-}y &= {}^{-}1 \\ y &= 1 \end{aligned}$$

(4, 1) ________ (is, is not) a solution for $2x - y = 7$. — is

104 To complete the solution (4, __) for $2x - 3y = 6$, the following steps are used:

$$\begin{aligned} 2x - 3y &= 6 \quad (4, \_\_) \\ 2 \cdot 4 - 3y &= 6 \\ 8 - 3y &= 6 \\ ^{-}3y &= {}^{-}2 \\ \frac{^{-}1}{3} \cdot {}^{-}3y &= \frac{^{-}1}{3} \cdot {}^{-}2 \\ y &= \frac{2}{3} \end{aligned}$$

$\left(4, \frac{2}{3}\right)$ ________ (is, is not) a solution for $2x - 3y = 6$. — is

105 (6, ________) is a solution for $2x - y = 7$. (Solve $12 - y = 7$.) — (6, 5)

106 (2, ________) is a solution for $3x + 2y = 13$. (Solve $6 + 2y = 13$ for y.) — $\left(2, \frac{7}{2}\right)$

107 The ordered pair (7, 3) is a solution for $2x + y = 17$ because $14 + 3 = 17$ is a true statement. Complete the ordered pair (________, 9) to obtain another solution for $2x + y = 17$. (Solve $2x + 9 = 17$.) — (4, 9)

108 (________, 7) is a solution for $3x + y = 13$. (Solve $3x + 7 = 13$.) — (2, 7)

109 (________, 7) is a solution for $2x + y = 12$.
(Solve $2x + 7 = 12$ for x.)

$\left(\frac{5}{2}, 7\right)$

110 The ordered pair (3, 4) is a solution for $x + y = 7$ because $3 + 4 = 7$ is a true statement. Complete the ordered pair (5, ________) to obtain another solution for $x + y = 7$. (Solve $5 + y = 7$.)

(5, 2)

111 The ordered pair (5, 4) is a solution for $x + 2y = 13$ because $5 + 8 = 13$ is a true statement. Complete the ordered pair (________, 5) to obtain another solution for $x + 2y = 13$. (Solve $x + 10 = 13$.)

(3, 5)

112 (5, ________) is a solution for $2x + 3y = 31$.
(Solve $10 + 3y = 31$.)

(5, 7)

113 (________, 6) is a solution for $5x + y = 12$.
(Solve $5x + 6 = 12$.)

$\left(\frac{6}{5}, 6\right)$

114 (________, 4) is a solution for $x + 3y = 20$.
(Solve $x + 12 = 20$.)

(8, 4)

115 (________, 2) is a solution for $2x + 5y = 13$.
(Solve $2x + 10 = 13$.)

$\left(\frac{3}{2}, 2\right)$

116 ($^{-}8$, ________) is a solution for $2x + y = 0$.
(Solve $^{-}16 + y = 0$.)

($^{-}8$, 16)

117 (0, ________) is a solution for $x + y = 11$.
(Solve $0 + y = 11$.)

(0, 11)

118 (________, $^{-}1$) is a solution for $3x + y = 13$.
(Solve $3x - 1 = 13$.)

$\left(\frac{14}{3}, {}^{-}1\right)$

119 (________, 4) is a solution for $3x + y = 13$.
(Solve $3x + 4 = 13$.)

(3, 4)

120 (5, ________) is a solution for $x + y = 11$.

(5, 6)

121 $(^-2,$ ______ $)$ is a solution for $5x + 3y = 2$.

$(^-2, \underline{4})$

122 $(5,$ ______ $)$ is a solution for $2x - 3y = {^-1}$.

$\left(5, \underline{\frac{11}{3}}\right)$

123 $($ ______ $, 4)$ is a solution for $2x - y = 6$.

$(\underline{5}, 4)$

124 $(2,$ ______ $)$ is a solution for $3x + 2y = {^-8}$.

$(2, \underline{^-7})$

125 $($ ______ $, {^-3})$ is a solution for $2x - 2y = 8$.

$(\underline{1}, {^-3})$

126 $(9,$ ______ $)$ is a solution for $x + y = 15$.

$(9, \underline{6})$

127 $(5,$ ______ $)$ is a solution for $x + 3y = 7$.

$\left(5, \underline{\frac{2}{3}}\right)$

128 $(^-3,$ ______ $)$ is a solution for $2x + y = 4$.

$(^-3, \underline{10})$

129 $($ ______ $, {^-2})$ is a solution for $5x + y = 17$.

$\left(\underline{\frac{19}{5}}, {^-2}\right)$

130 $($ ______ $, 1)$ is a solution for $2x + y = 9$.

$(\underline{4}, 1)$

131 Use the equation $x + 2y = 18$ to complete each of the following ten solutions. Check each solution as you complete it.

(a) $(2,$ ______ $)$. — $(2, \underline{8})$

(b) $($ ______ $, 5)$. — $(\underline{8}, 5)$

(c) $(6,$ ______ $)$. — $(6, \underline{6})$

(d) $($ ______ $, 9)$. — $(\underline{0}, 9)$

(e) $(14,$ ______ $)$. — $(14, \underline{2})$

(f) $($ ______ $, {^-3})$. — $(\underline{24}, {^-3})$

(g) $(^-2,$ ______ $)$. — $(^-2, \underline{10})$

(h) $($ ______ $, 1)$. — $(\underline{16}, 1)$

(i) $(^-4,$ ______ $)$. — $(^-4, \underline{11})$

(j) $($ ______ $, 0)$. — $(\underline{18}, 0)$

132 Use the equation $3x + y = 11$ to complete each of the following ten solutions. Check each solution as you complete it.

(a) (______, 2).
(b) (1, ______).
(c) (______, 17).
(d) (4, ______).
(e) (______, 11).
(f) ($^{-}3$, ______).
(g) (6, ______).
(h) ($^{-}1$, ______).
(i) (______, 5).
(j) (______, $^{-}4$).

$(\underline{3}, 2)$
$(1, \underline{8})$
$(\underline{^{-}2}, 17)$
$(4, \underline{^{-}1})$
$(\underline{0}, 11)$
$(^{-}3, \underline{20})$
$(6, \underline{^{-}7})$
$(^{-}1, \underline{14})$
$(\underline{2}, 5)$
$(\underline{5}, ^{-}4)$

There is no limit to the number of solutions for the equation $x + 5y = 9$. Any rational number may be used as a replacement for x to obtain an ordered-pair solution. Since there is no limit to the possible replacements for x, there is also no limit to the ordered-pair solutions of $x + 5y = 9$.

Self-Quiz # 2

The following questions test the objectives of the preceding section. 100% mastery is desired.

1. Find the solution for $x + 2y = 16$ when y is replaced by $^{-}5$. ______.

2. Find the solution for $3x - y = 10$ when x is replaced by 2. ______.

3. Complete the following solutions for $2x + y = 15$:
(a) (3, ___).
(b) (___, 7).
(c) ($^{-}1$, ___).
(d) (___, $^{-}2$).

There is no limit to the number of ordered-pair solutions of the equation $2x + y = 4$. One way to indicate all of the solutions is by graphing the equation on a rectangular set of axes such as those shown below.

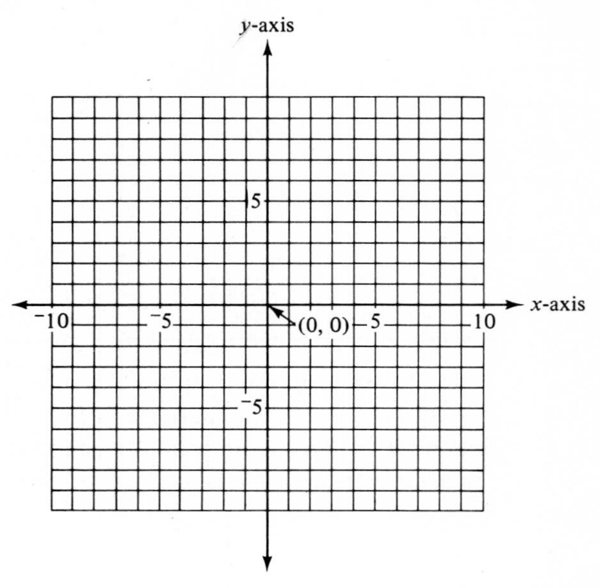

133 Every ordered pair (x, y) can be shown as a *point* on the graph. (4, 3) can be shown as a ________ on the graph.

point

134 Every ordered pair can be shown as a point on the graph. (2, $^-3$) can be shown as a point on the graph. Can (3, 5) be shown as a point on the graph? ________ .

Yes

135 Can the ordered pair ($^-7$, 3) be shown as a point on the graph? ________ .

Yes

136 (0, 0) can be shown on the graph as the point where the x-axis and y-axis intersect (see above figure). What ordered pair can be shown by the point where the x-axis and y-axis intersect? ________ .

(0, 0)

137 The point where the x-axis and y-axis intersect is called the origin (see above figure). The ordered pair for the point of origin is ________ .

(0, 0)

138 Every point on the x-axis represents an ordered pair (x, y) in which $y = 0$. [*Note:* y is always the second number in an ordered pair.] The ordered pair (________, 0) represents the point indicated by A.

(2, 0)

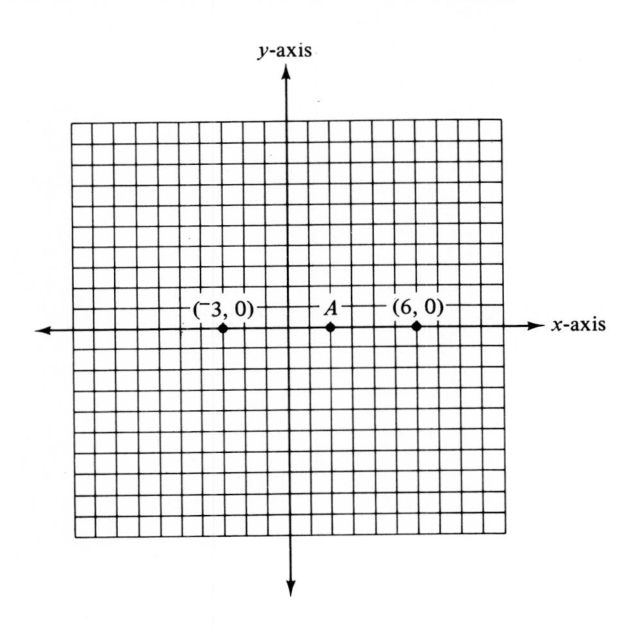

139 Every point on the x-axis represents an ordered pair (x, y) in which $y = 0$. The point marked K represents the ordered pair ______ . (1, 0)

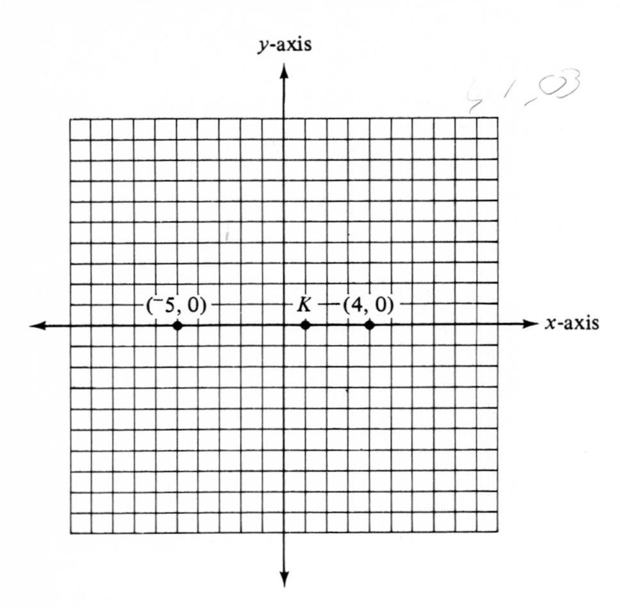

140 Point P represents the ordered pair ______ . (0, 0)

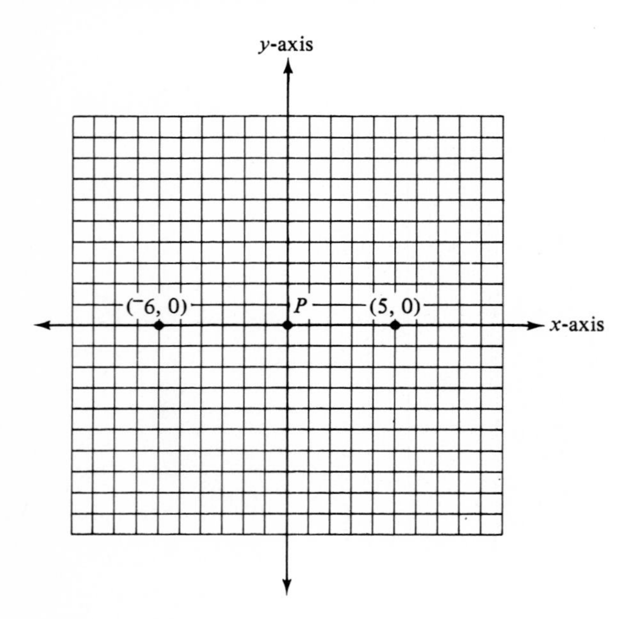

141 Point M is represented by the ordered pair (3, 0); point N by (⁻2, 0); and Point P by (⁻5, 0). What ordered pair does B represent? ________ . (1, 0)

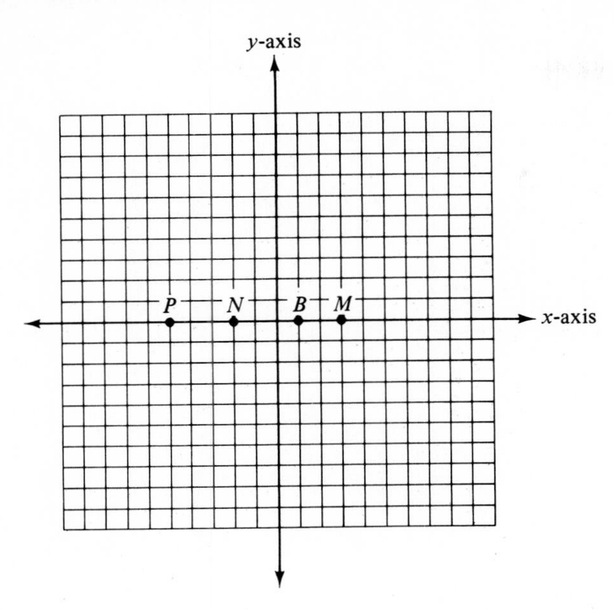

142 Point A represents the ordered pair (⁻4, 0). What ordered pair is represented by point B? ________ . (⁻1, 0)

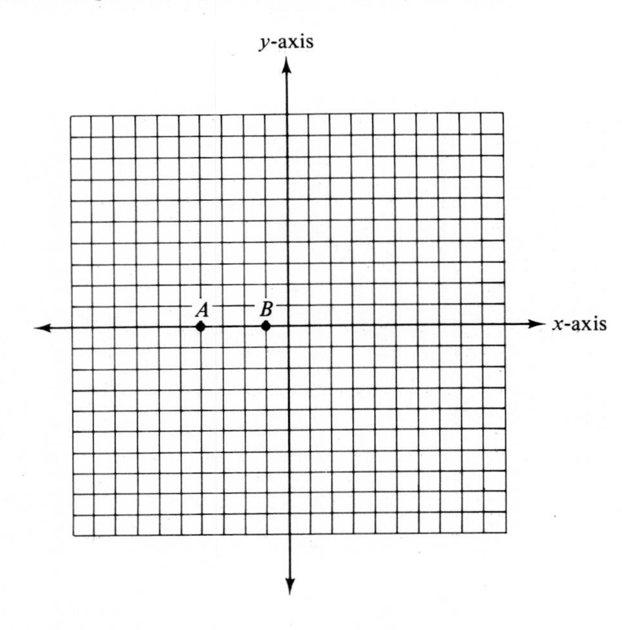

143 Point M is represented by $(5, 0)$. Point L is represented by ________ .

$(2, 0)$

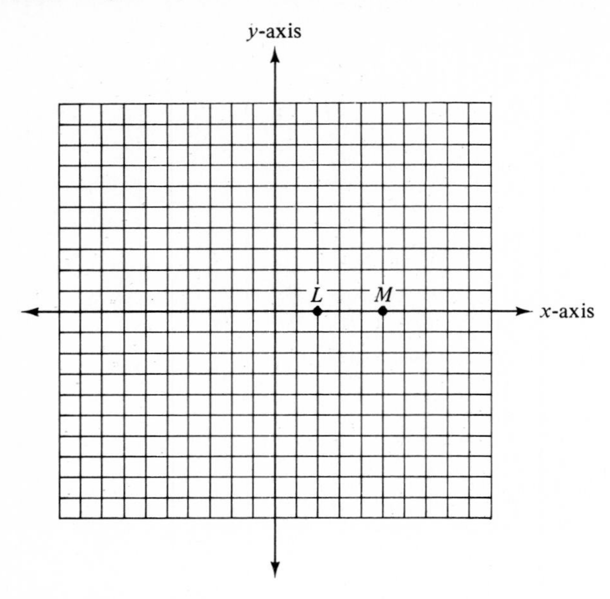

144 Every point on the y-axis represents an ordered pair (x, y) in which $x = 0$. [*Note:* x is always the first number of the ordered pair.] The point marked A represents the ordered pair (0, ________).

(0, 1)

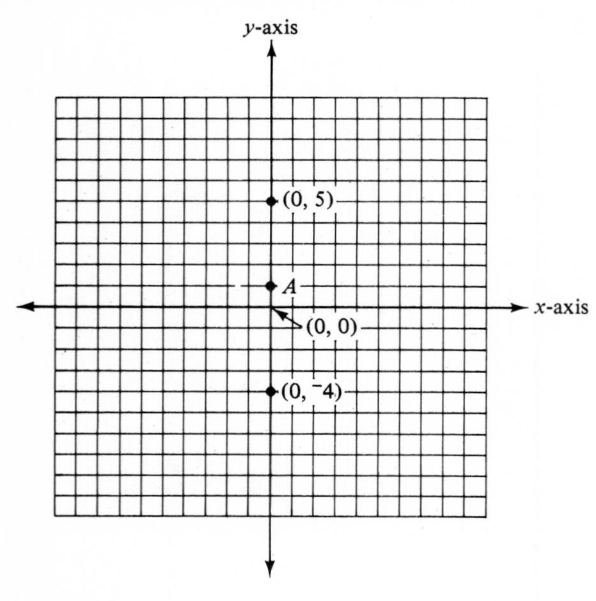

145 Every point on the y-axis represents an ordered pair (x, y) in which $x = 0$. The point marked B represents the ordered pair ______ .

(0, ⁻2)

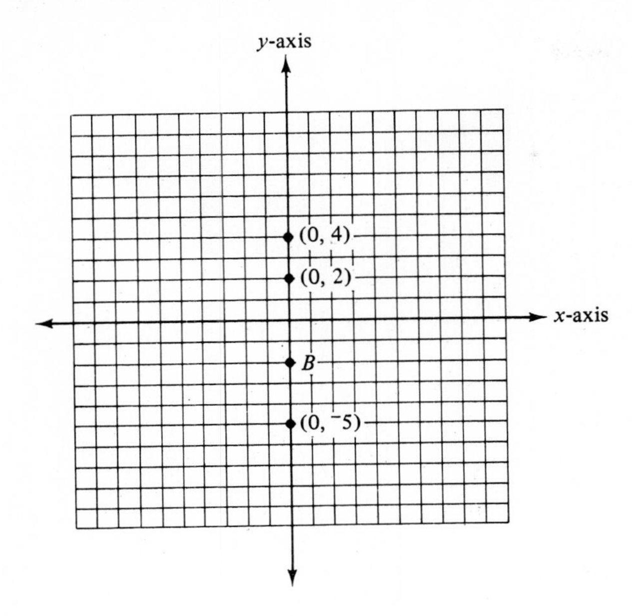

146 Point C represents the ordered pair ______ .

(0, 0)

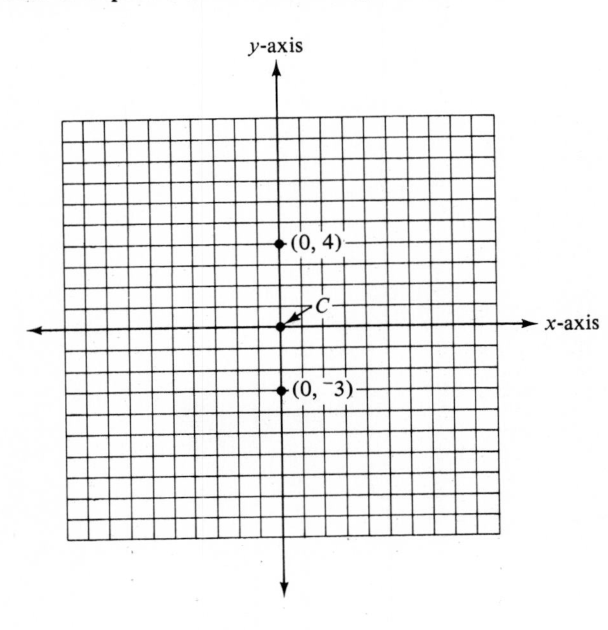

147 Point A is represented by the ordered pair (0, 4); point B by $(0, {}^{-}3)$; point C by $(0, {}^{-}5)$. What ordered pair does point D represent? ________ .

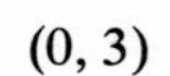
(0, 3)

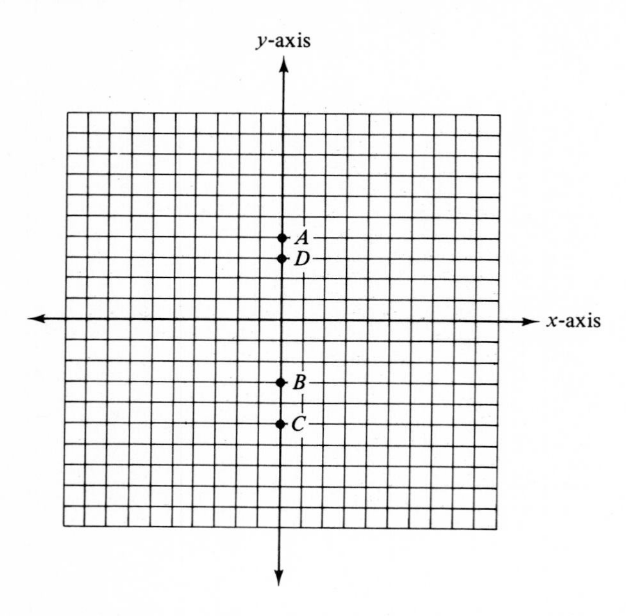

148 Point A represents the ordered pair $(0, {}^{-}4)$. What ordered pair is represented by B? ________ .

$(0, {}^{-}1)$

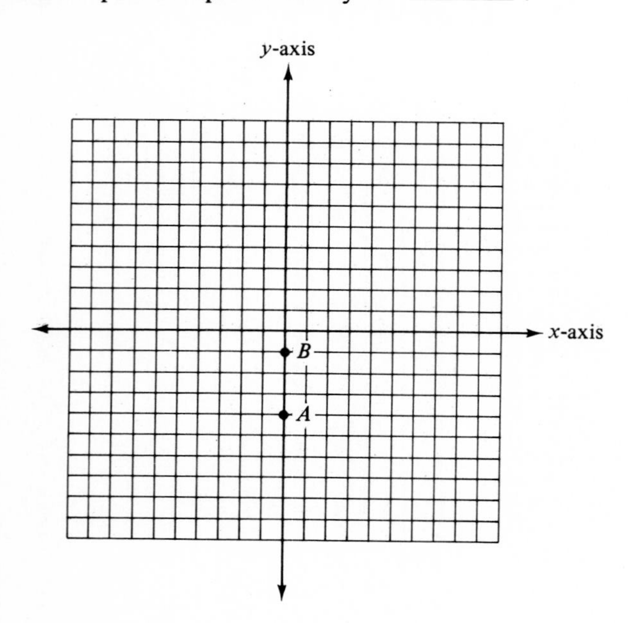

149 Point M is represented by $(0, {}^-5)$. Point L is represented by ______ .

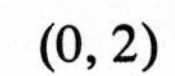

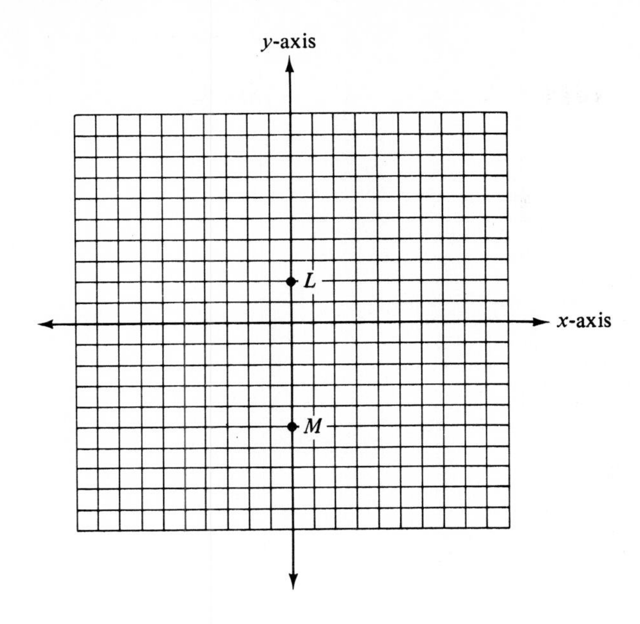

150 Point A is represented by ______ .

$(0, {}^-4)$

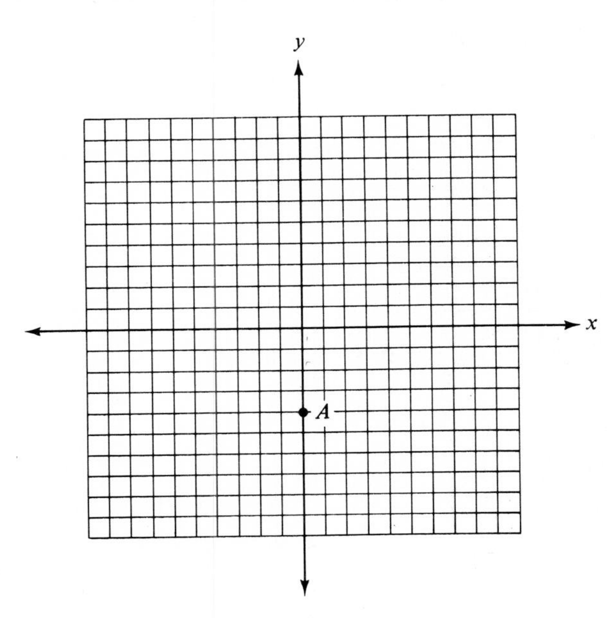

151 Point Q represents the ordered pair (0, 4). Point R represents the ordered pair ________ .

(0, ⁻5)

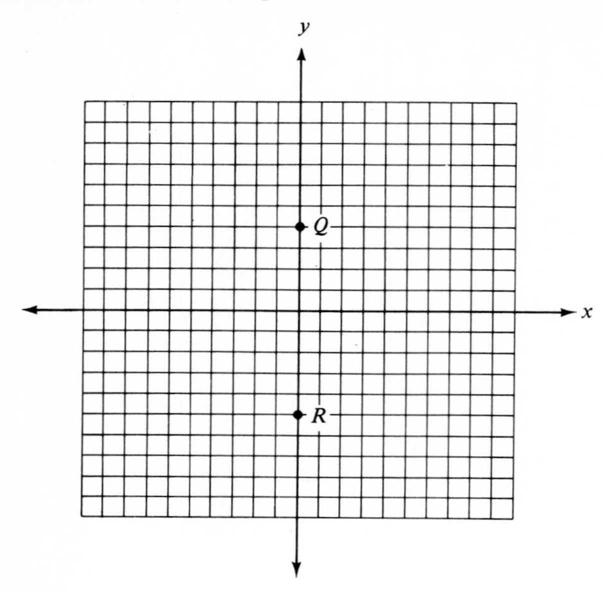

152 The ordered pair (2, 1) shows that $x = 2$ and $y = 1$. To find the location of (2, 1), start at the origin; go 2 to the right and up 1. To find the location of (5, 3), start at the origin; go 5 to the right and up ________ .

3

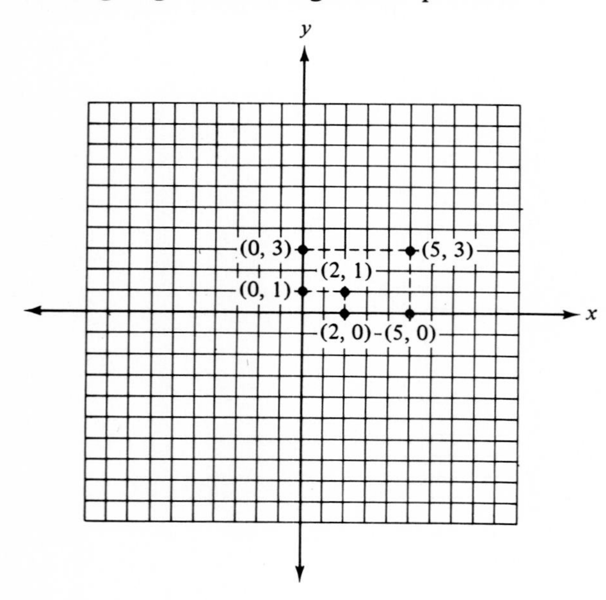

153 The ordered pair $(3, 2)$ is located by starting at the origin, going 3 to the right, and up ______ .

2

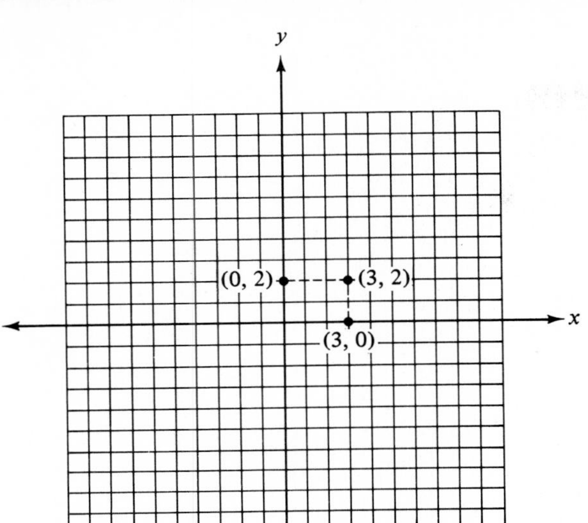

154 The ordered pair $(^-5, 1)$ is located by going 5 to the left of the origin and up ______ .

1

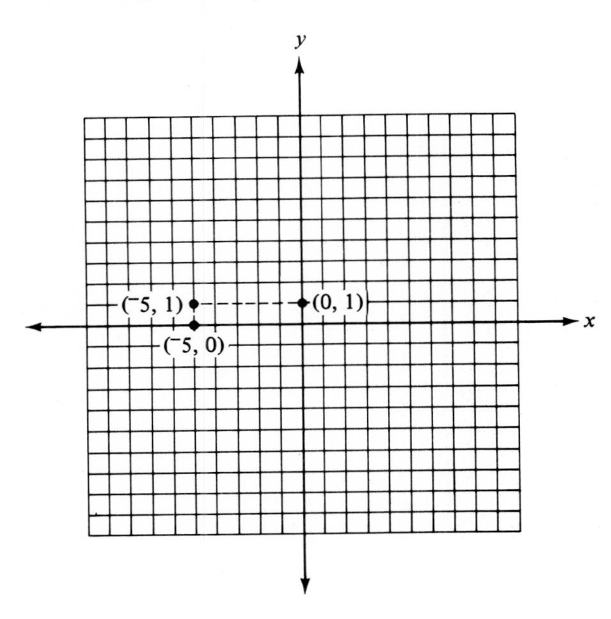

155 The ordered pair ($^-4$, $^-3$) is located 4 to the left of the origin and down ________ .

3

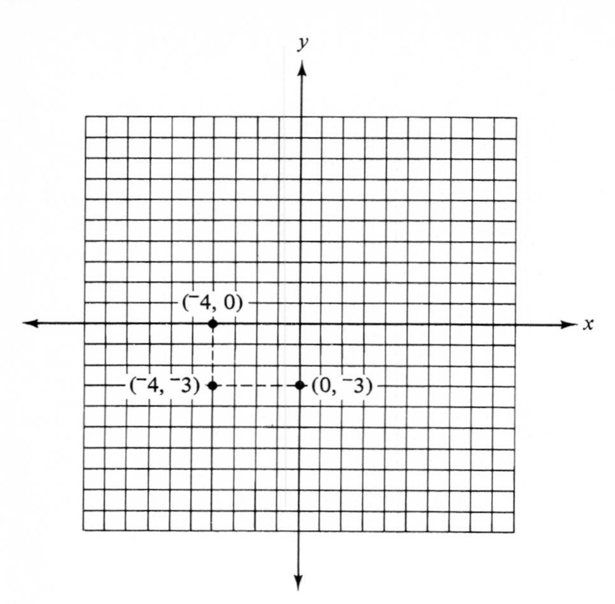

156 The ordered pair (5, $^-4$) is located 5 to the right of the origin and ________ 4.

down

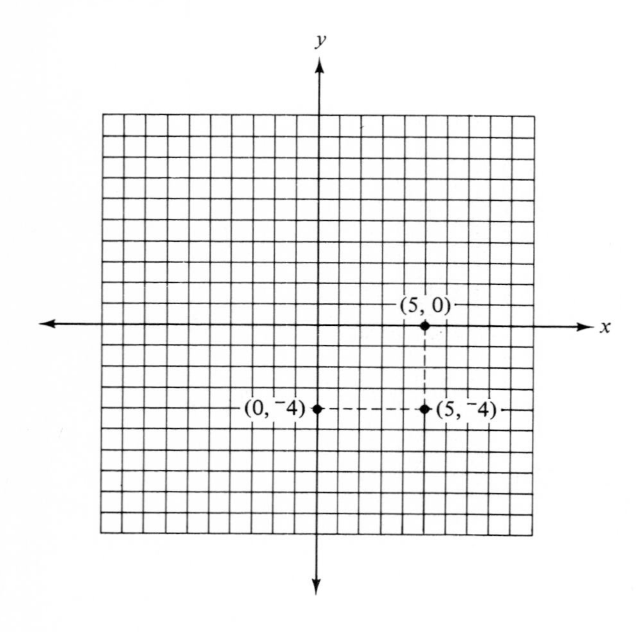

157 The ordered pair $(7, 5)$ is 7 to the ______ of the origin and up ______ .

right, 5

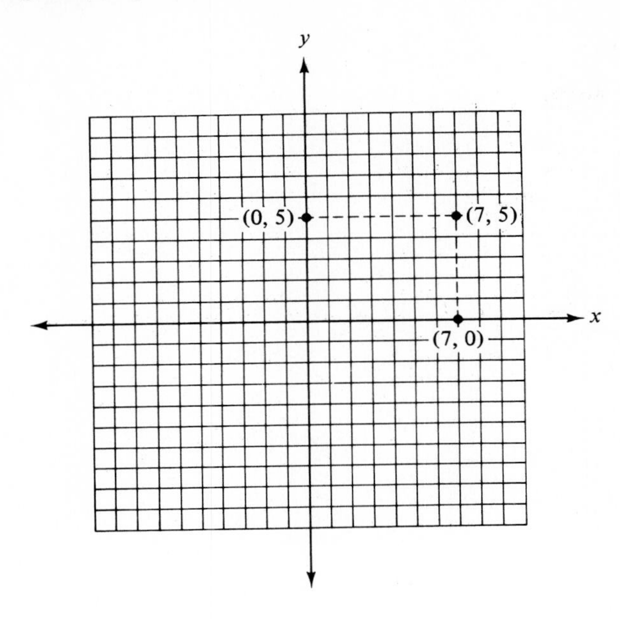

158 The ordered pair $(^-4, ^-6)$ is 4 to the ______ of the origin and down ______ .

left, 6

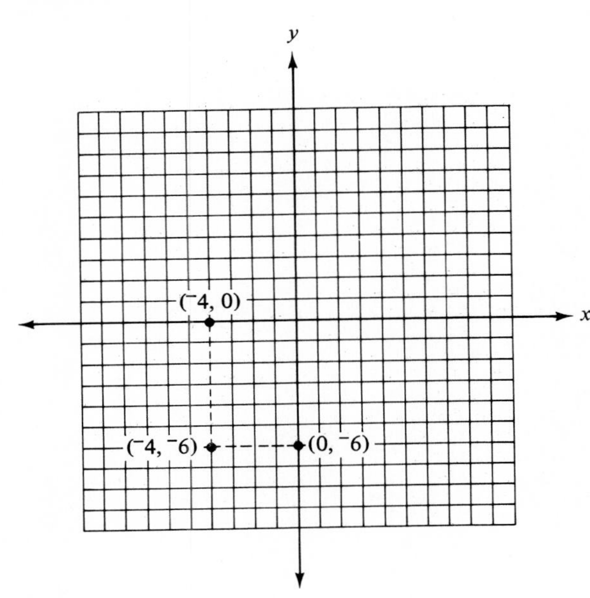

159 On the graph below what ordered pair is located 5 to the right of the origin and down 2? ________ (5, ⁻2)

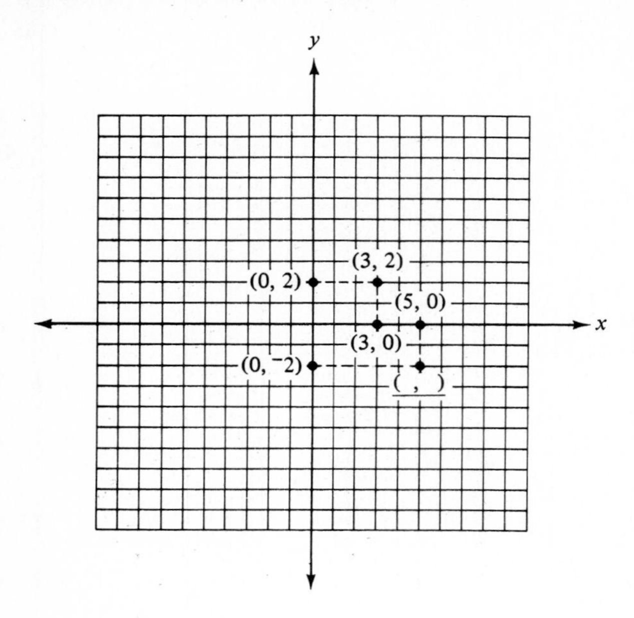

160 The graph below shows three points. The ordered pairs are (⁻5, 0), (0, ⁻6), and ________ . (⁻5, ⁻6)

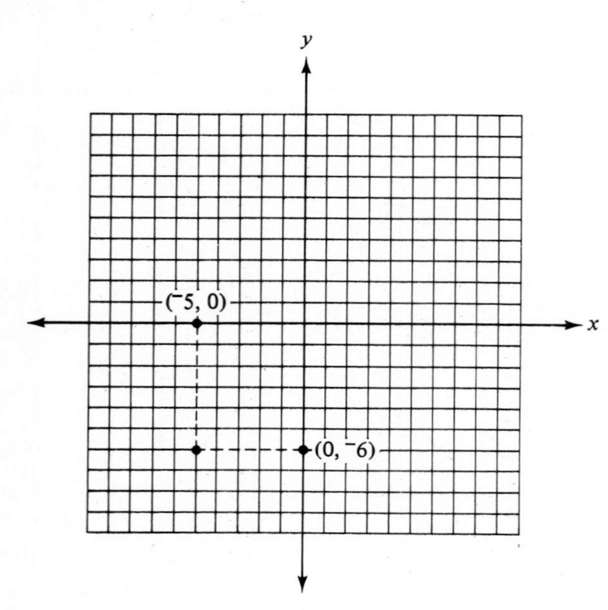

The ordered pair (3, 7) is not the same as the ordered pair (7, 3), as shown by the graph below.

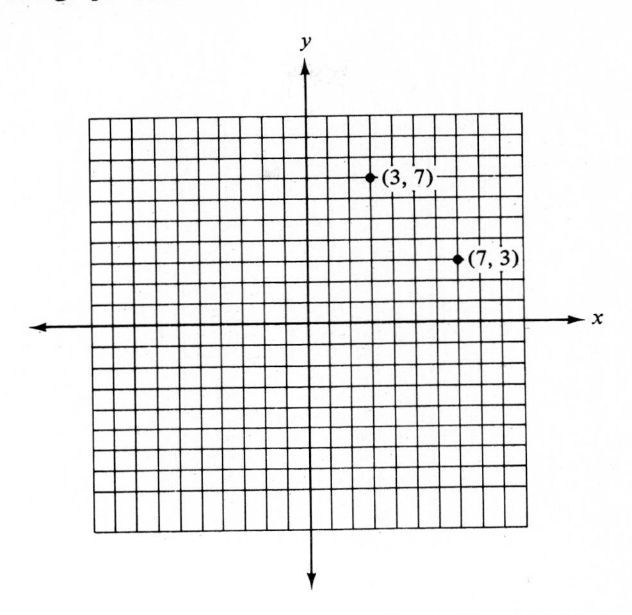

Since (7, 3) is different from (3, 7) be sure that answers for the following frames show the numbers in the same order as the answers in the text. The x-value is always first and the y-value second in an ordered pair (x, y).

161 Write the ordered pair (x, y) for point A. ________ . (3, 2)

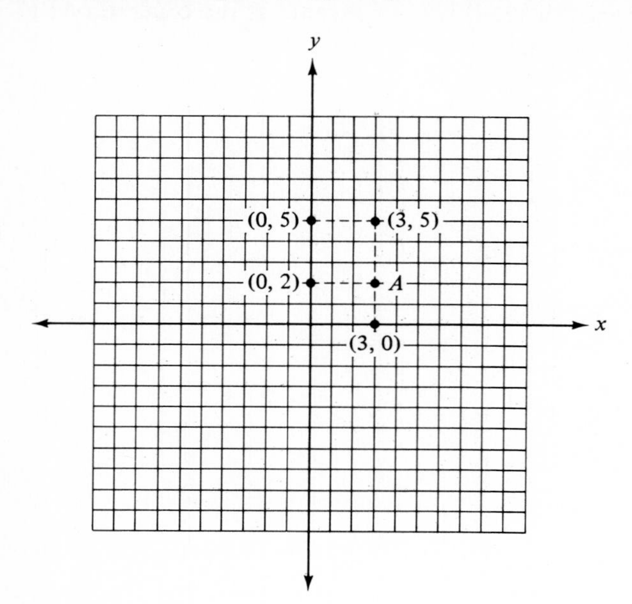

162 Write the ordered pair (x, y) for point B. ________ . (2, 3)

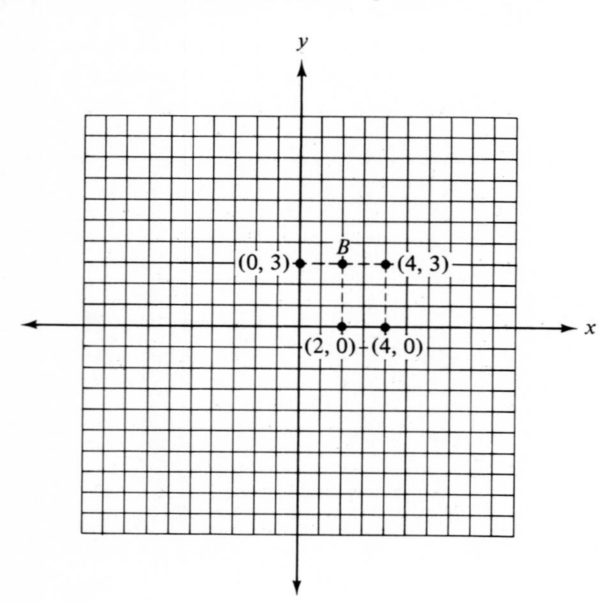

163 Write the ordered pair (x, y) for point C. ________ .

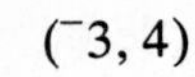
$(^{-}3, 4)$

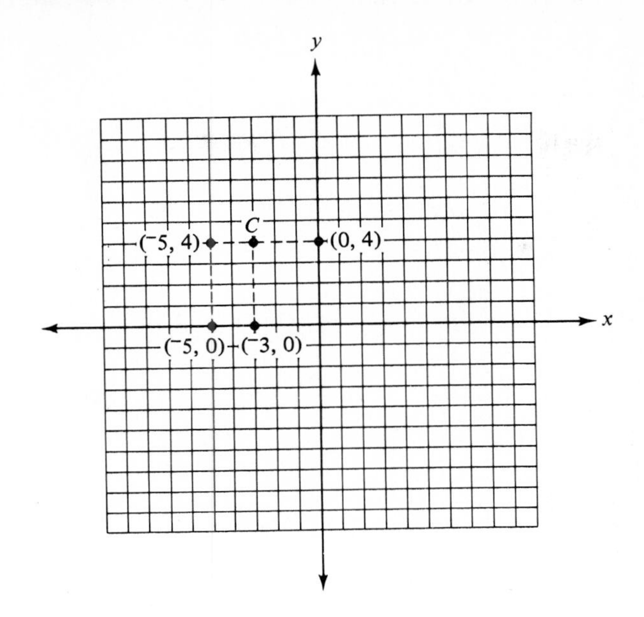

164 Write the ordered pair (x, y) for point D. ________ .

$(1, ^{-}1)$

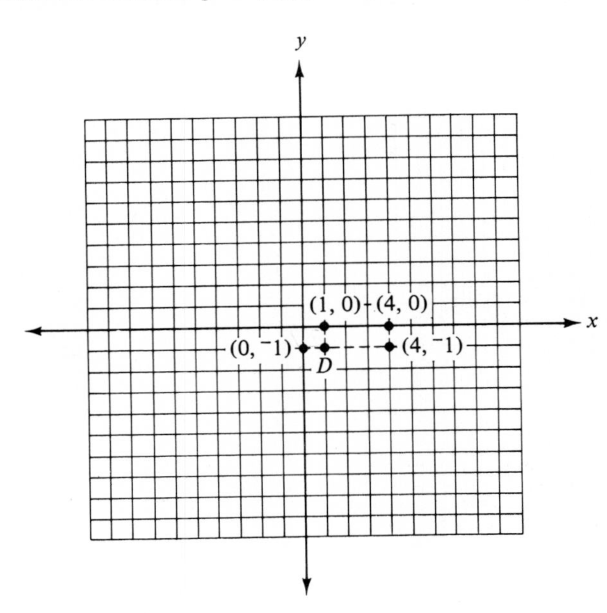

165 Write the ordered pair (x, y) for point P. ______ .

(¯2, ¯3)

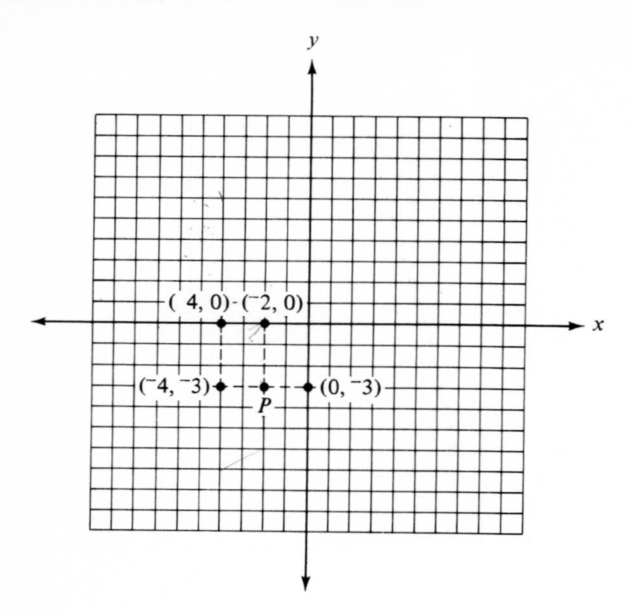

166 Write the ordered pair (x, y) for point A. ______ .

(6, ¯2)

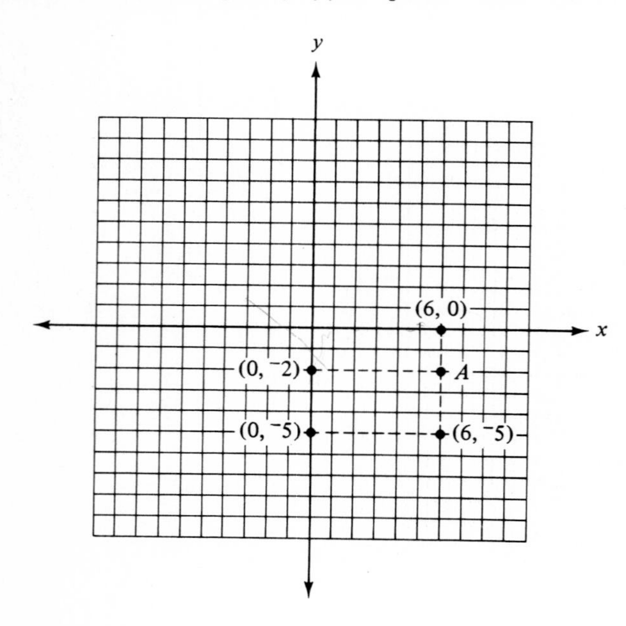

167 Write the ordered pair (x, y) for point D. ________ . $(^{-}3, 4)$

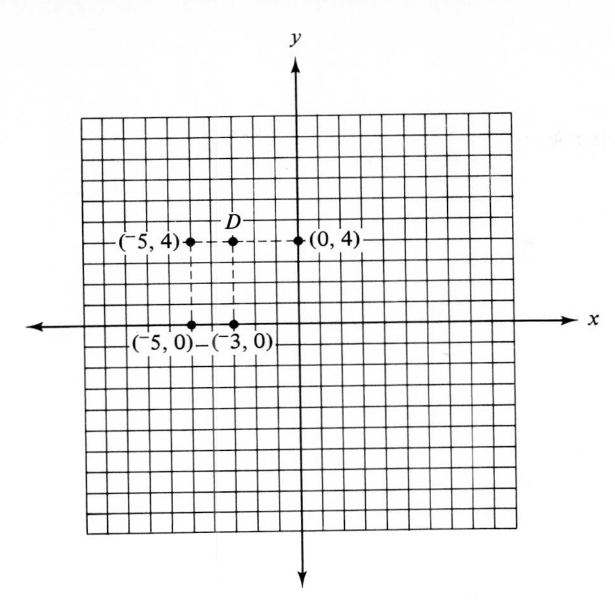

168 Mark and label the point to represent (4, 3).

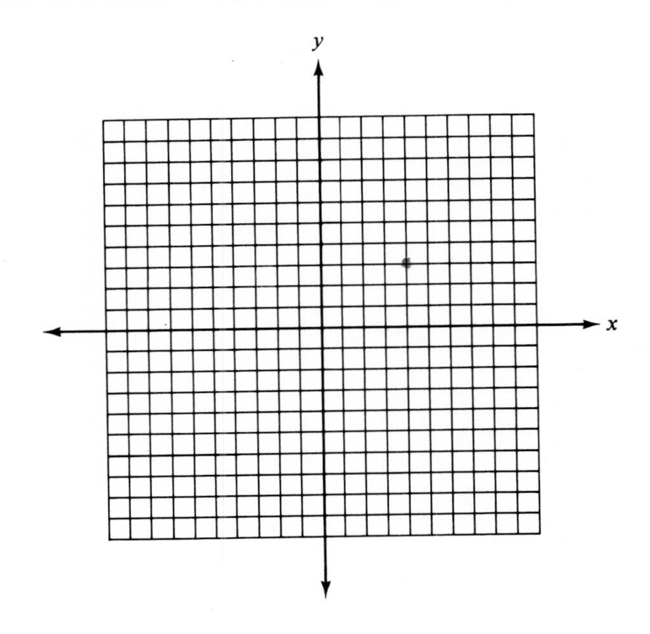

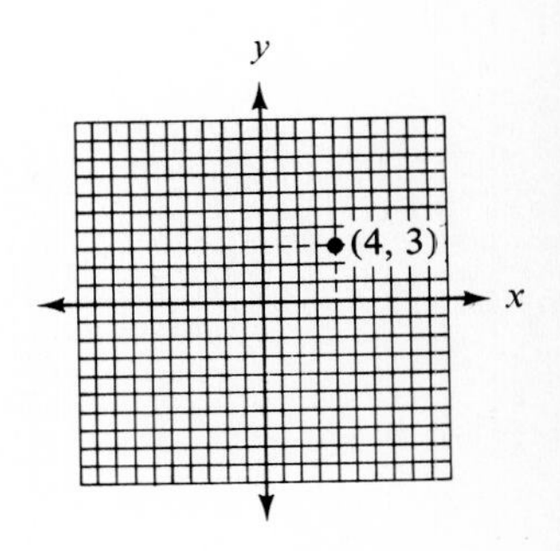

169 Mark and label the point to represent (5, ⁻4).

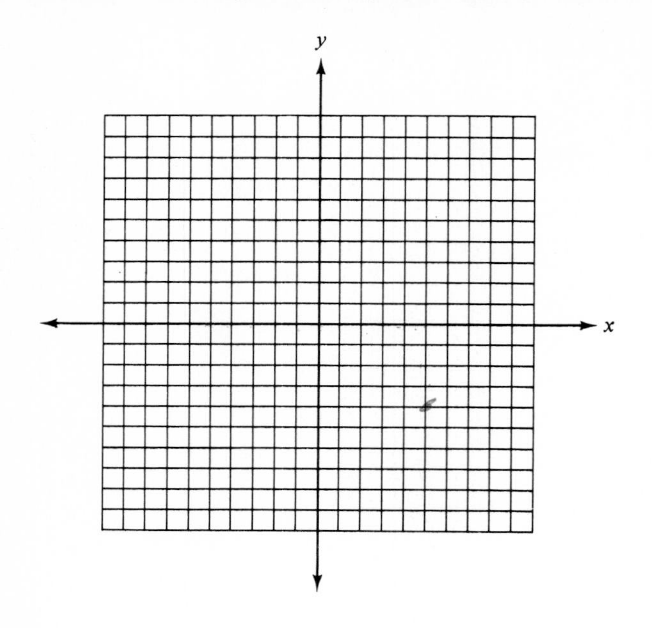

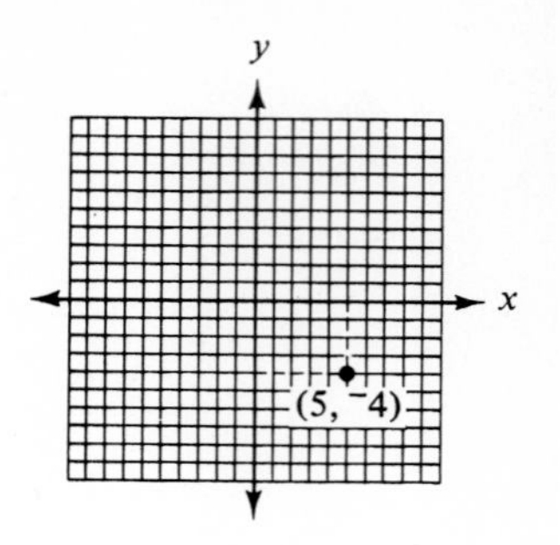

170 Mark and label the point to represent (⁻5, 2).

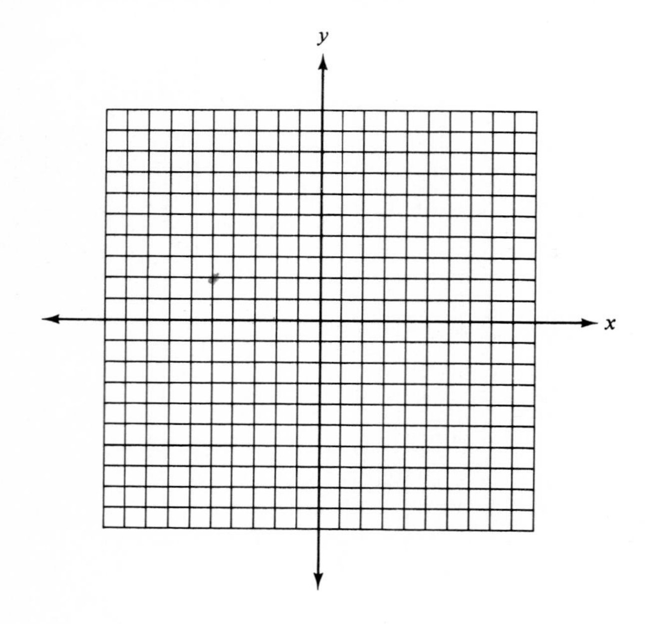

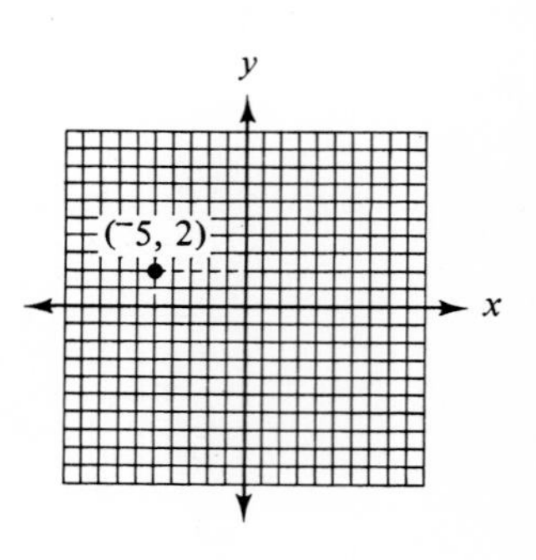

171 Mark and label the point to represent $(^{-}3, {}^{-}3)$.

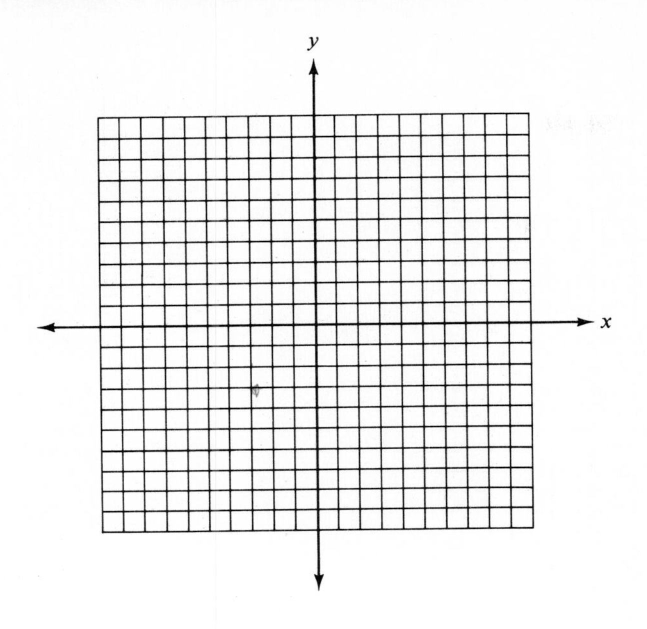

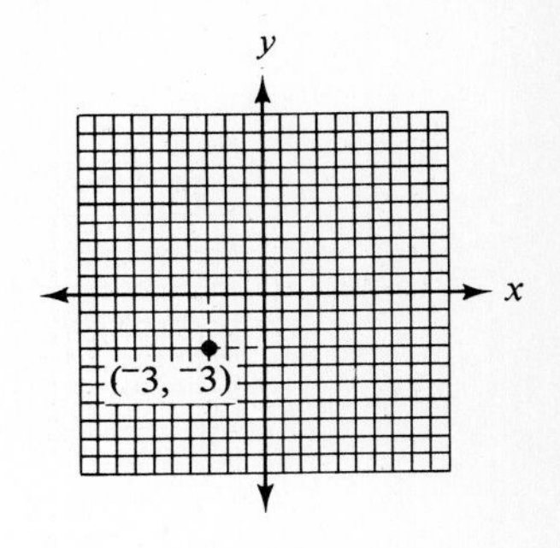

172 Mark and label the point to represent $(^{-}4, 3)$.

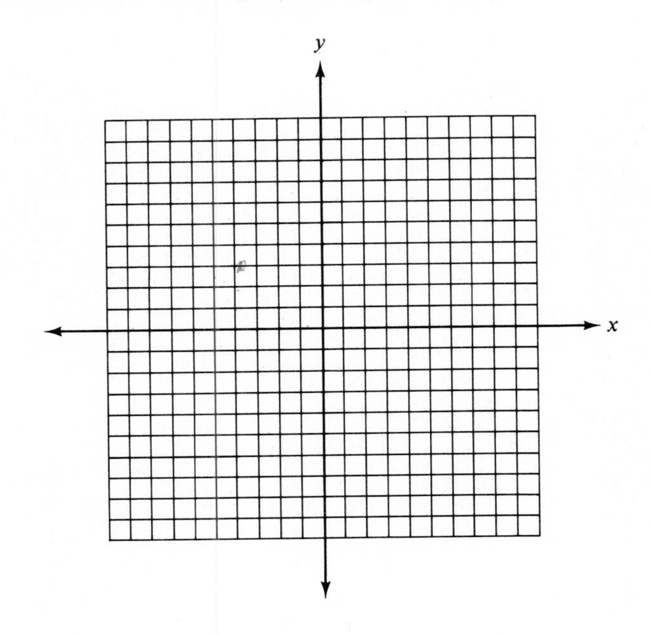

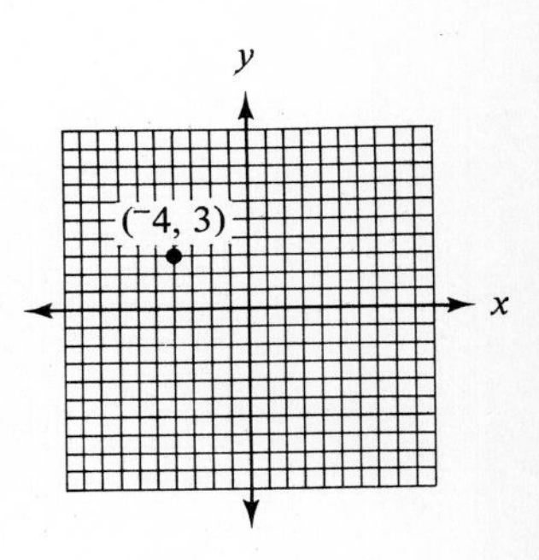

173 Mark and label the point to represent (7, 0).

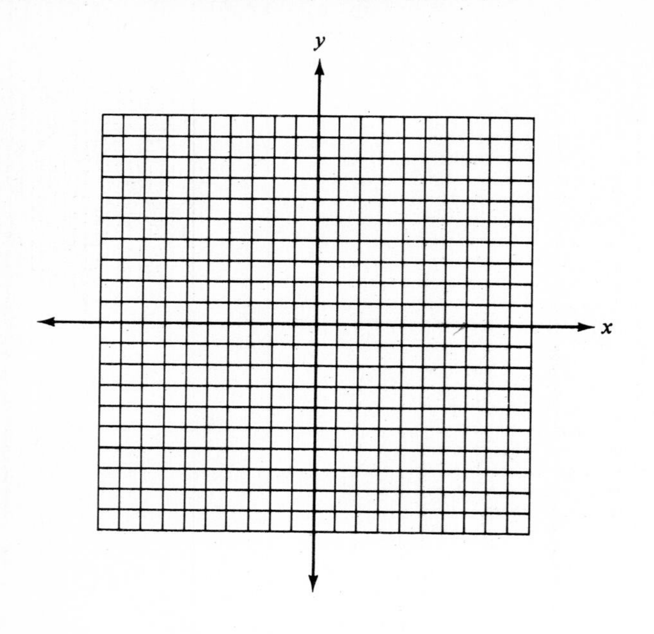

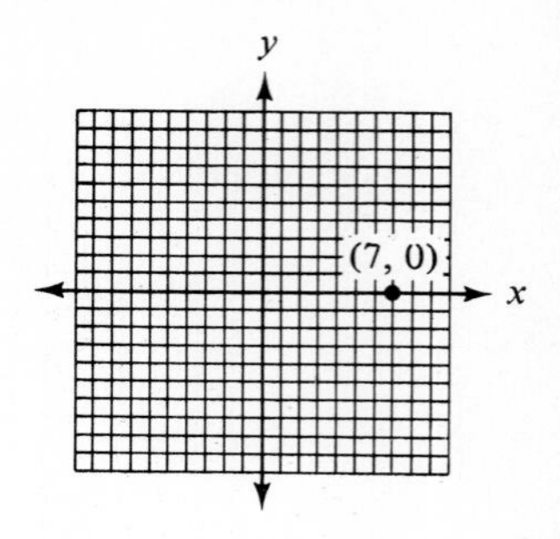

174 Mark and label the point to represent ($^-5$, 2).

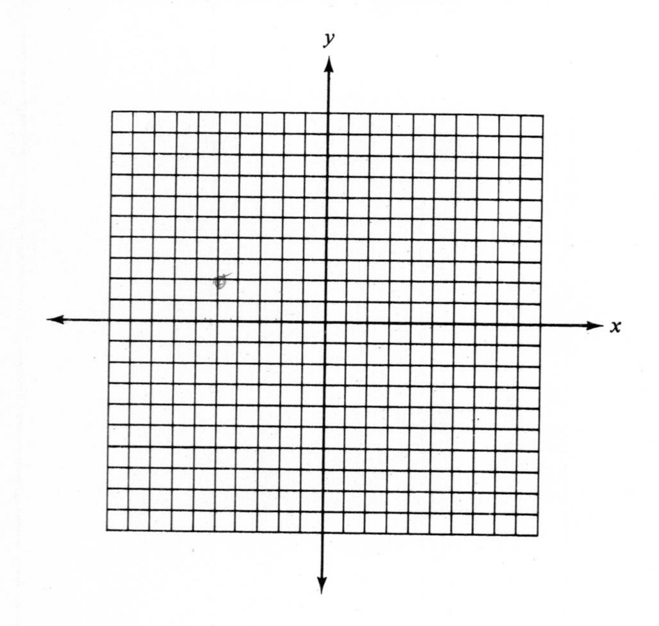

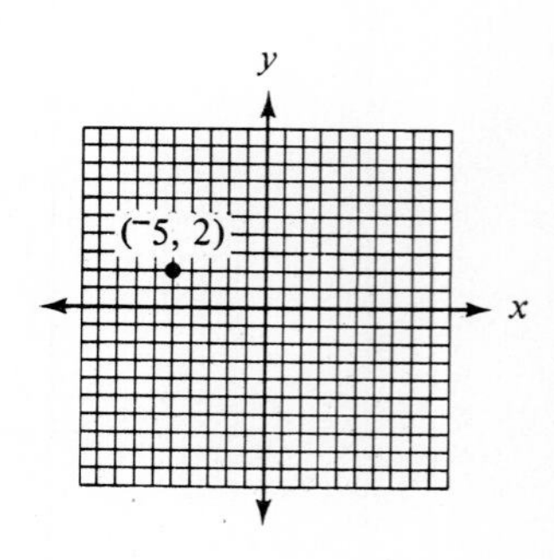

175 Mark and label the point to represent ($^-5, 0$).

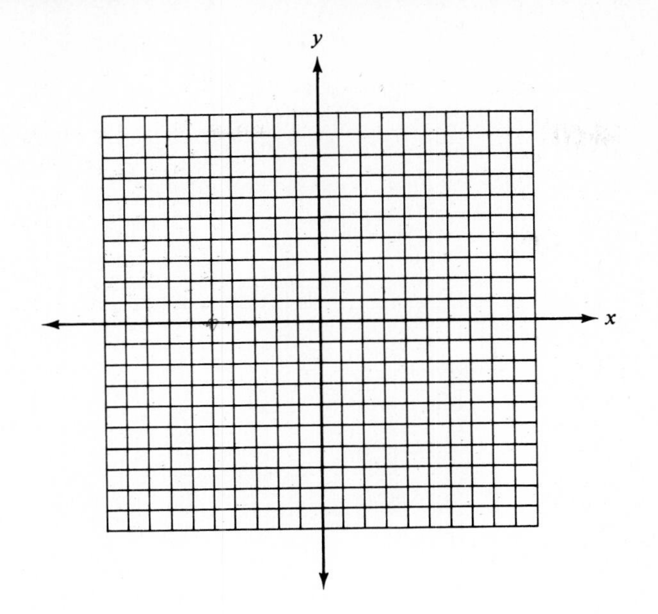

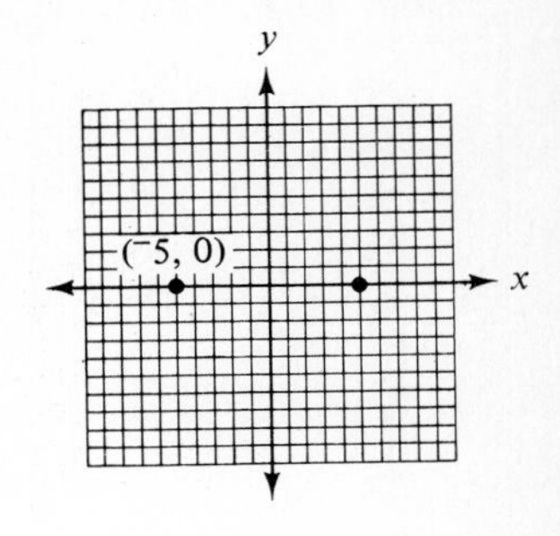

Self-Quiz # 3

The following questions test the objectives of the preceding section. 100% mastery is desired.

1. Write the ordered pairs for each point on the graph.

A ________ .
B ________ .
C ________ .
D ________ .
E ________ .
F ________ .

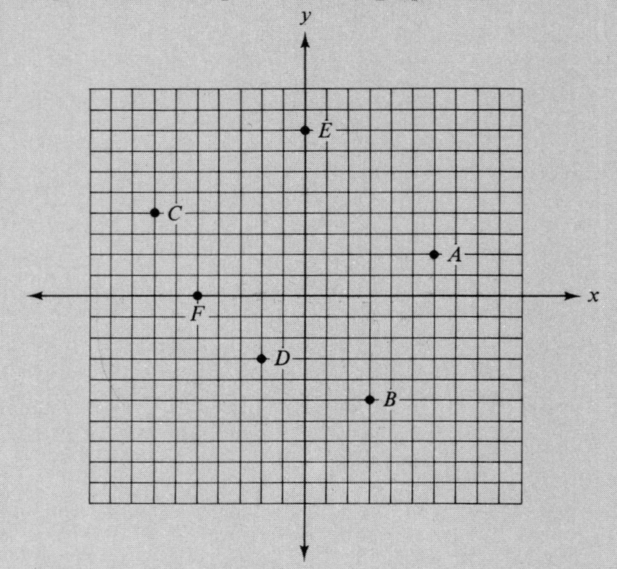

2. Mark the following points on the graph:

A (1, 5).
B ($^-3$, 4).
C ($^-2$, $^-5$).
D (4, 0).
E (3, $^-4$).
F (0, 0).

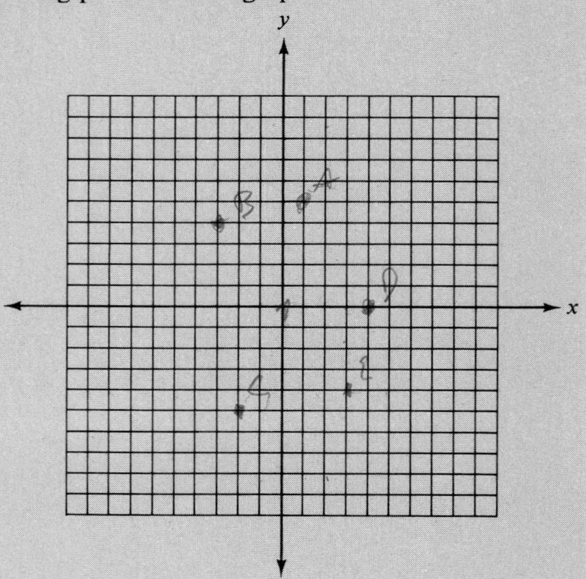

Every solution for the equation $2x + y = 8$ is an ordered pair (x, y). (3, 2) is a solution for $2x + y = 8$ because $2 \cdot 3 + 2 = 8$ is a true statement. (4, 0) is another solution for $2x + y = 8$ because $2 \cdot 4 + 0 = 8$ is a true statement.

In the following frames, solutions for equations such as $2x + y = 8$ will be graphed as a straight line representing *all* the solutions.

---

176 Are (1, $^-3$) and (4, 6) both solutions for $3x - y = 6$? ________ .

Yes

177 (1, $^-3$) and (4, 6) are both solutions for $3x - y = 6$. Is ($^-1$, $^-9$) also a solution for $3x - y = 6$? ________ .

Yes

178 Are (4, 1) and (2, 5) both solutions for $2x + y = 9$? ________ .

Yes

179 (4, 1) and (2, 5) are both solutions for $2x + y = 9$. Is (3, 3) also a solution for $2x + y = 9$? ________ .

Yes

180 (4, 1), (2, 5), and (3, 3) are all solutions for $2x + y = 9$. Mark the points to represent these three ordered pairs on the graph below.

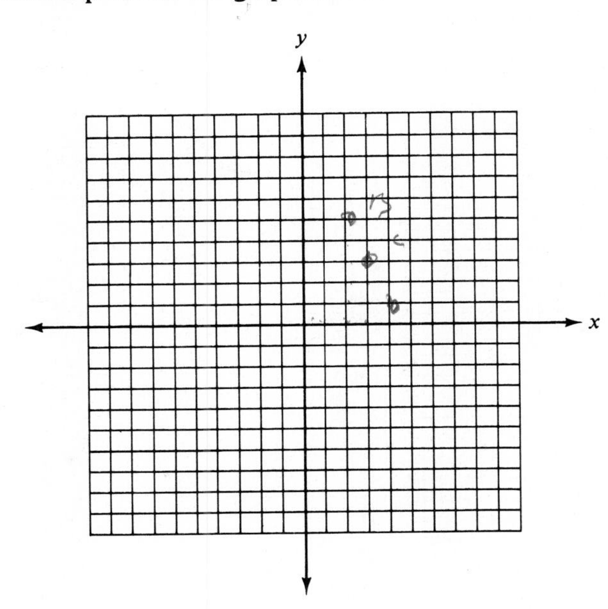

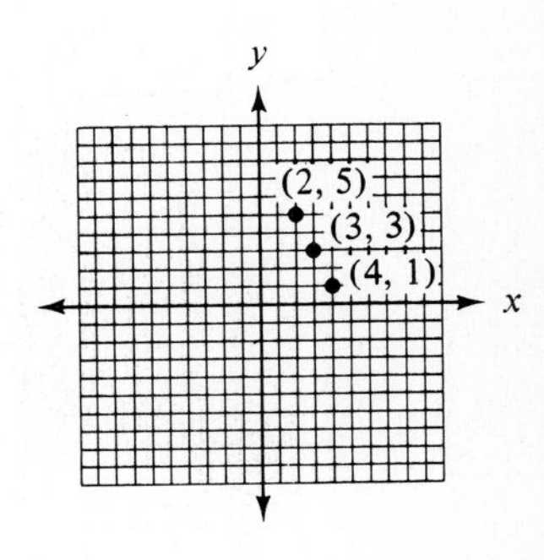

181 The three solutions (4, 1), (2, 5), and (3, 3) for $2x + y = 9$ are shown on the graph below. Can one straight line be drawn through all three points? ________ .

Yes

182 (0, 5), ($^{-}2$, 3), and ($^{-}5$, 0) are all solutions for $x - y = {}^{-}5$ and are shown on the following graph. Can one straight line be drawn through all three points? ________ .

Yes

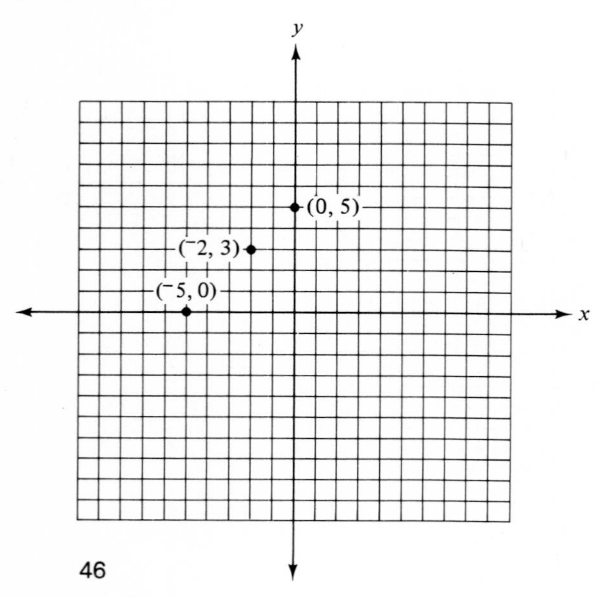

183 (1, 1), $(3, {}^-3)$, and (0, 3) are all solutions for $2x + y = 3$ and are shown on the following graph. Can one straight line be drawn through all three points? ________ .

Yes

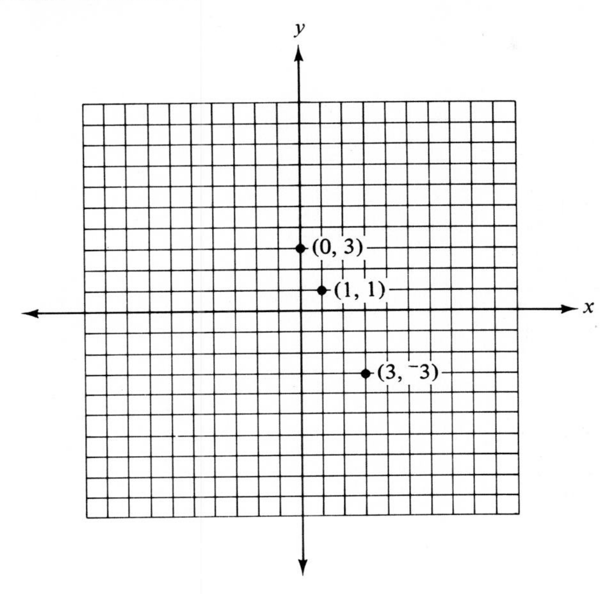

184 (2, 6), (8, 0), and $({}^-2, 10)$ are all solutions for $x + y = 8$ and are shown on the following graph. Can one straight line be drawn through all three points? ________ .

Yes

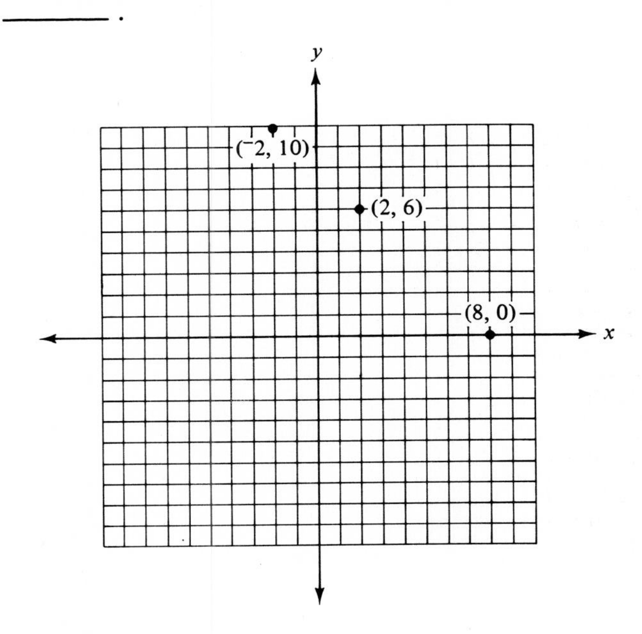

185 (0, 1), ($^-2$, $^-5$), and (2, 7) are all solutions for $3x - y = {}^-1$ and are shown on the following graph. Can one straight line be drawn through all three points? ________.

Yes

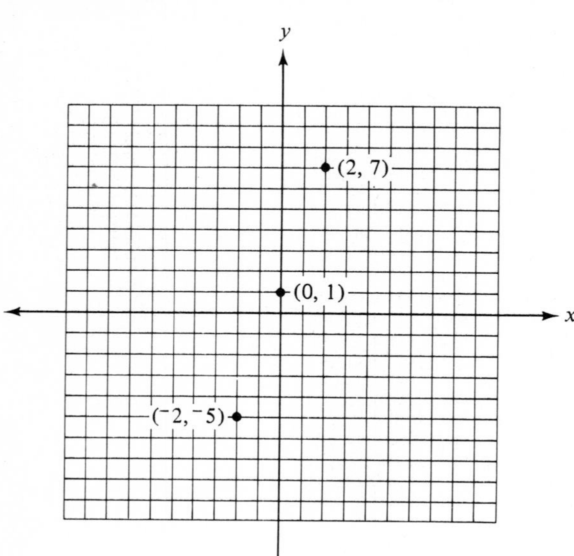

186 (6, 1), (2, $^-3$), and (8, 3) are all solutions for $x - y = 5$ and are shown on the following graph. Can one straight line be drawn through all three points? ________.

Yes

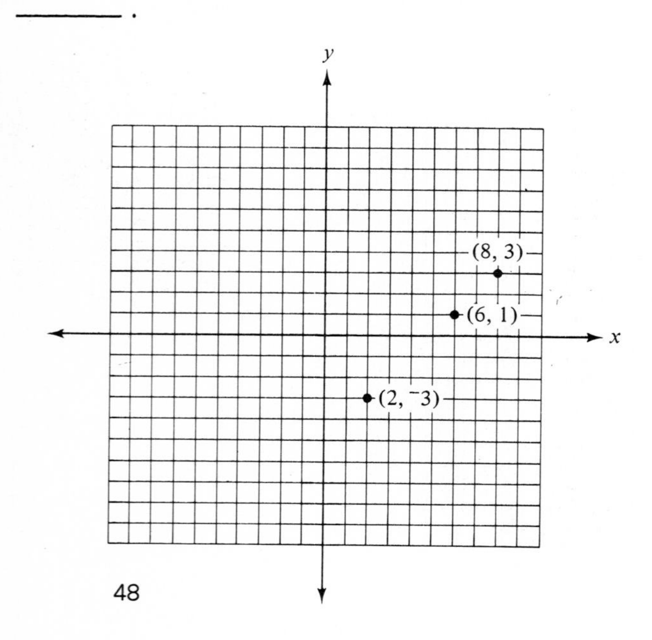

187 (5, 3) and (10, 8) are both solutions for $x - y = 2$. Locate the points on the following graph and draw a straight line through them. [*Note:* Extend the line so that it intersects both the x and the y axes.]

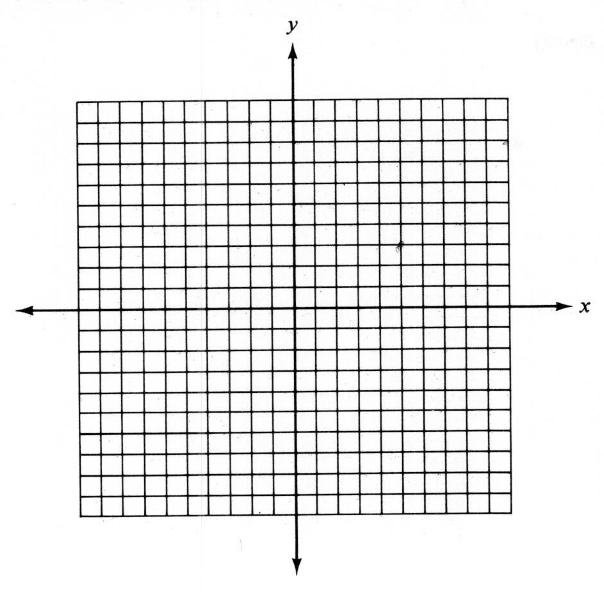

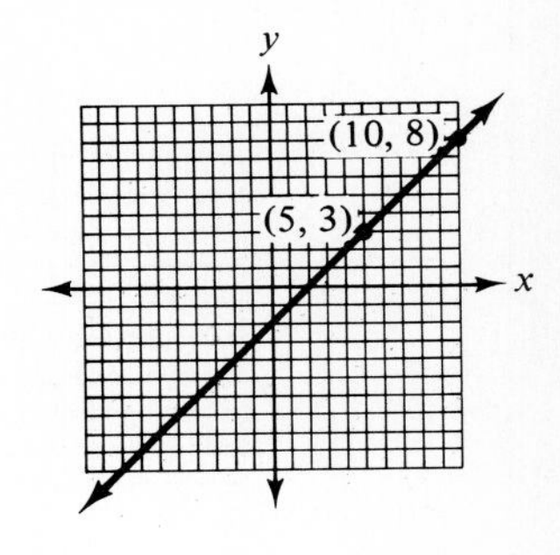

188 (0, 1) and (2, 5) are both solutions for $2x - y = {}^{-}1$. Locate the points on the following graph and draw a straight line through them.

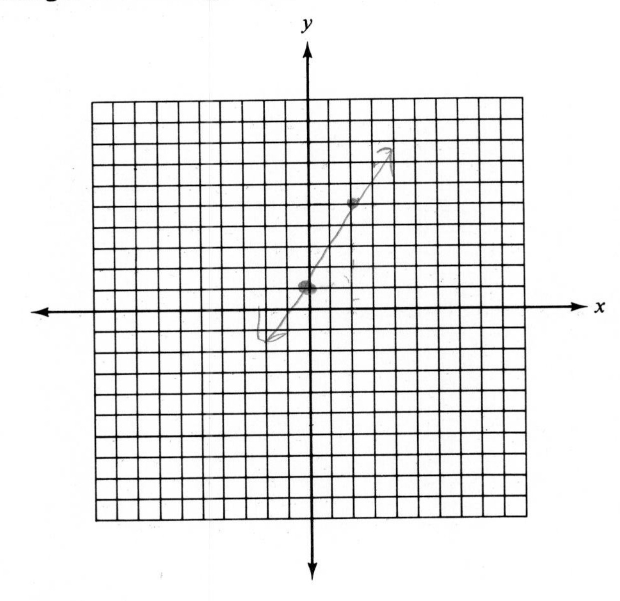

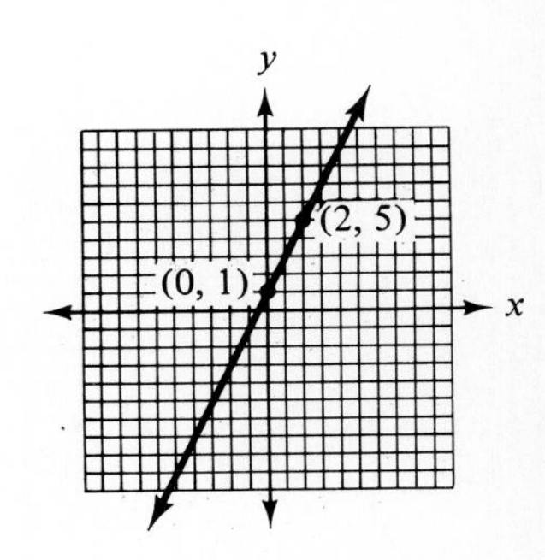

189 (4, 5) and (1, 8) are both solutions for $x + y = 9$. Locate the points on the following graph and draw a straight line through them.

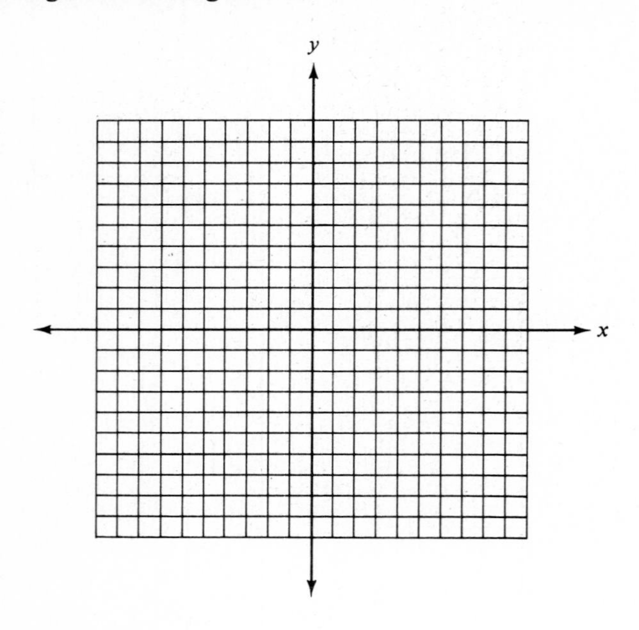

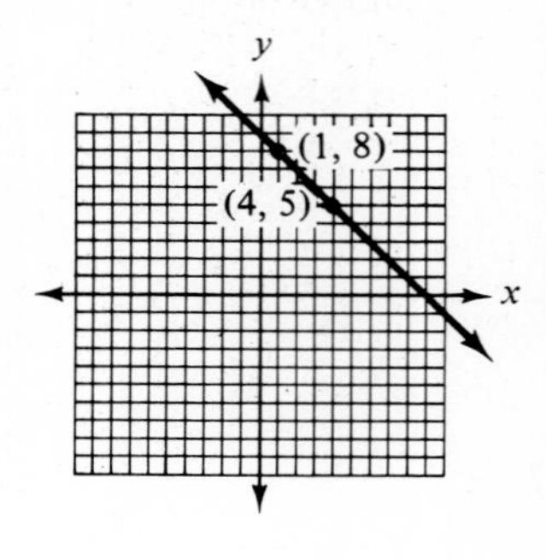

190 (1, 2) and (⁻2, 5) are both solutions for $x + y = 3$. Locate the points on the following graph and draw a straight line through them.

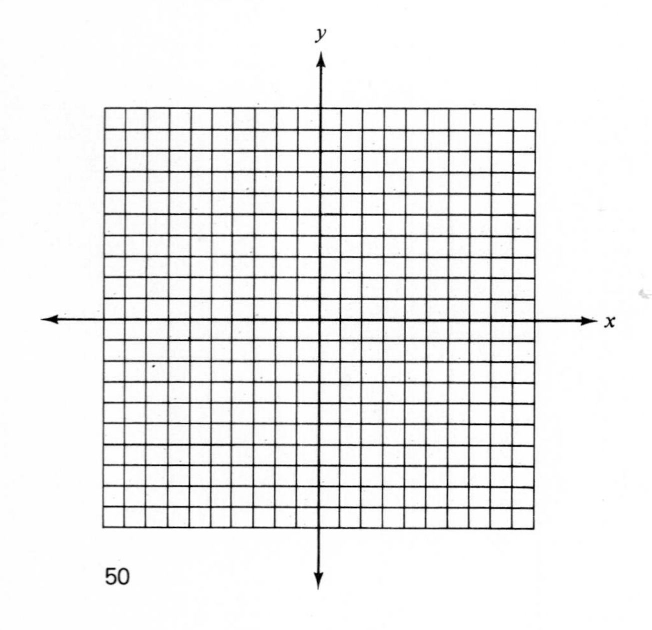

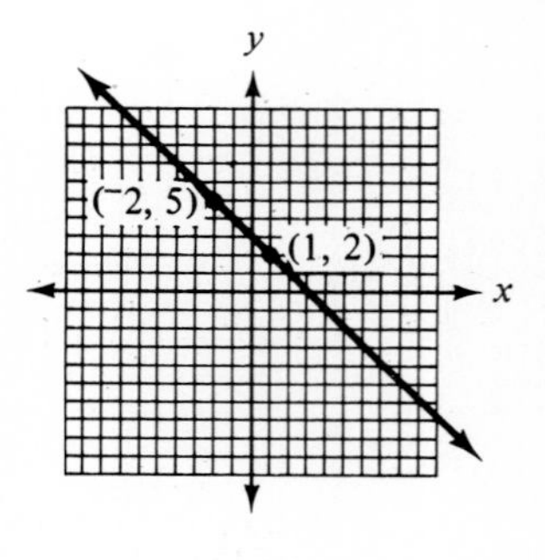

191 (5, 1) and (2, $^{-}2$) are both solutions for $x - y = 4$. Locate the points on the following graph and draw a straight line through them.

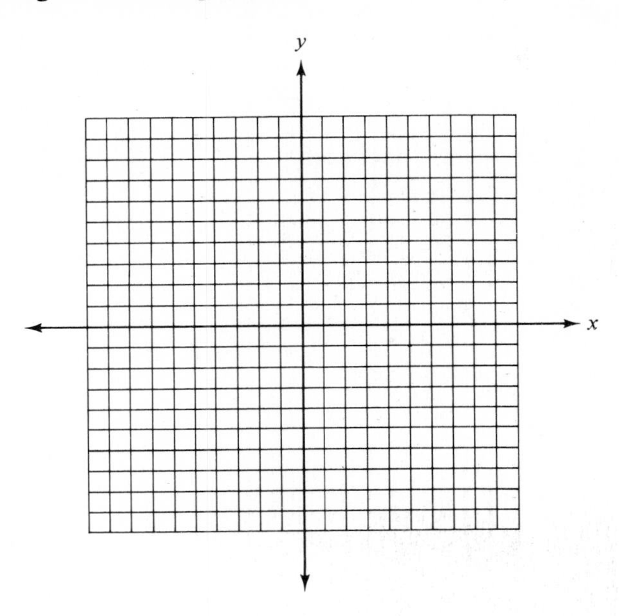

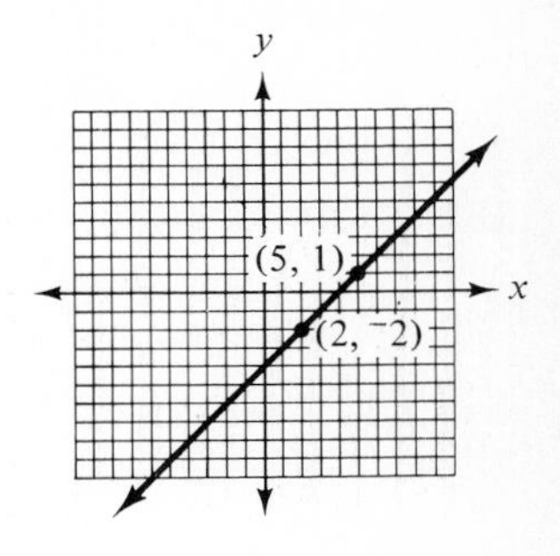

192 The line below was obtained by graphing ($^{-}8$, $^{-}6$) and (6, 8) and drawing the line through them. Label each of the points marked by a capital letter with its ordered pair (x, y).

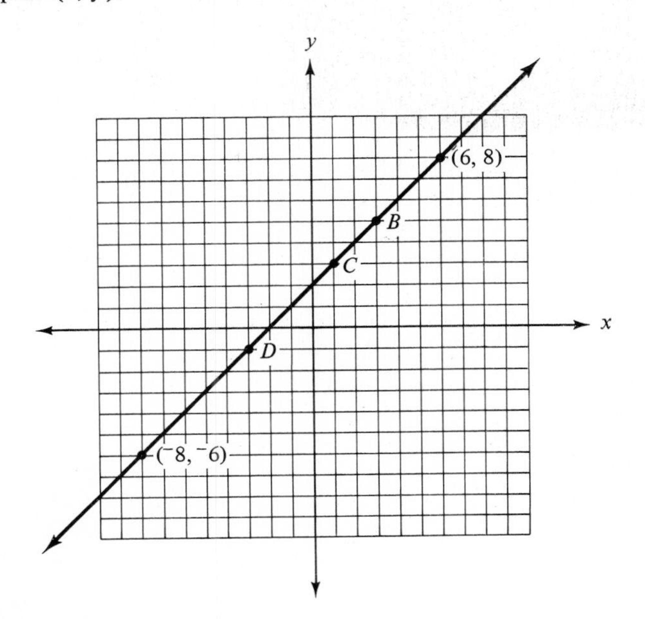

B is (3, 5),
C is (1, 3),
D is ($^{-}3$, $^{-}1$)

193 $(^-8, ^-6)$ and $(6, 8)$ are both solutions for $x - y = ^-2$. $(3, 5)$, $(1, 3)$, and $(^-3, ^-1)$ are the ordered pairs representing B, C, and D of the previous frame. Are $(3, 5)$, $(1, 3)$, and $(^-3, ^-1)$ solutions for $x - y = ^-2$?

________ .

Yes

194 The line below was obtained by graphing $(5, 7)$ and $(^-2, ^-7)$ and drawing the line through them. Label each of the points marked by a capital letter with its ordered pair (x, y).

A is $(4, 5)$,
B is $(1, ^-1)$,
C is $(0, ^-3)$

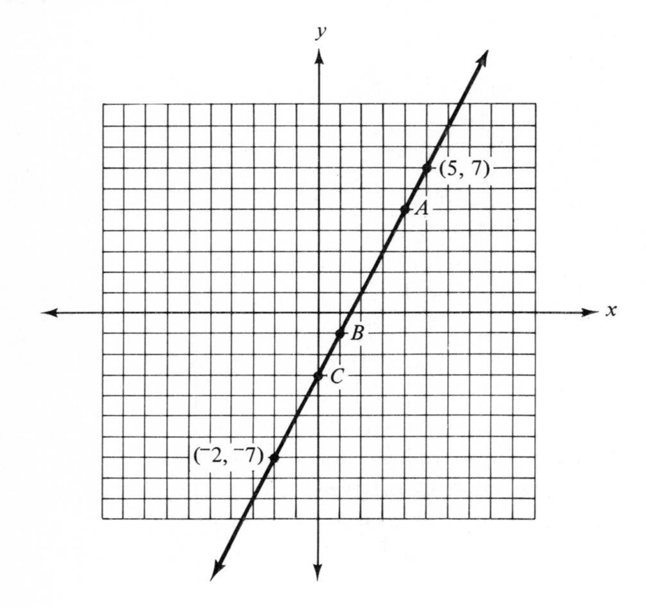

195 $(5, 7)$ and $(^-2, ^-7)$ are both solutions for $2x - y = 3$. $(4, 5)$, $(1, ^-1)$, and $(0, ^-3)$ are ordered pairs representing A, B, and C of the previous frame. Are $(4, 5)$, $(1, ^-1)$, and $(0, ^-3)$ solutions for $2x - y = 3$?

________ .

Yes

196 The line below was obtained by graphing (0, 4) and (8, 0) and drawing the line through them. Label each of the points marked by a capital letter with its ordered pair (x, y).

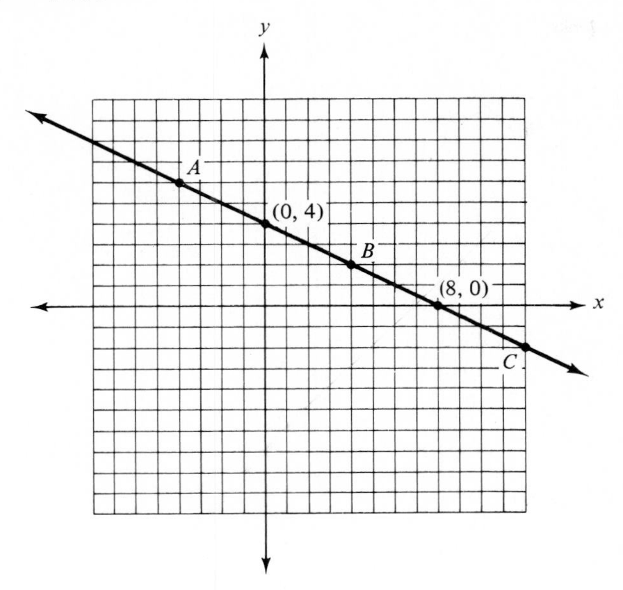

A is $(^-4, 6)$,
B is $(4, 2)$,
C is $(12, ^-2)$

197 (0, 4) and (8, 0) are both solutions for $x + 2y = 8$. $(^-4, 6)$, $(4, 2)$, and $(12, ^-2)$ are the ordered pairs representing A, B, and C of the previous frame. Are $(^-4, 6)$, $(4, 2)$, and $(12, ^-2)$ solutions for $x + 2y = 8$? ________.

Yes

198 (0, 3) and (2, 5) are solutions for $x - y = {}^{-}3$ and determine the following line. $({}^{-}3, 0)$ and $({}^{-}5, {}^{-}2)$ are also solutions for $x - y = {}^{-}3$. Plot $({}^{-}3, 0)$ and $({}^{-}5, {}^{-}2)$ on the graph.

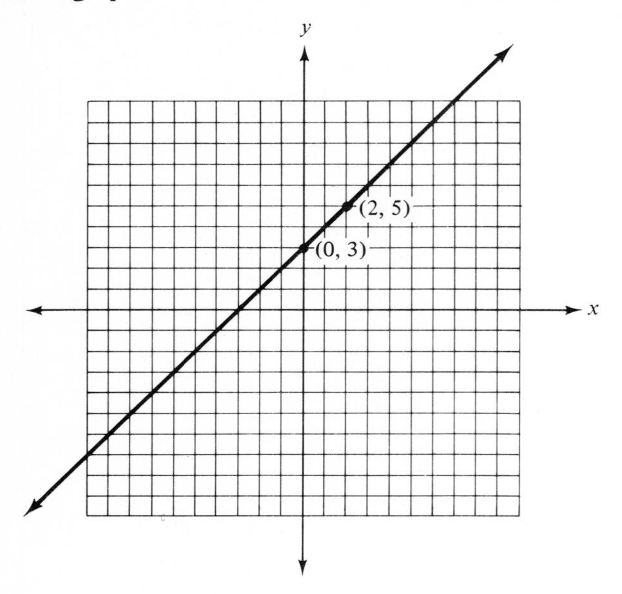

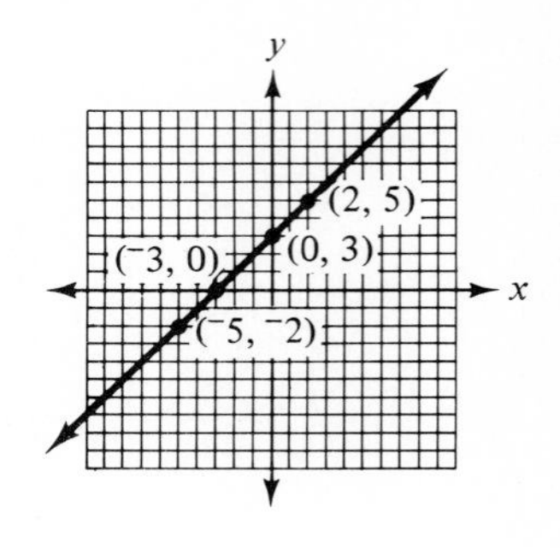

199 (2, 3) and $(0, {}^{-}1)$ are solutions for $y = 2x - 1$ and determine the following line. (1, 1) and (4, 7) are also solutions for $y = 2x - 1$. Plot (1, 1) and (4, 7) on the graph.

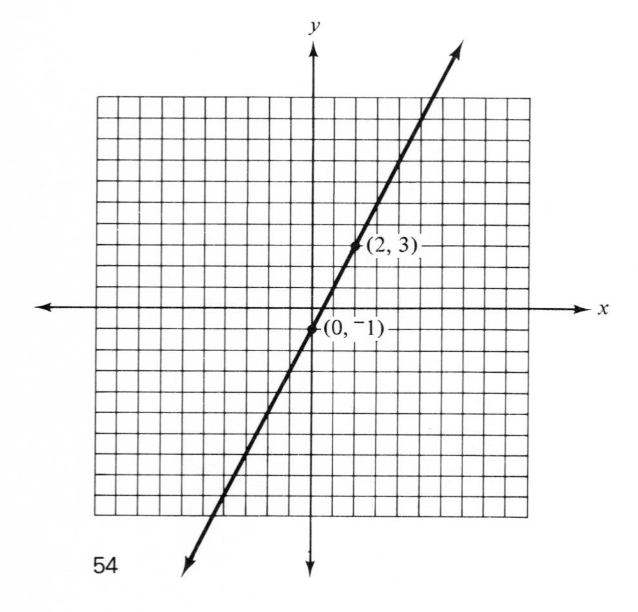

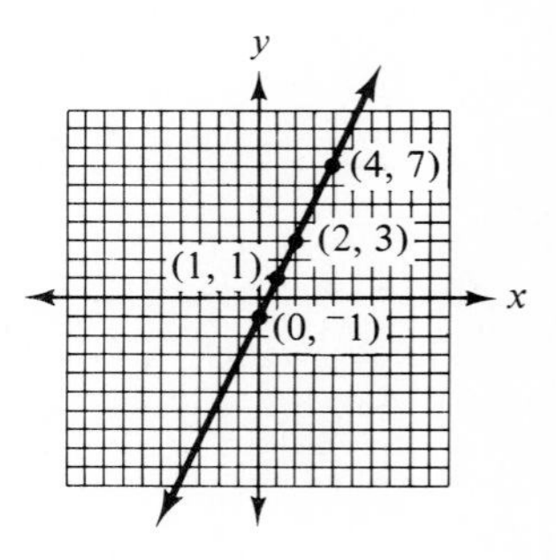

The line shown below was obtained by finding two ordered-pair solutions for $2x + y = 3$: $(0, 3)$ and $(2, {}^{-}1)$. This line is called the graph of the equation $2x + y = 3$ because every solution of the equation is on the line and every point of the line represents an ordered-pair solution of the equation.

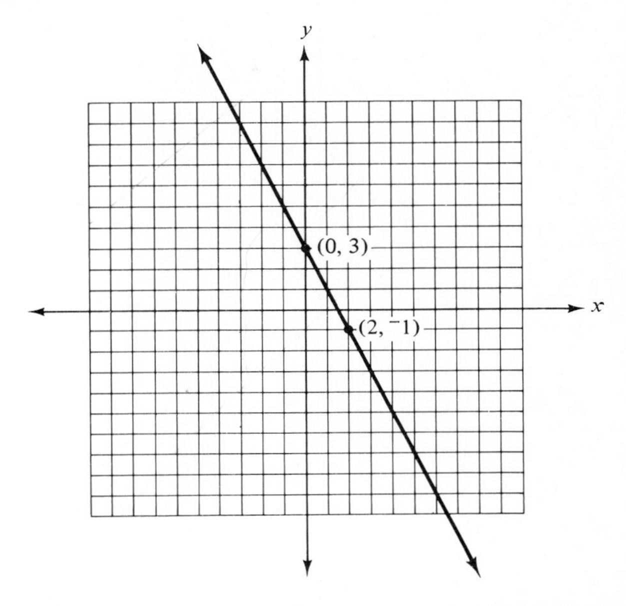

200 Every solution of the equation $2x + y = 3$ is on the line of the graph of the equation. $(3, {}^{-}3)$ is a solution for $2x + y = 3$. Is $(3, {}^{-}3)$ on the line of the equation $2x + y = 3$? ________ .

Yes

201 Every solution of the equation $x + y = 5$ is on the line of the graph of the equation. $(2, 3)$ is a solution for $x + y = 5$. Is $(2, 3)$ on the line of the equation $x + y = 5$? ________ .

Yes

202 Every solution of $x + y = 7$ is on the line of $x + y = 7$ shown below. Which of the points (5, 2) or ($^-3$, 2) is *not* a solution for $x + y = 7$? ________ .

($^-3$, 2)

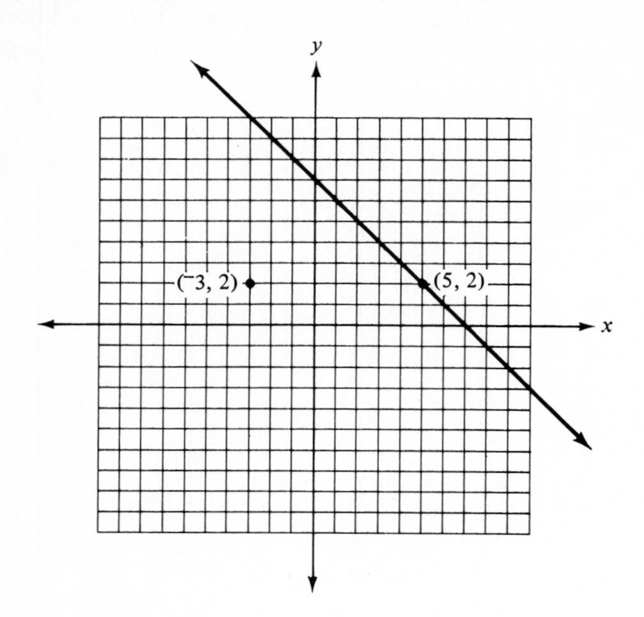

203 ($^-3$, 2) is not a solution for the equation $x + y = 7$ because $^-3 + 2 = 7$ is a false statement. Is (8, 1) a solution for $x + y = 7$? ________ .

No

204 Every solution of $2x + y = 5$ is on the line of $2x + y = 5$ shown below. Which of the points (2, 1) or (1, 8) is *not* a solution for $2x + y = 5$ because it does not lie on the line? ________ .

(1, 8)

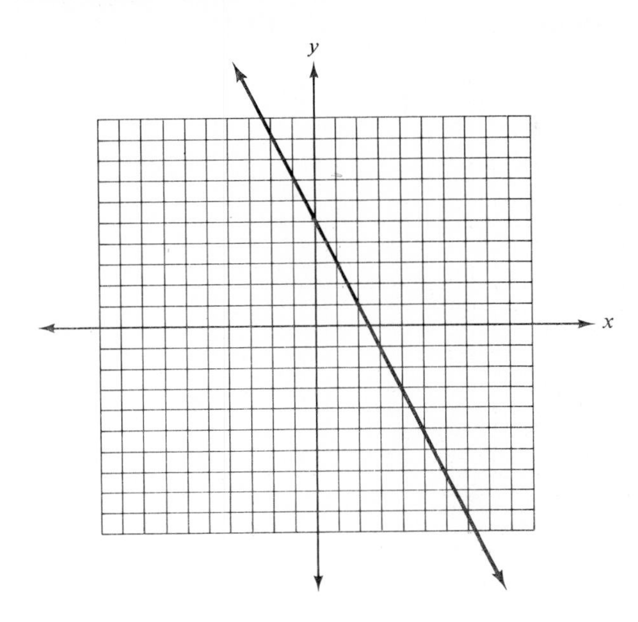

205 Which of the points (8, 2), (1, 5), (⁻1, ⁻5), or (0, 0) is *not* a solution for $y = 5x$ (shown below) because it does not lie on the line? ________ .

(8, 2)

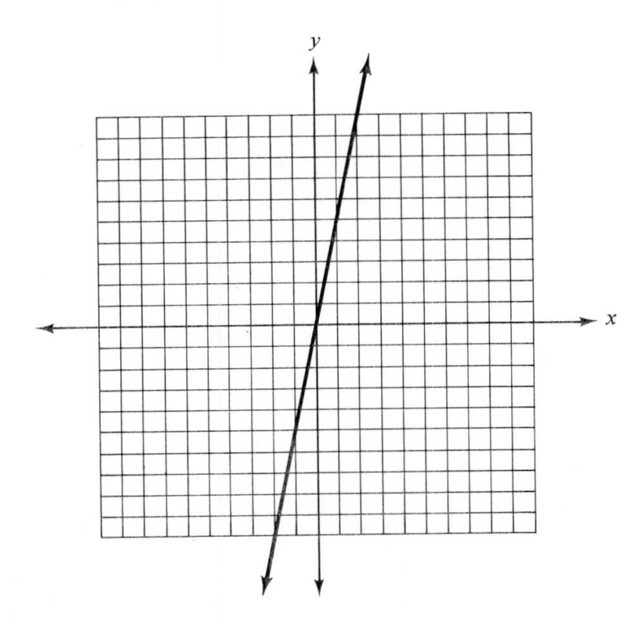

206 The graph of the equation $3x - y = {}^{-}2$ is shown below. Which of the ordered pairs (1, 5), (2, 2), or (⁻2, ⁻4) is *not* a solution for $3x - y = {}^{-}2$ because it is *not* on the line? ________ .

(2, 2)

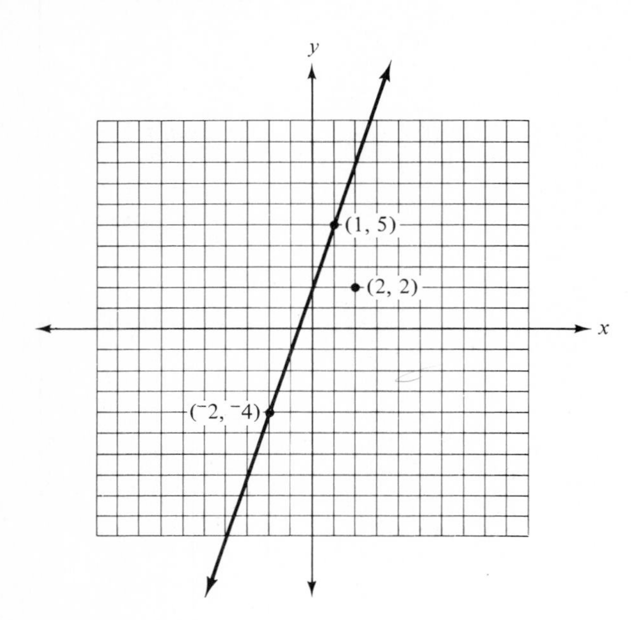

207 The graph below shows the line of $2x - y = 6$. Is every ordered-pair solution of $2x - y = 6$ represented by a point on the line? ________ .

Yes

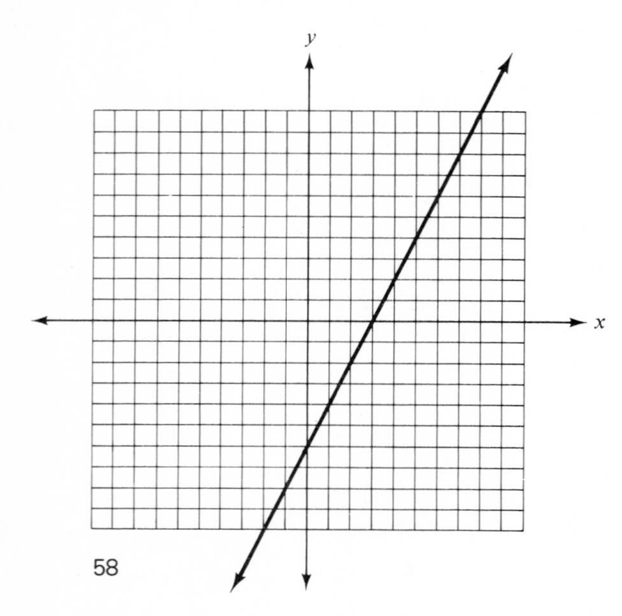

Self-Quiz # 4

The following questions test the objectives of the preceding section. 100% mastery is desired.

1. The line below was obtained by graphing $(^-4, ^-9)$ and $(5, 9)$ and drawing the line through them. Label each point marked with a capital letter with its ordered pair (x, y).

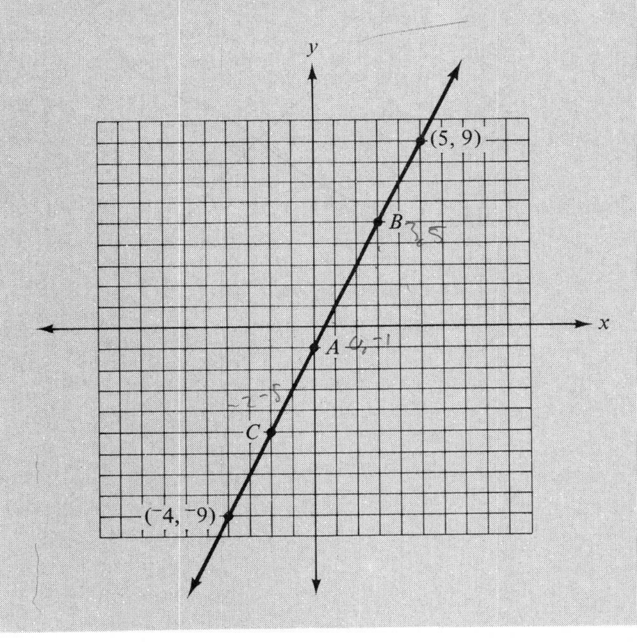

2. The graph of the equation $2x - 3y = 6$ is shown below. Which of the ordered pairs is *not* a solution for $2x - 3y = 6$: $(0, ^-2)$, $(^-2, 3)$, or $(3, 0)$? ______ .

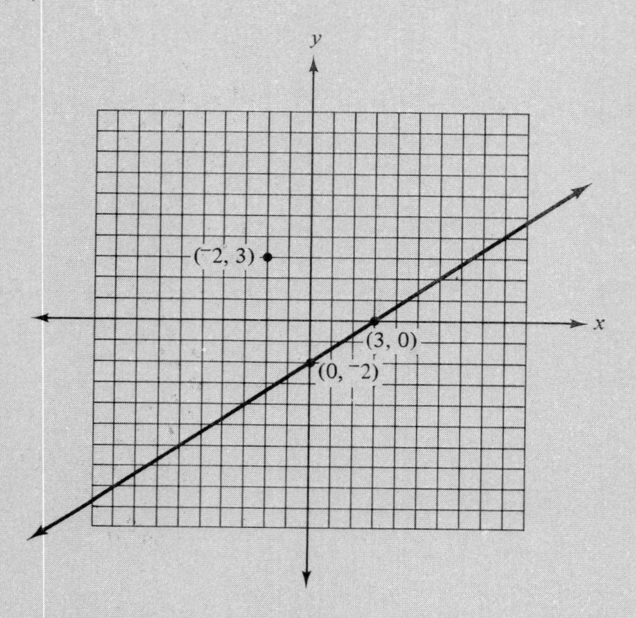

3. Is $(2, ^-5)$ a solution for the equation $5x + 2y = {}^-10$? ________ .

4. Is $(3, ^-2)$ a solution for the equation $x - 3y = 9$? ________ .

5. Is $(7, ^-2)$ a solution for the equation $x - y = 5$? ________ .

Any two ordered-pair solutions of an equation that are plotted on a graph will determine the line that represents all solutions of the equation. For example, the line passing through $(1, 2)$ and $(5, 6)$ will show all solutions for $y = x + 1$.

---

208 To graph the line of $3x + y = 8$, the first step is to find two ordered-pair solutions of the equation. Is $(1, 5)$ a solution for $3x + y = 8$? ________________ .

Yes, $3 \cdot 1 + 5 = 8$

209 Replace x by 2 and complete the ordered-pair solution (2, ____) for $2x + y = 7$. (2, 3)

210 Replace x by 0 in $2x + y = 7$ and complete the ordered pair (0, ____) as a solution for the equation. (0, 7)

The three steps in graphing an equation such as $4x + y = 6$ are:

1. Find two ordered-pair solutions for $4x + y = 6$.
2. Plot the solutions on the graph.
3. Draw the line through the two points.

The following steps are used to graph $4x + y = 6$:

1. Find the solutions (0, __) and (__, 0).
2. Plot (0, 6) and $\left(\frac{3}{2}, 0\right)$ on the graph.
3. Draw the line through the points (0, 6) and $\left(\frac{3}{2}, 0\right)$.

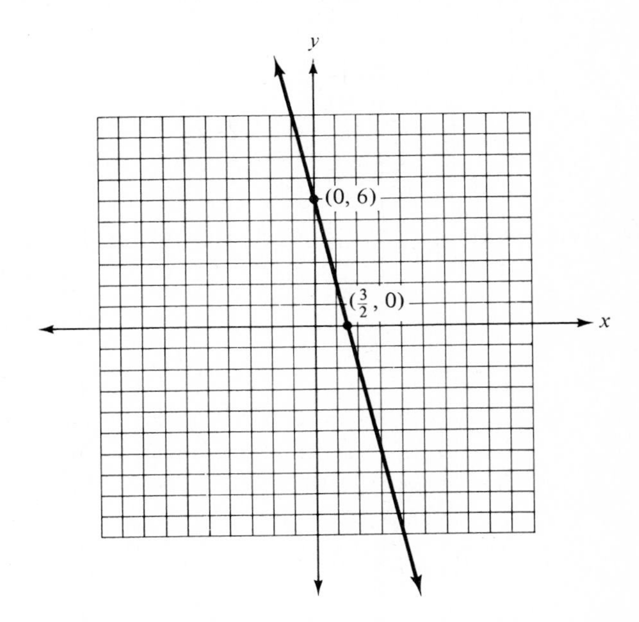

---

211 The first step in graphing $2x + y = 5$ is to find two ordered-pair solutions of the equation. The pair (0, __) is completed by replacing x by zero and solving the equation for y.

$$2x + y = 5$$
$$2 \cdot 0 + y = 5$$
$$y = 5$$

(0, ________) is a solution for $2x + y = 5$.

(0, 5)

212 (0, 5) is a solution for $2x + y = 5$. To find a second solution for $2x + y = 5$, replace y by zero.

$$2x + y = 5$$
$$2x + 0 = 5$$
$$2x = 5$$
$$x = \frac{5}{2}$$

(________, 0) is a solution for $2x + y = 5$.

$\left(\frac{5}{2}, 0\right)$

213 $(0, 5)$ and $\left(\frac{5}{2}, 0\right)$ are two solutions for the equation $2x + y = 5$. Graph $(0, 5)$ and $\left(\frac{5}{2}, 0\right)$ on the axes below. $\left[\textit{Note: } \frac{5}{2} = 2\frac{1}{2}.\right]$

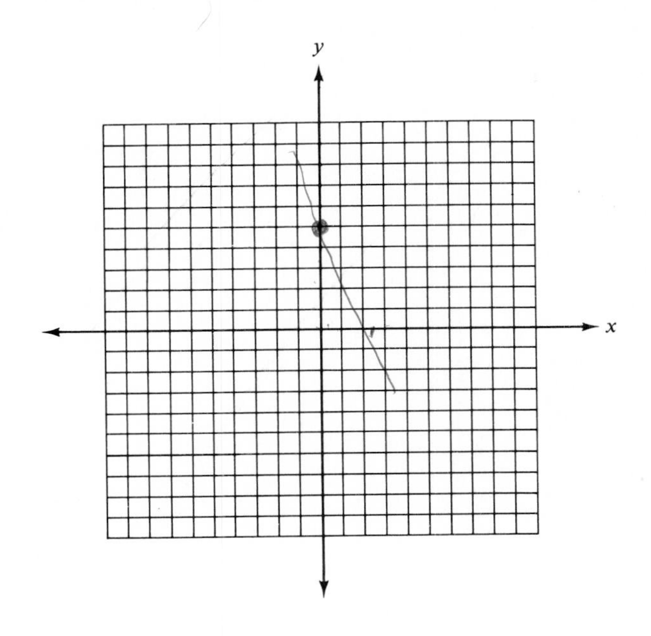

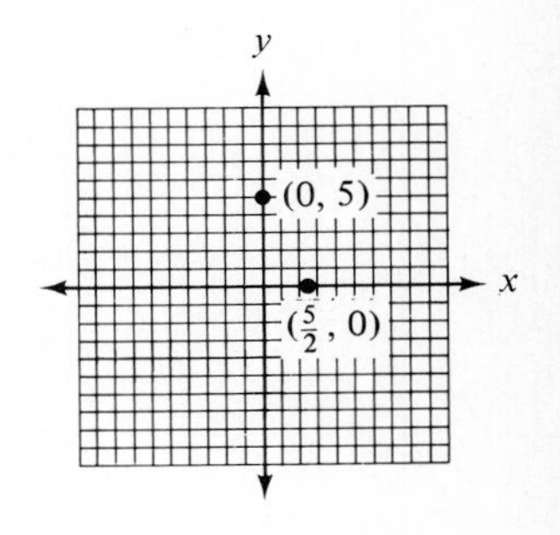

214 The three steps in graphing $2x + y = 5$ are:

(a) Find the solutions (0, ________) and (________, 0).

(b) Locate the solutions (0, 5) and $\left(\frac{5}{2}, 0\right)$ on the graph.

(c) Draw the line through the two points.

Complete the third step in graphing $2x + y = 5$.

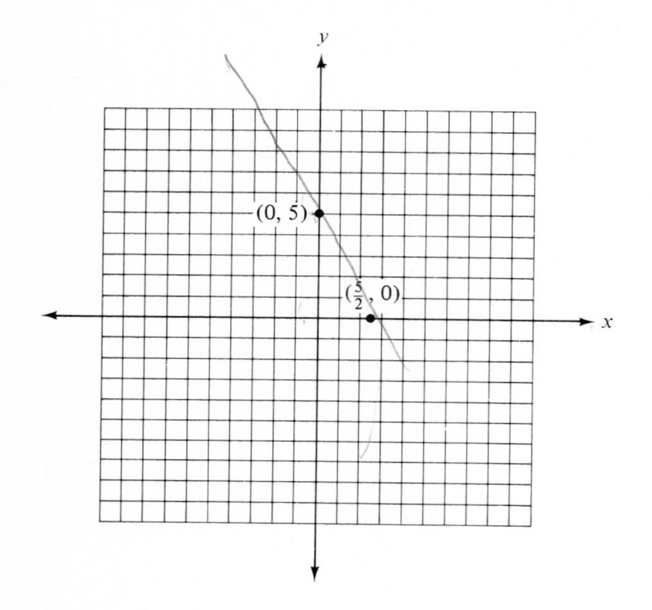

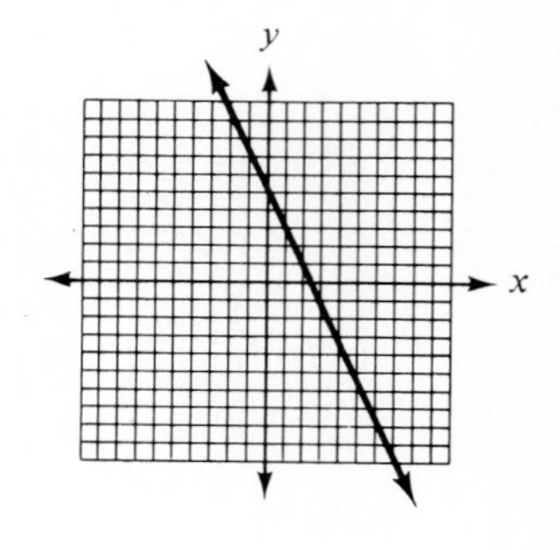

215 The first step in graphing $3x - 2y = 8$ is to find two solutions for the equation. To complete (0, ________), replace x by zero in the equation $3x - 2y = 8$.

$$3x - 2y = 8$$
$$3 \cdot 0 - 2y = 8$$
$$0 - 2y = 8$$
$$^{-}2y = 8$$
$$y = {}^{-}4$$

(0, ________) is a solution of $3x - 2y = 8$.

(0, $^{-}4$)

216 $(0, {}^{-}4)$ is one solution of $3x - 2y = 8$. To find another solution, replace y by zero.

$$3x - 2y = 8$$

$$3x - 2 \cdot 0 = 8$$

$$3x - 0 = 8$$

$$3x = 8$$

$$x = \frac{8}{3}$$

(________, 0) is a solution of $3x - 2y = 8$.

$\left(\frac{8}{3}, 0\right)$

217 $(0, {}^{-}4)$ and $\left(\frac{8}{3}, 0\right)$ are two solutions of $3x - 2y = 8$.

Graph them on the axes below. $\left[\textit{Note: } \frac{8}{3} = 2\frac{2}{3}.\right]$

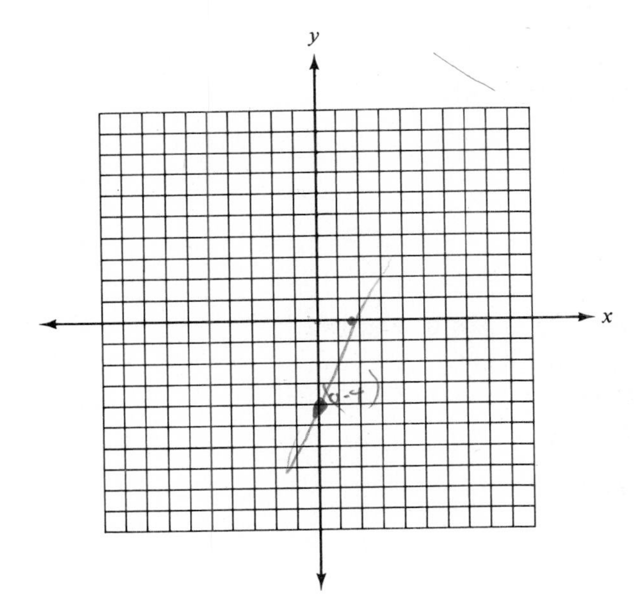

y

$(2\frac{2}{3}, 0)$

x

$(0, {}^{-}4)$

218 To graph $3x - 2y = 8$,

(a) Find two solutions (0, ________) and (________, 0).

(b) Graph $(0, {}^-4)$ and $\left(\frac{8}{3}, 0\right)$.

(c) Draw the line through the two points.

Complete the third step in graphing $3x - 2y = 8$.

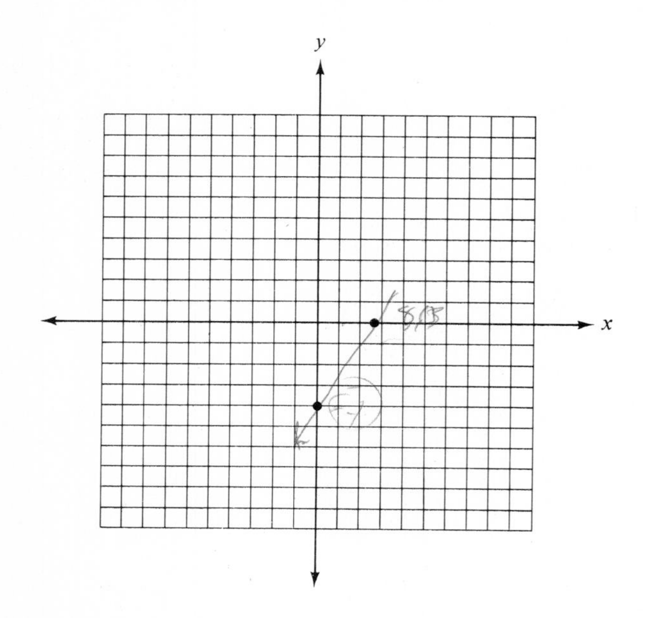

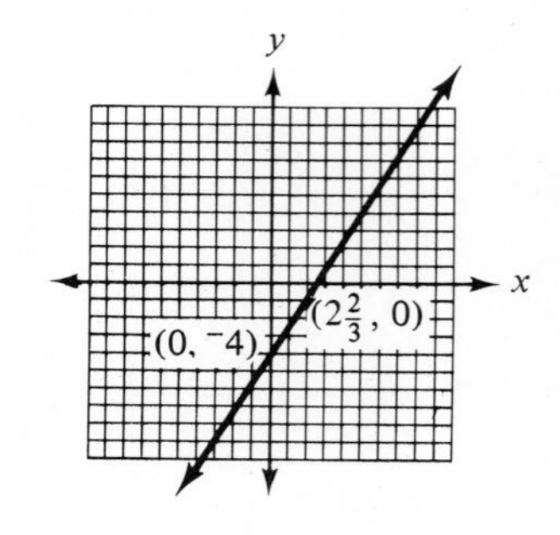

219 The first step in graphing $5x - y = 3$ is to find two solutions of the equation. (0, ________) and (________, 0) are solutions of $5x - y = 3$.

$(0, {}^-3)$, $\left(\frac{3}{5}, 0\right)$

220 $(0, {}^{-}3)$ and $\left(\frac{3}{5}, 0\right)$ are solutions of $5x - y = 3$. Locate the two points on the axes below and draw the line of $5x - y = 3$.

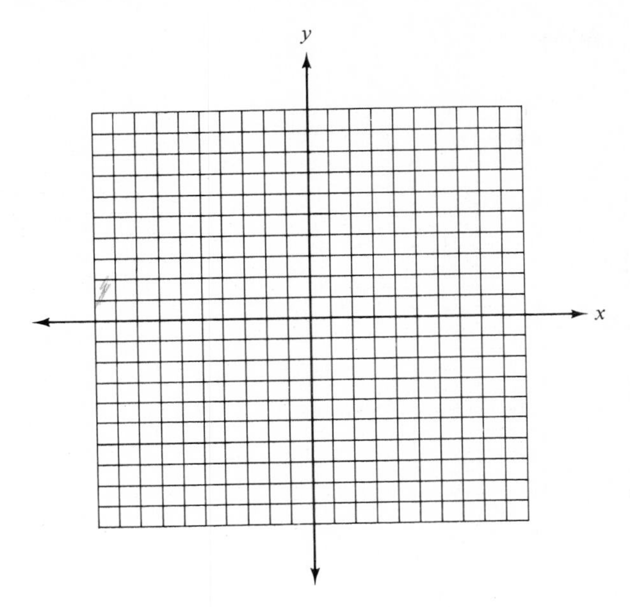

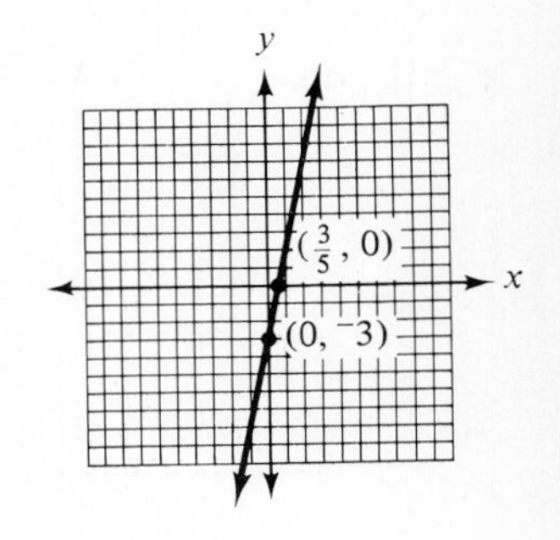

221 To graph $4x + 3y = 5$, first find two solutions of the equation. (0, ________) and (________, 0) are solutions of $4x + 3y = 5$.

$\left(0, \frac{5}{3}\right)$, $\left(\frac{5}{4}, 0\right)$

222 $\left(0, \frac{5}{3}\right)$ and $\left(\frac{5}{4}, 0\right)$ are two solutions of $4x + 3y = 5$. Graph the line of $4x + 3y = 5$.

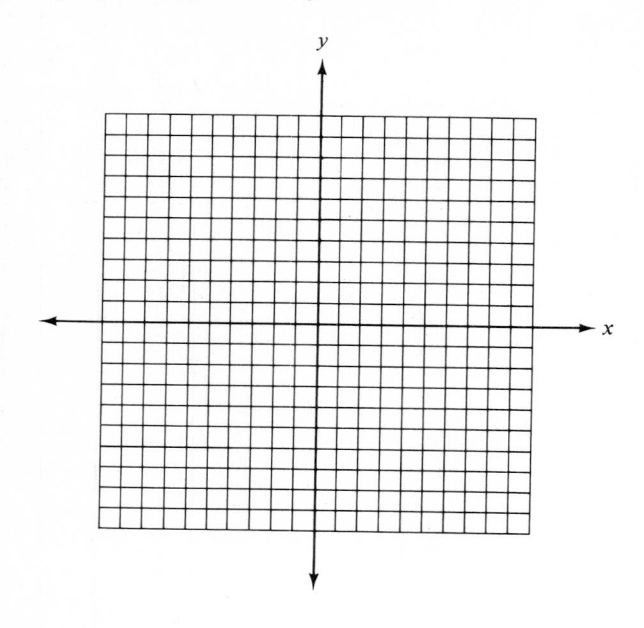

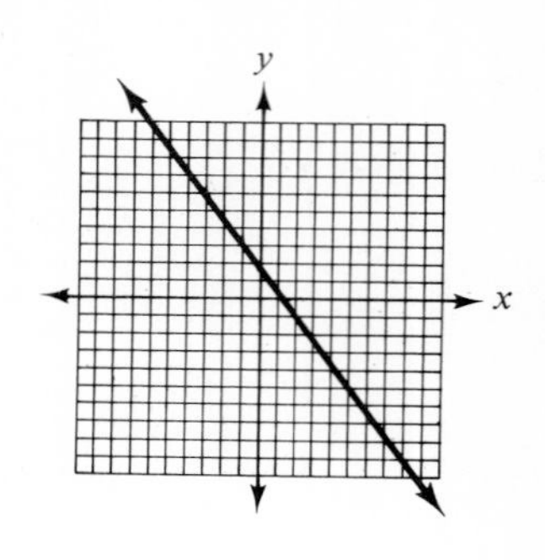

223 Graph $x + y = 5$ by first finding the solutions (0, ________) and (________, 0).

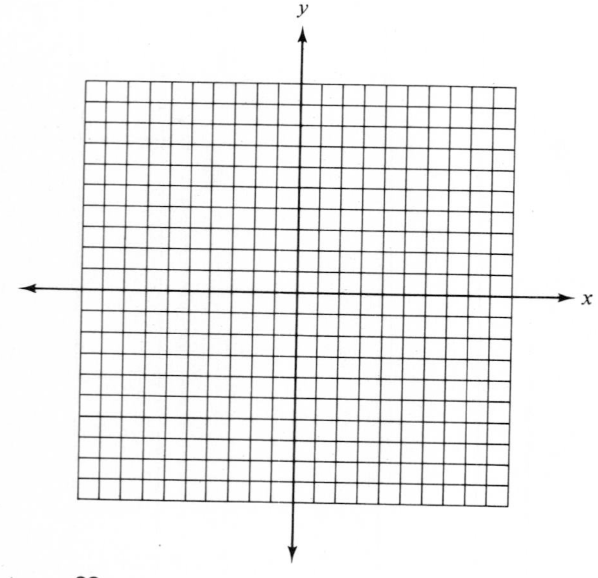

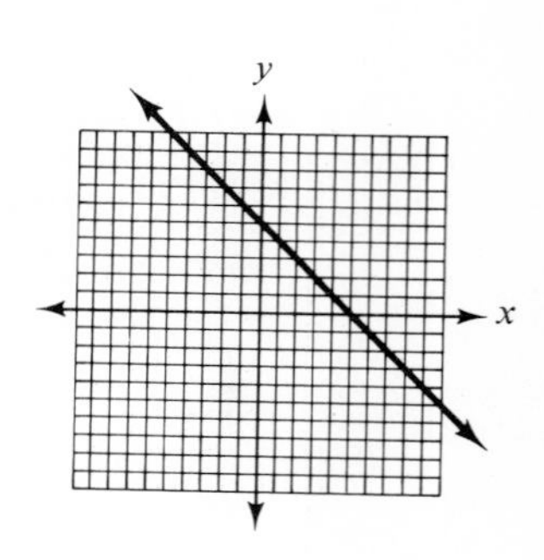

224 Graph $2x - 3y = 12$ by first finding the solutions (0, ________) and (________, 0).

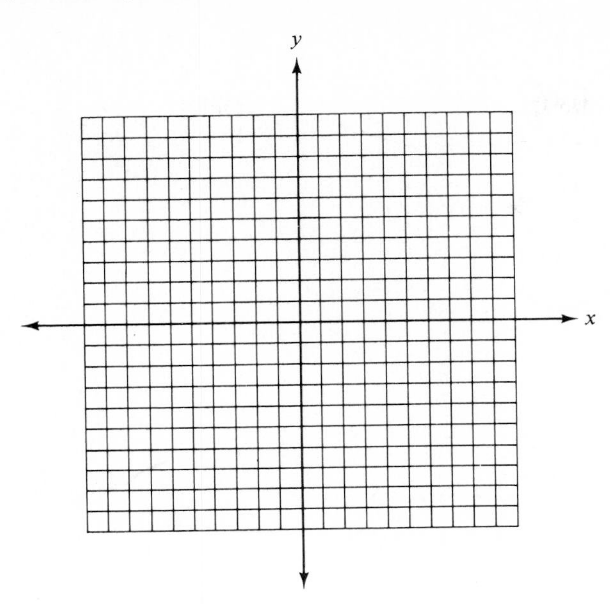

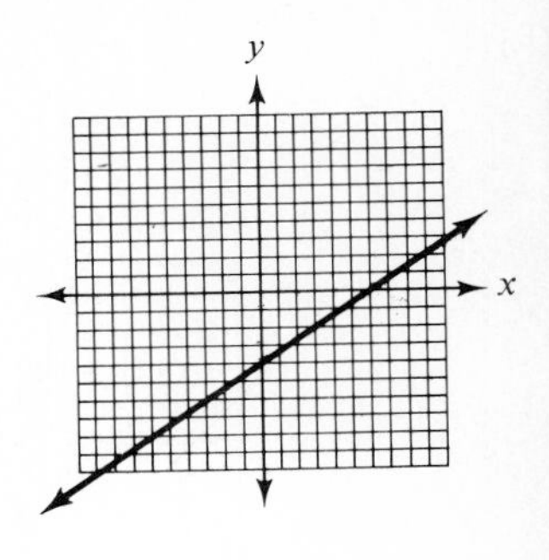

225 Graph $4x - y = 3$ by first finding the solutions (0, ________) and (________, 0).

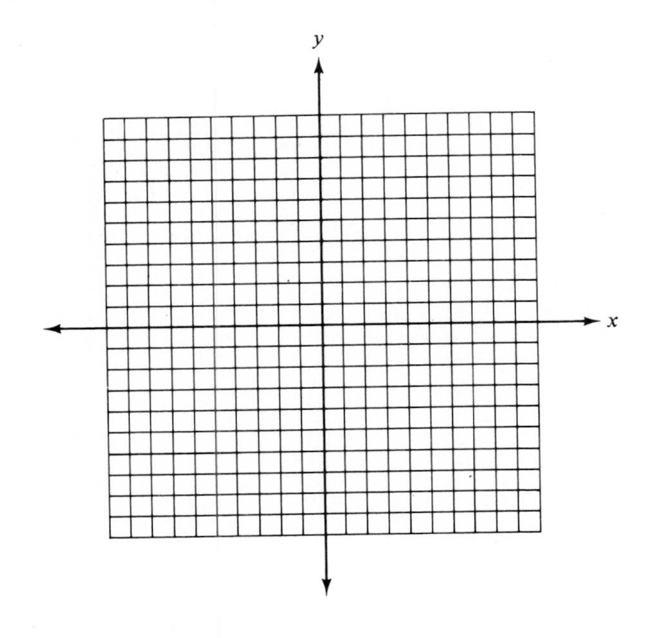

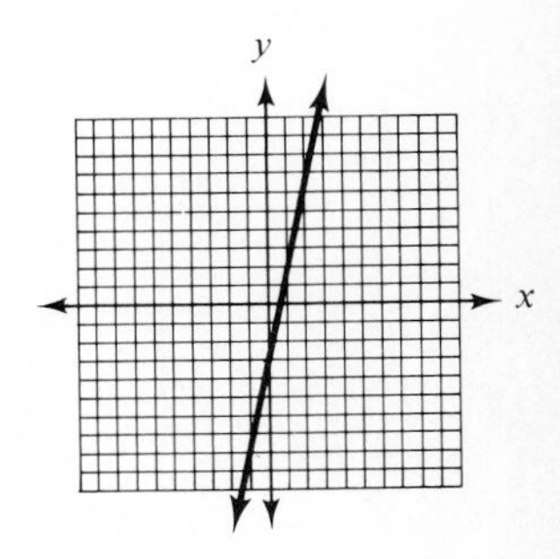

226 Graph $3x - 5y = {}^{-}10$.

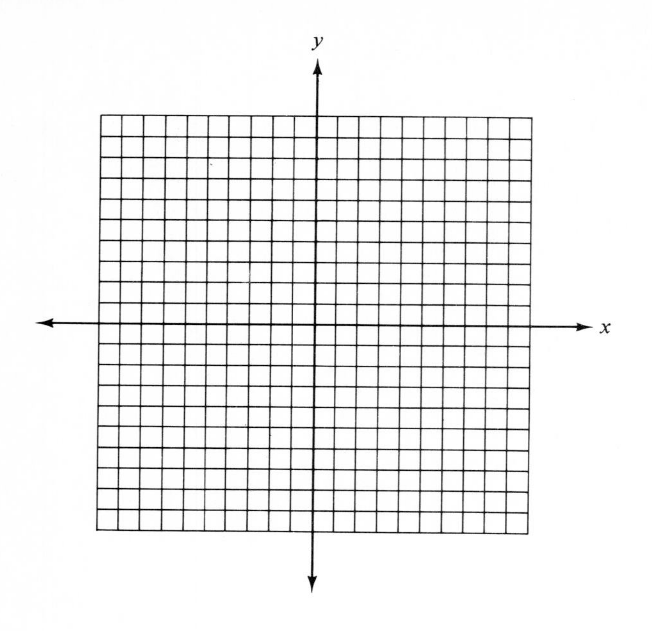

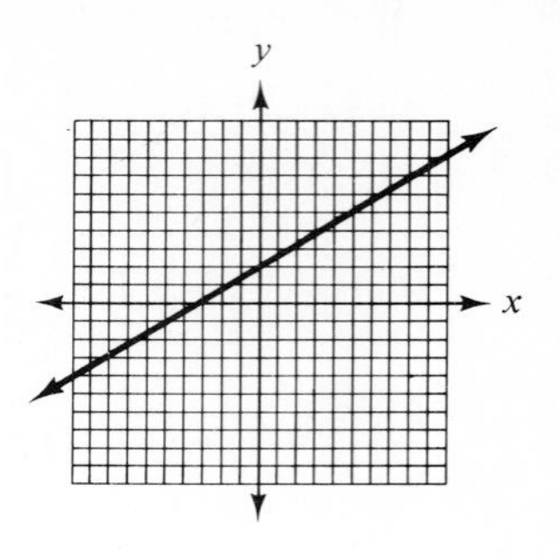

227 Graph ${}^{-}2x + y = 5$.

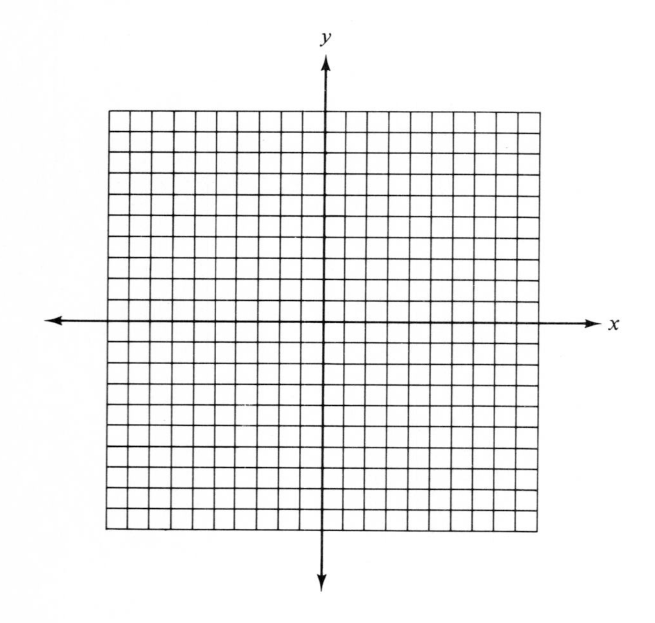

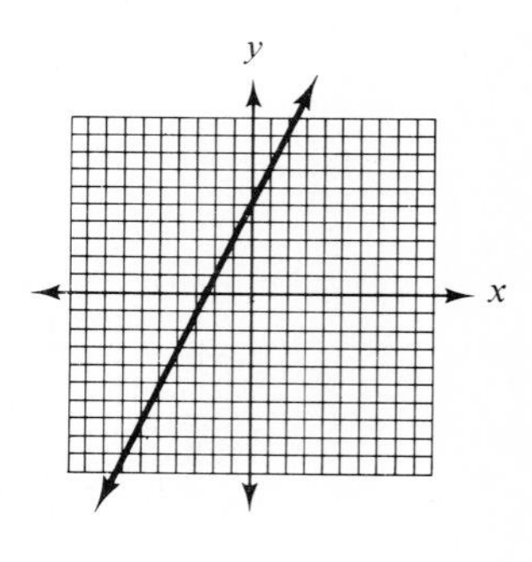

228 Graph $x - 2y = 6$.

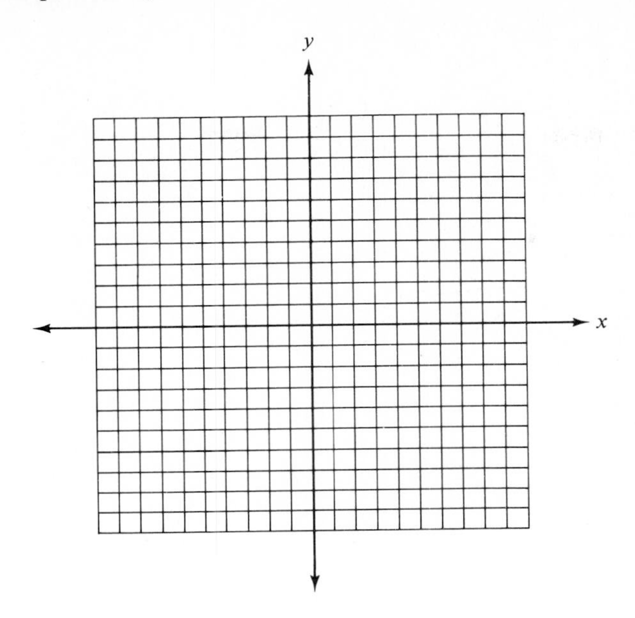

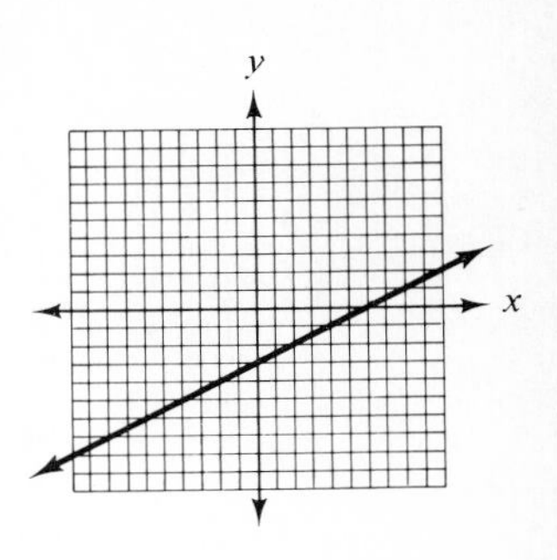

229 Graph $x + 3y = 6$.

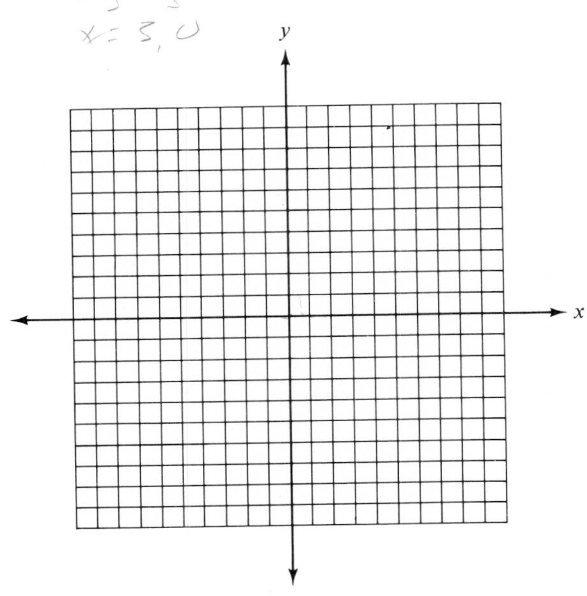

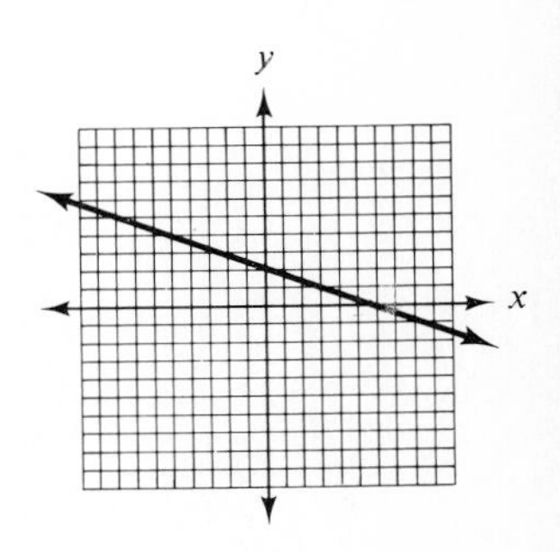

230 Graph $2x - 5y = {}^{-}10$.

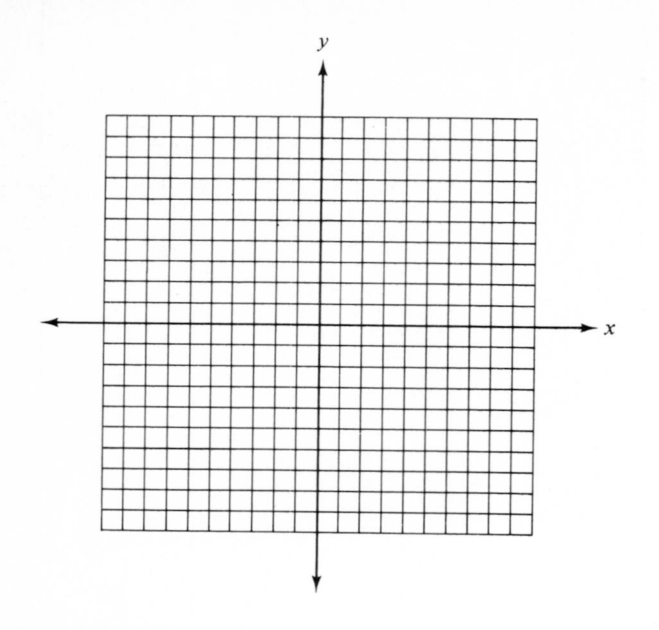

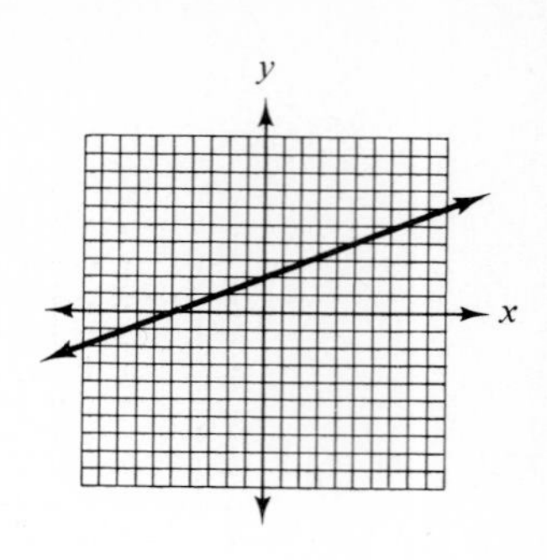

231 Graph $3x + y = 4$.

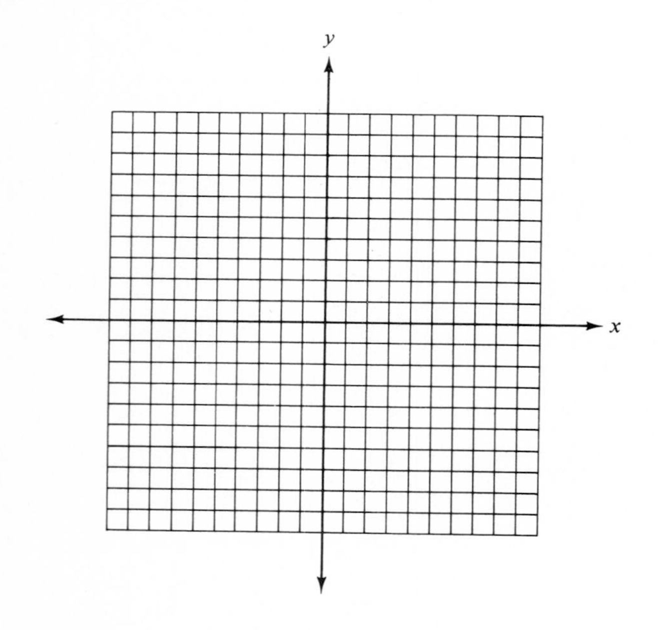

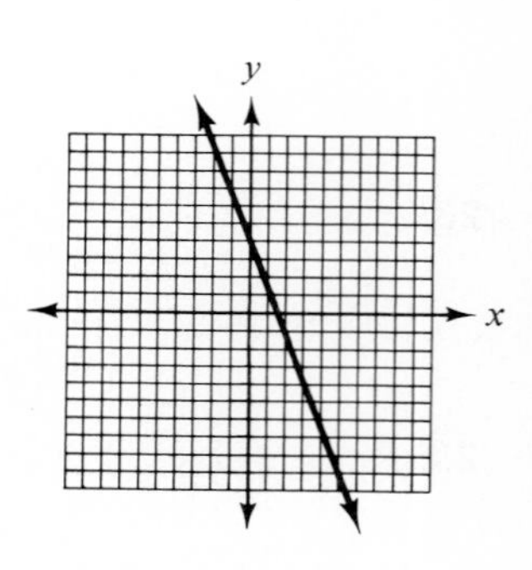

232 Graph $2x + y = 7$.

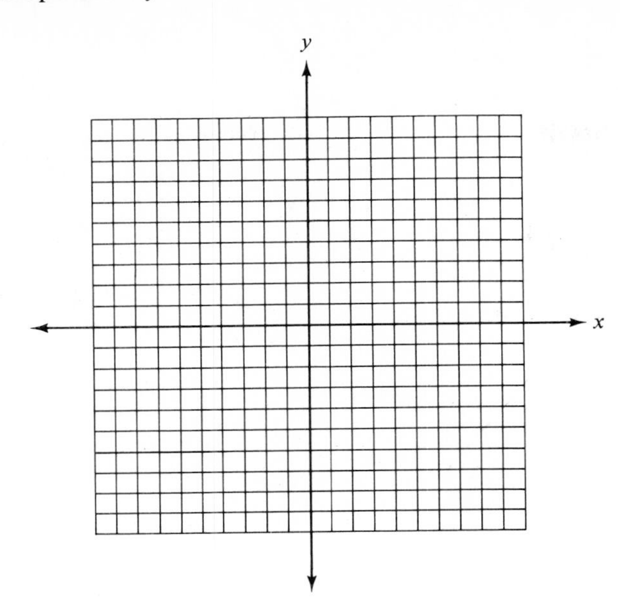

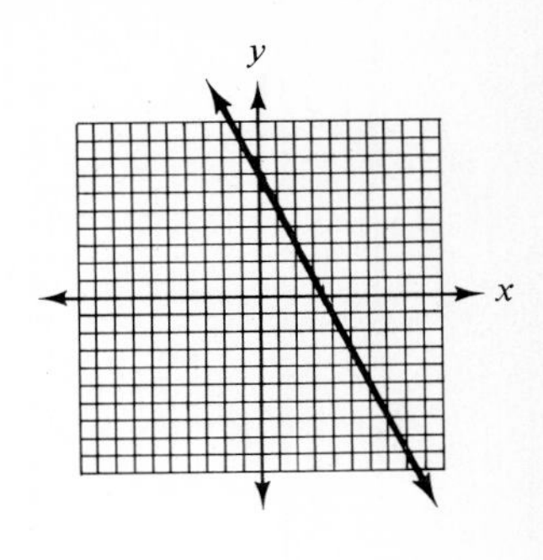

233 To graph $3x - y = 0$, any two solutions must be found. Complete the solutions (0, ________) and (________, 0).

$(0, \underline{0})$, $(\underline{0}, 0)$

234 $(0, \underline{0})$ and $(\underline{0}, 0)$ are the same solutions. To graph $3x - y = 0$, two solutions are necessary. Complete the solution (2, ________).

$(2, \underline{6})$

235 Whenever (0, 0) is a solution of an equation, one other solution must be found by replacing either x or y by some other number. Complete (________, 4) as a solution for $3x - y = 0$.

$\left(\underline{\frac{4}{3}}, 4\right)$

236 (0, 0), (2, 6), and $\left(\frac{4}{3}, 4\right)$ are three solutions of $3x - y = 0$. Locate the three points on the graph below and draw the line of $3x - y = 0$.

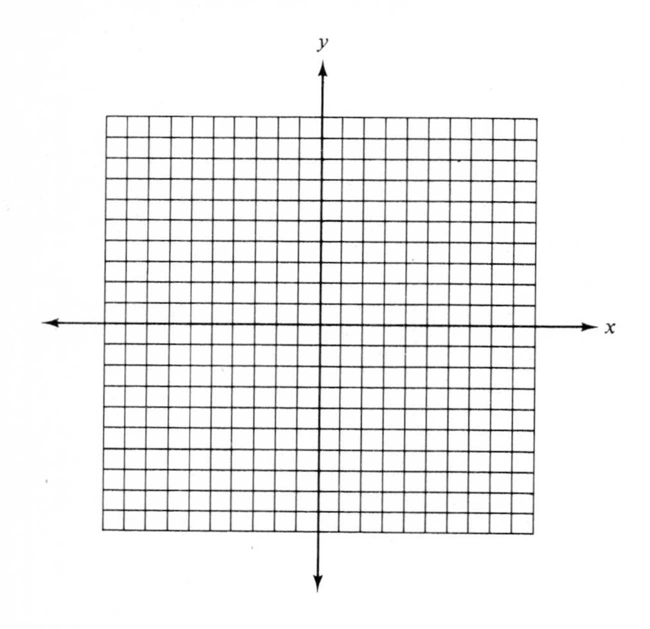

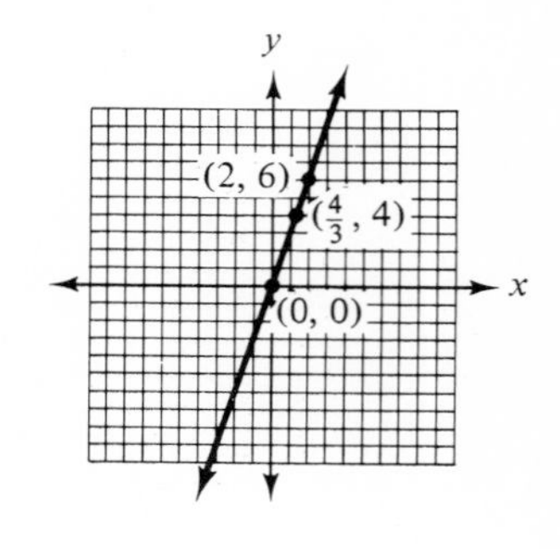

237 The graph of $5x + y = 0$ goes through (0, 0). Find another solution of $5x + y = 0$ and draw the line of the equation.

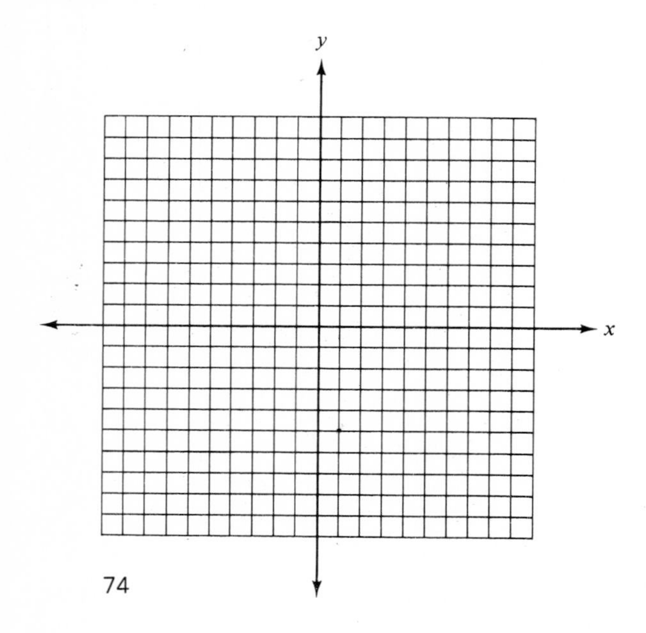

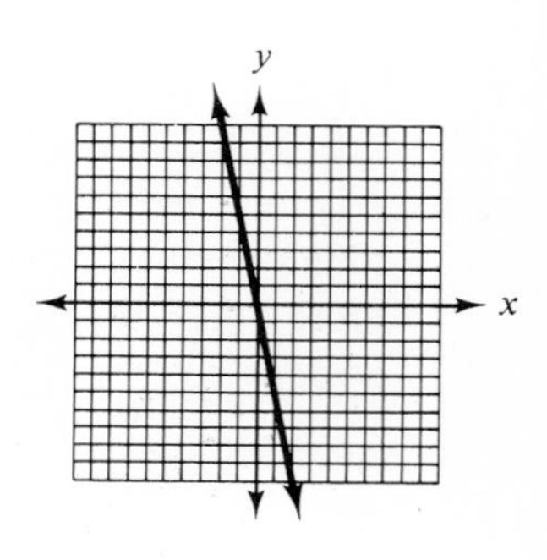

238 Graph $x + y = 0$ by finding any two solutions of the equation.

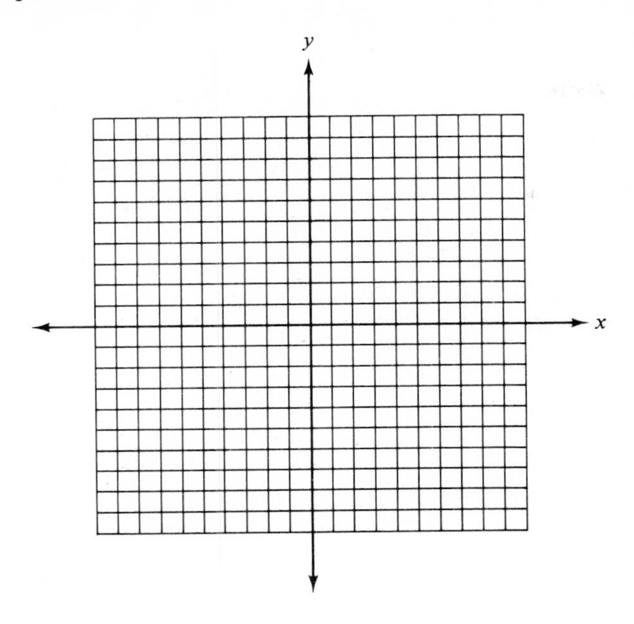

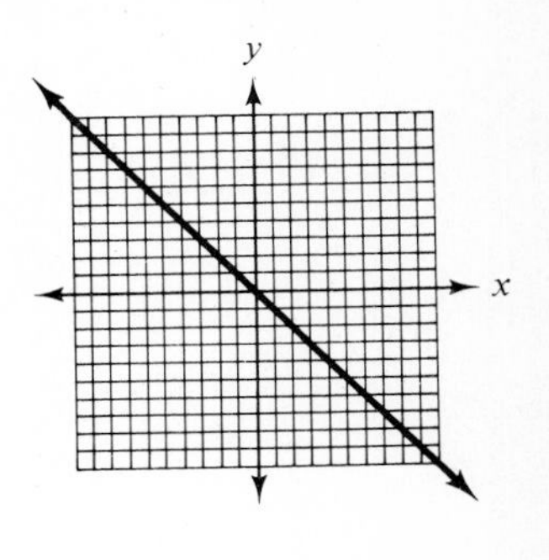

239 Graph $2x - y = {}^{-}3$.

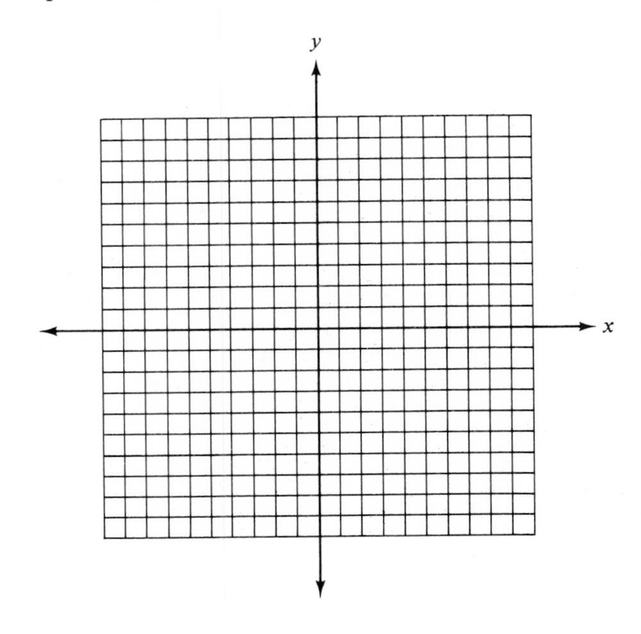

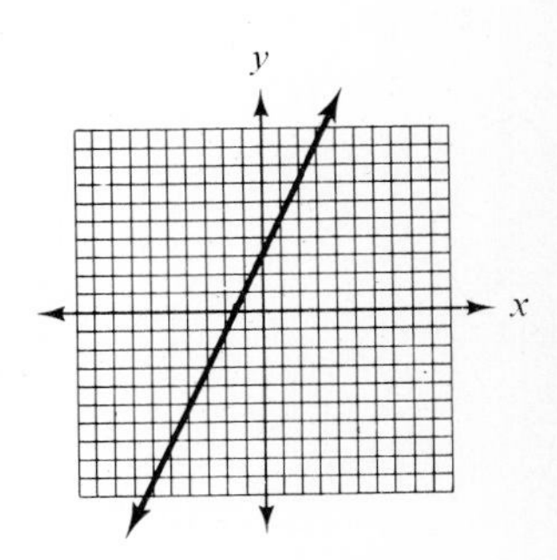

240 Graph $x - 2y = 2$.

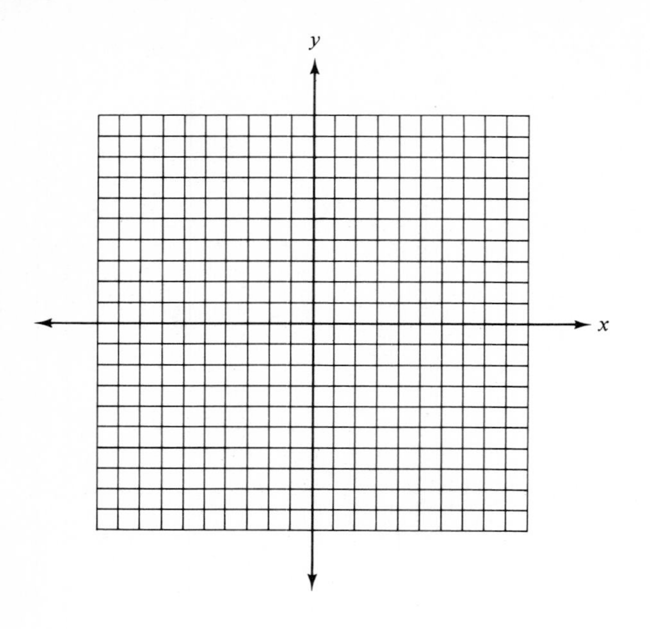

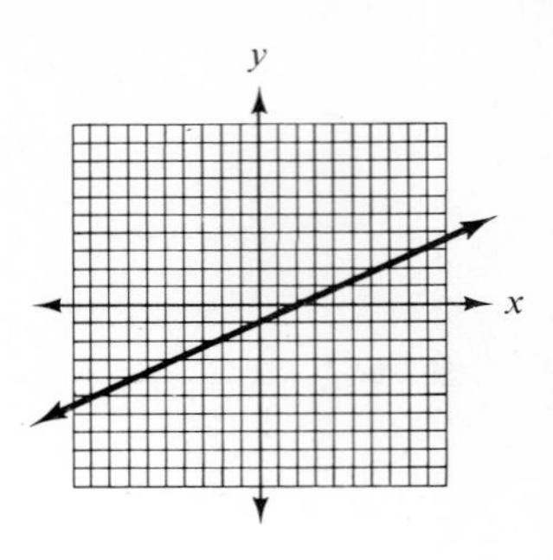

241 Graph $x - 4y = 8$.

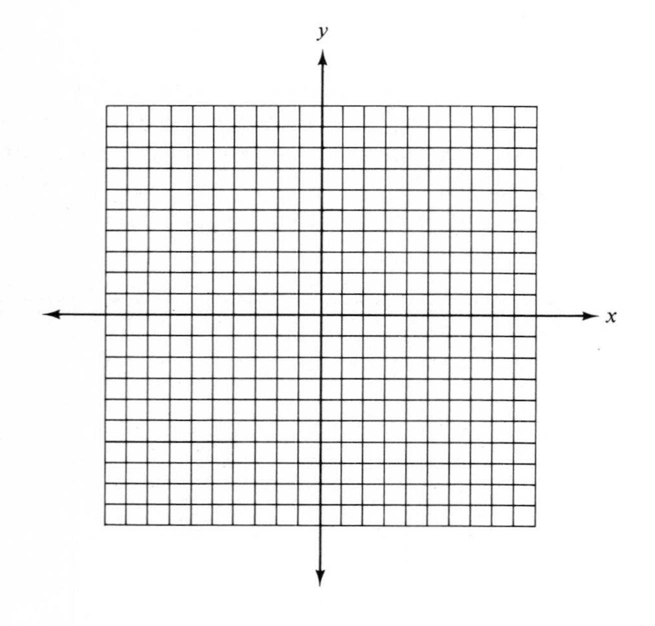

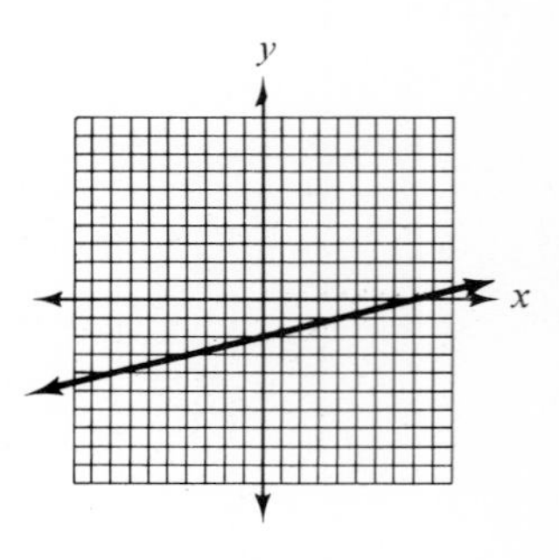

Another helpful and practical way of determining solutions for linear equations with two variables is developed in Appendix A.

Self-Quiz # 5

The following questions test the objectives of the preceding section. 100% mastery is desired.

Graph:

1. $x + y = 4$.

2. $x - y = 7$.

3. $2x - 3y = 6$.

4. $4x - y = 0$.

5. $5x + 3y = 15$.

If two equations are separately graphed on a single pair of axes, the point where the two lines intersect is called the *common solution.*

In graphing equations to find their common solution some drawing errors often occur. Therefore, your answers in the following section are acceptable if the lines and common solution are approximately the same as the answers shown in the program.

---

242 The lines of $2x + y = 7$ and $x - y = 2$ are shown below. What ordered pair represents the point that is on both lines? ________ . (3, 1)

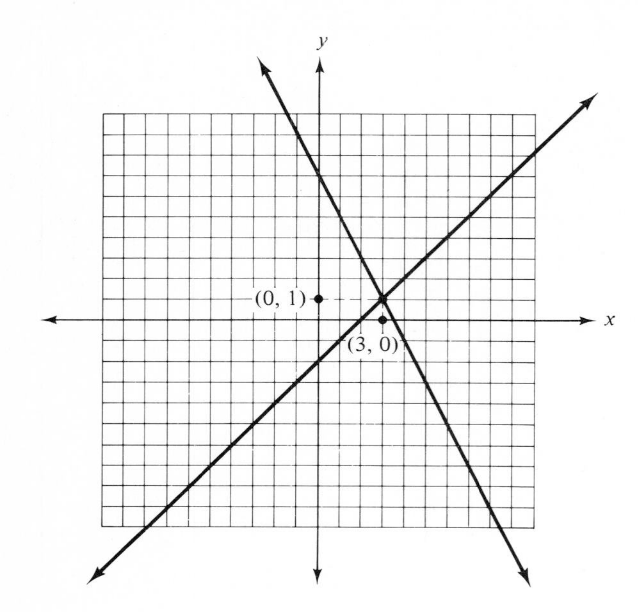

243 The graphs of $3x - y = 2$ and $2x + y = 8$ are shown below. What ordered pair represents the point that is common to both lines? ________ . (2, 4)

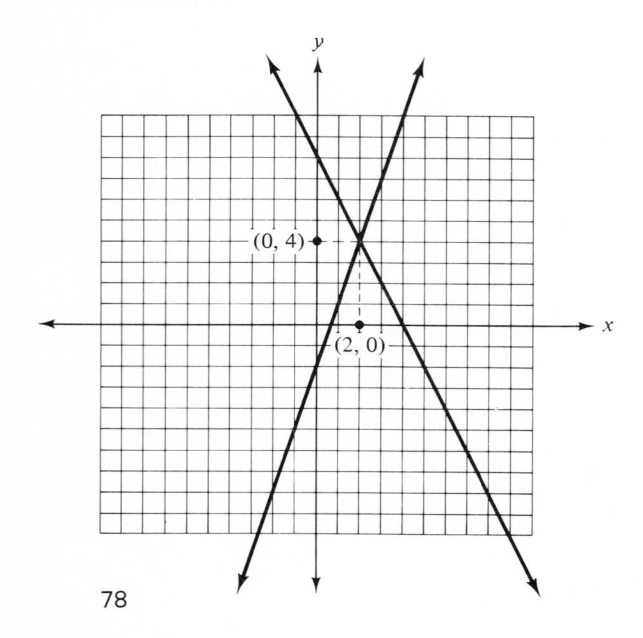

244 Graph both $2x - 5y = {}^-10$ and $x + y = {}^-5$ on the following axes and use an ordered pair to show the point that is common to both lines.

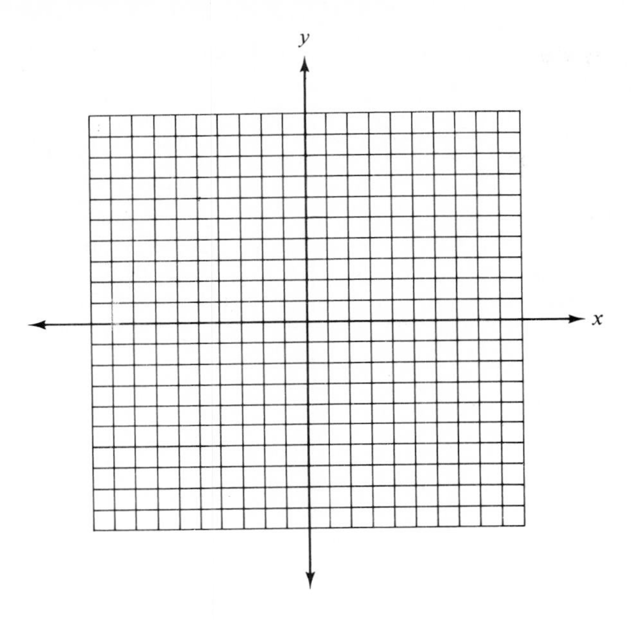

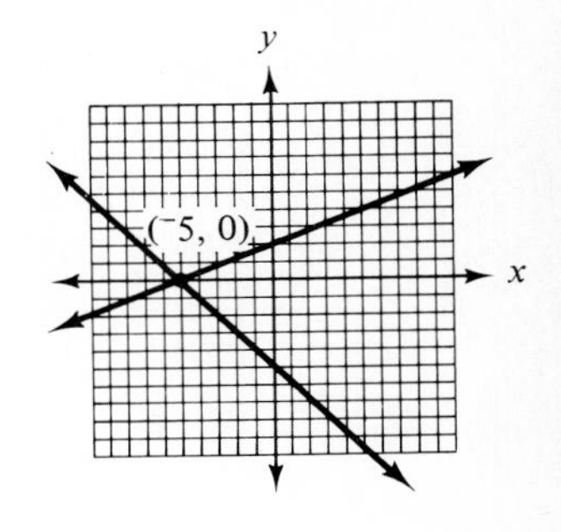

245 The graph of $2x + y = 5$ is shown below. Graph $x + y = 3$ on the same axes and use an ordered pair to show the point that is common to both lines.

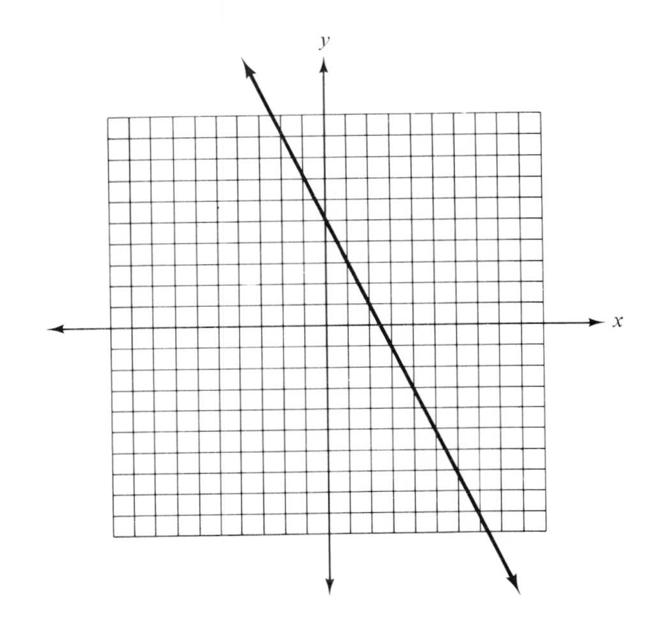

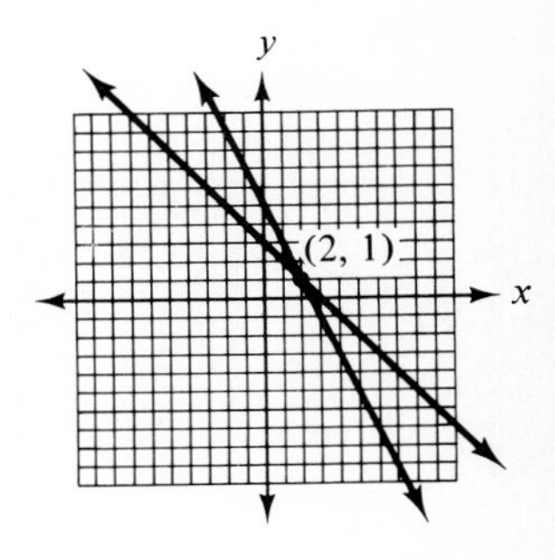

246 The graph of $x - 3y = 6$ is shown below. Graph $x - y = 4$ on the same set of axes and use an ordered pair to show the point that is common to both lines.

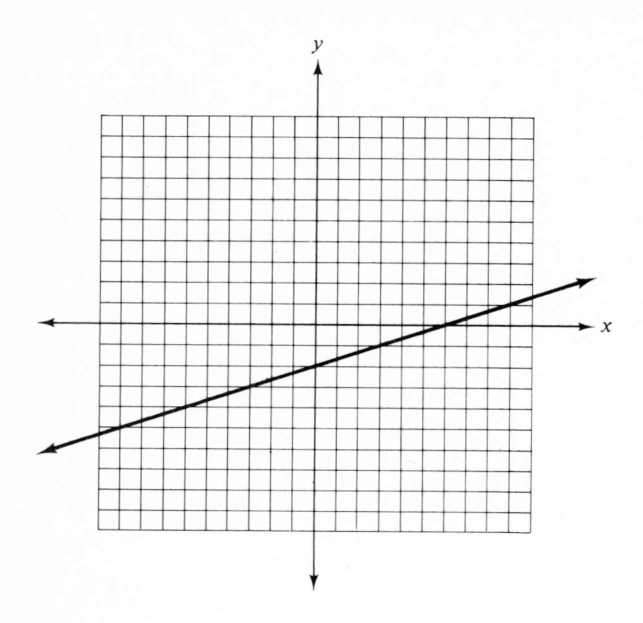

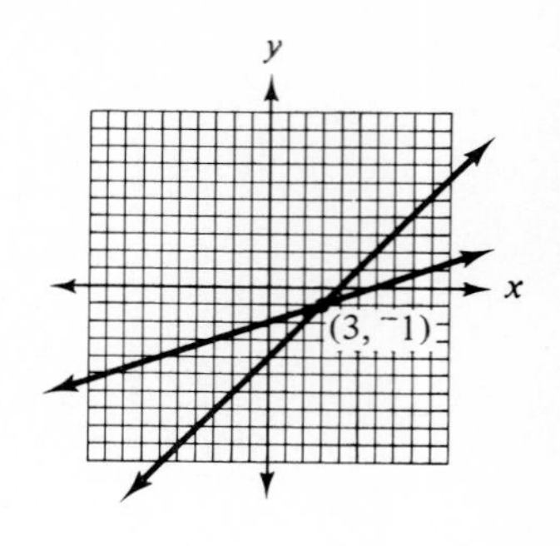

247 Graph both $x - 3y = 9$ and $x + y = 1$ on the following axes and use an ordered pair to show the point that is common to both lines.

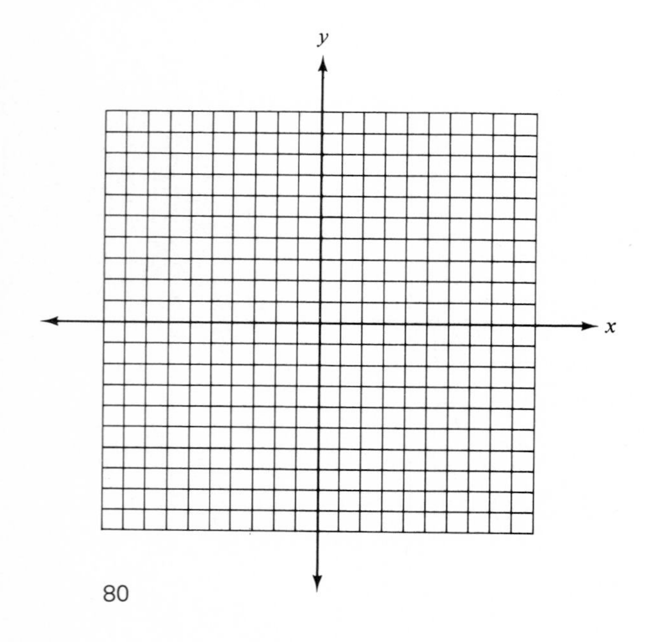

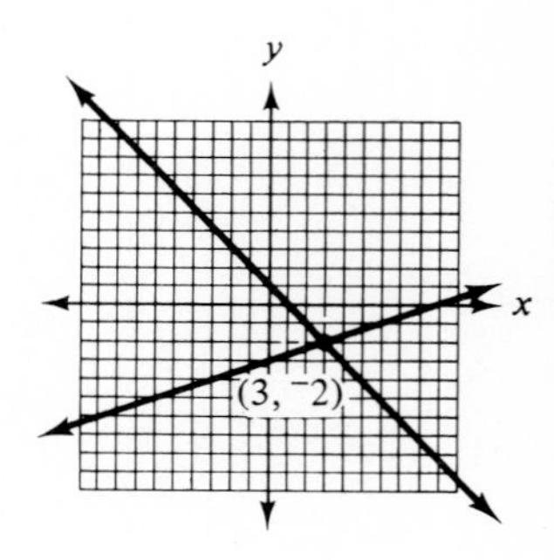

248 Graph both $4x - 3y = 12$ and $x + y = {}^{-}4$ on the following axes and use an ordered pair to show the point that is common to both lines.

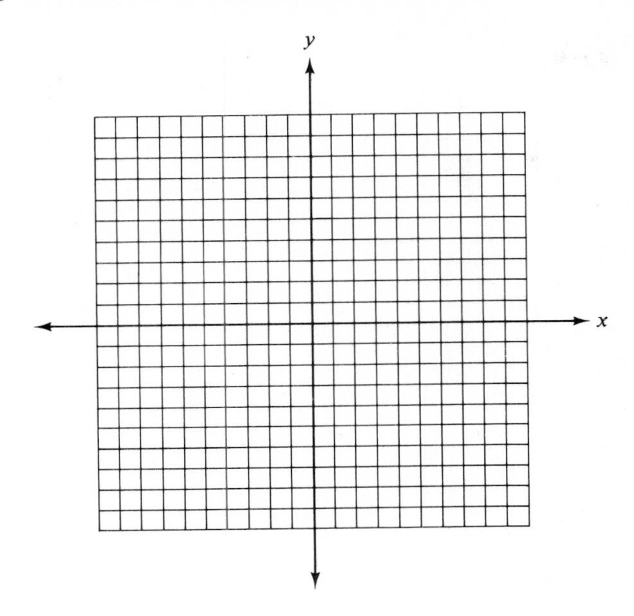

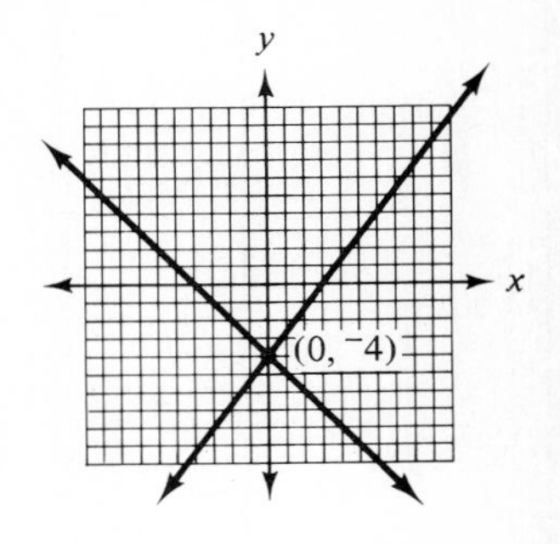

249 Graph both $2x - y = {}^{-}3$ and $x - y = 1$ on the following axes and use an ordered pair to show the point that is common to both lines.

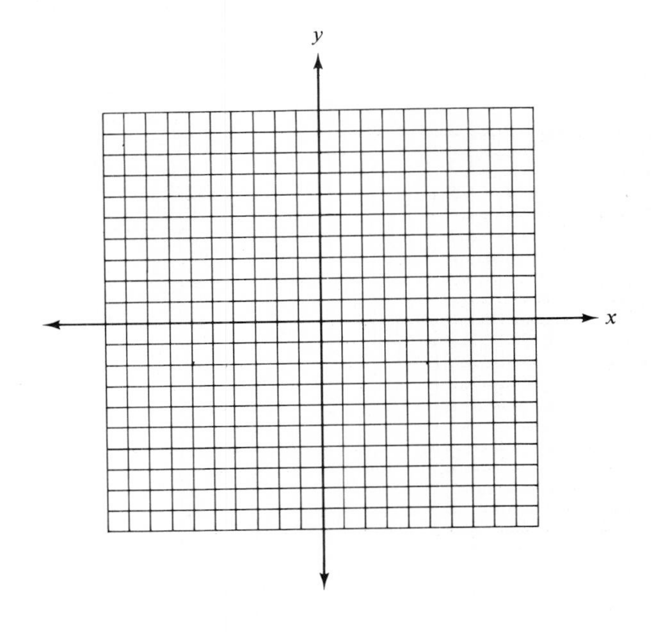

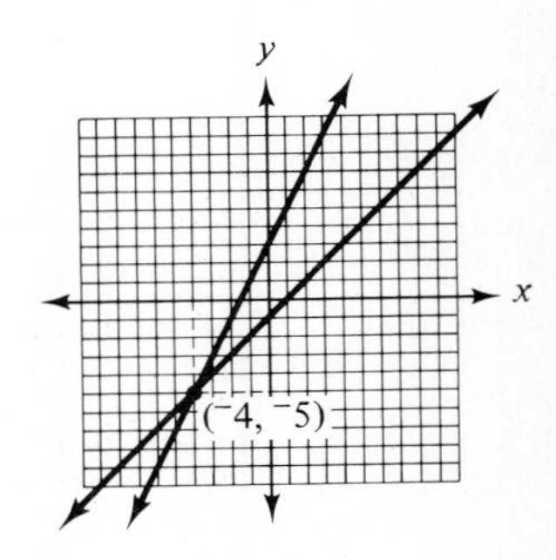

250 Graph both $x - 3y = 6$ and $4x - y = {}^{-}9$ on the following axes and use an ordered pair to show the point that is common to both lines.

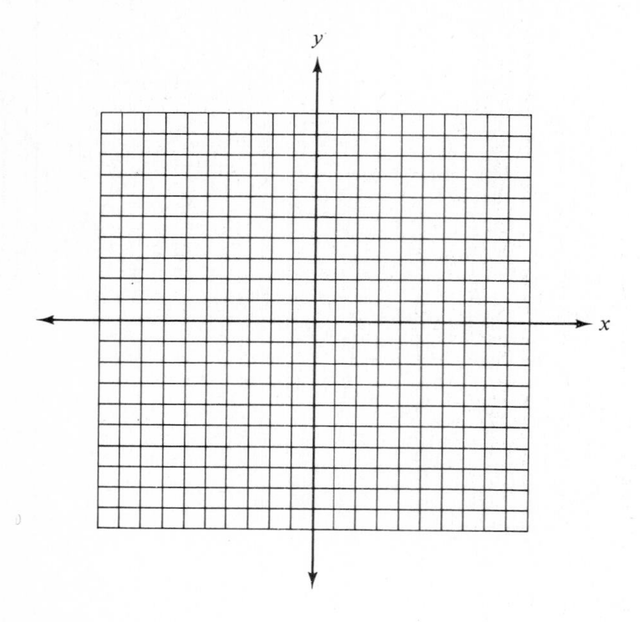

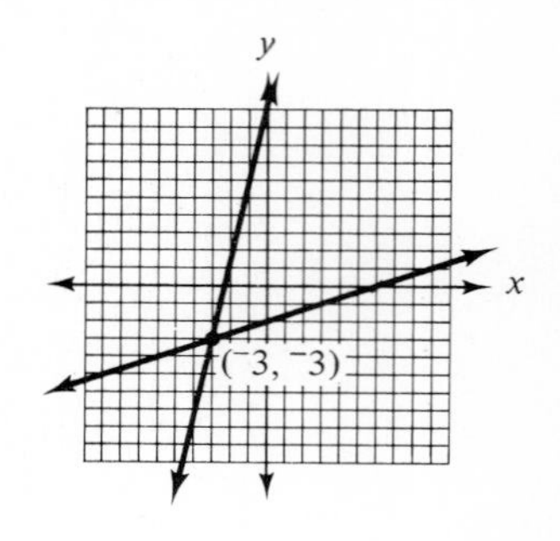

251 Graph both $2x + y = 5$ and $x - 3y = 6$ on the following axes and use an ordered pair to show the point that is common to both lines.

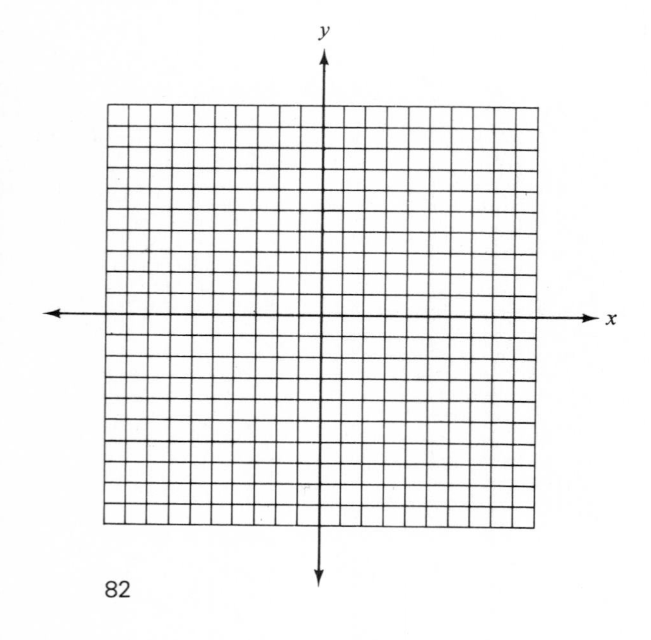

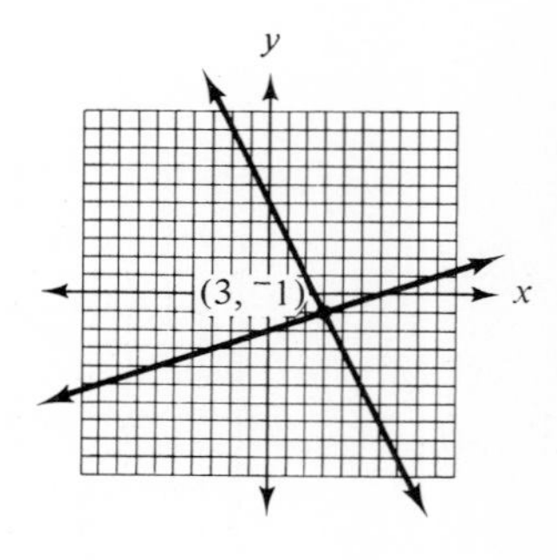

252 Graph both $2x + y = 6$ and $x + y = 2$ on the following axes and use an ordered pair to show the point that is common to both lines.

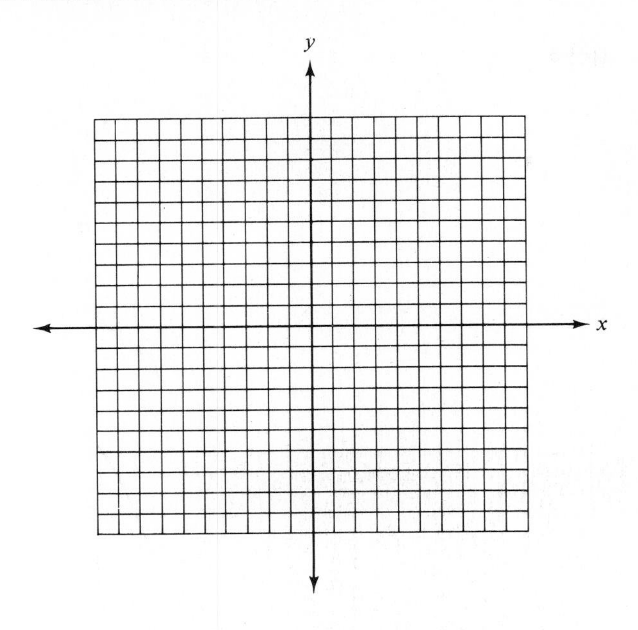

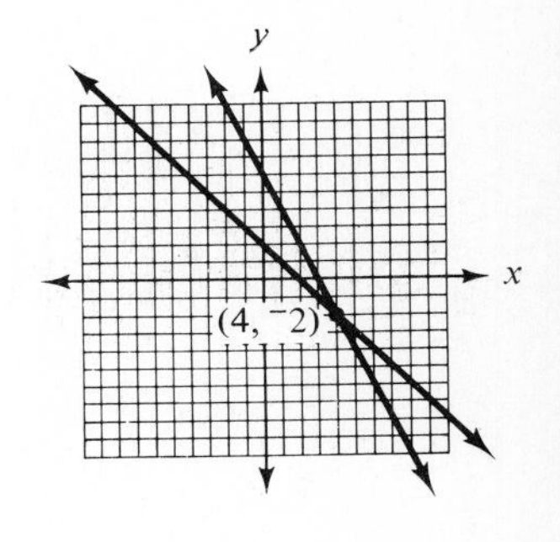

253 Graph both $3x - y = {}^{-}1$ and $x - y = 1$ on the following axes and use an ordered pair to show the point that is common to both lines.

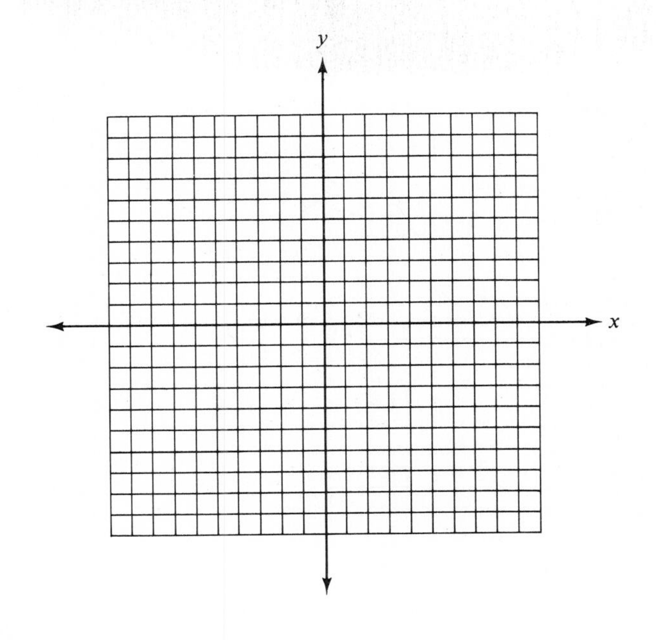

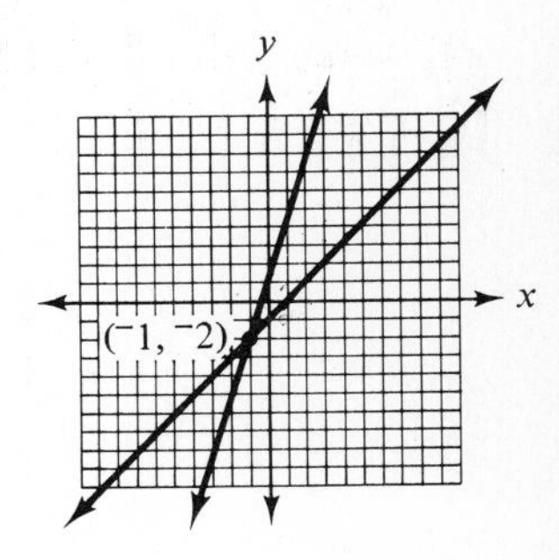

Self-Quiz # 6

The following questions test the objectives of the preceding section. 100% mastery is desired.

Graph both equations for each problem and use an ordered pair to show the point that is common to both lines.

1. $2x + y = 1$ and $x + y = {}^{-}1$.

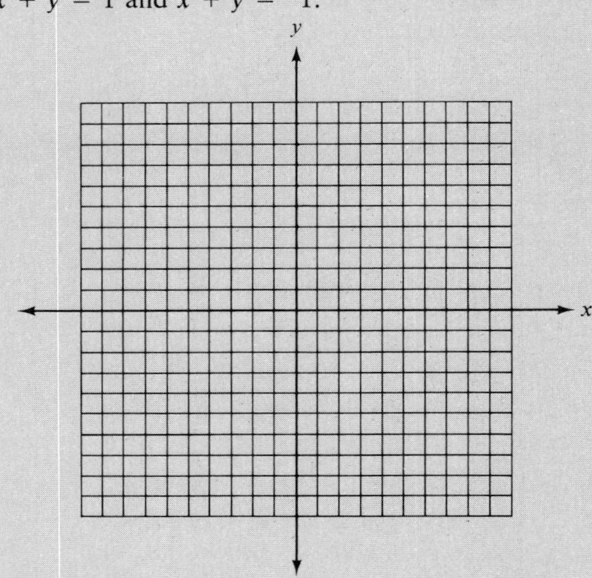

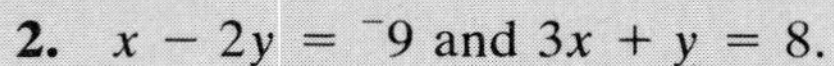

2. $x - 2y = {}^{-}9$ and $3x + y = 8$.

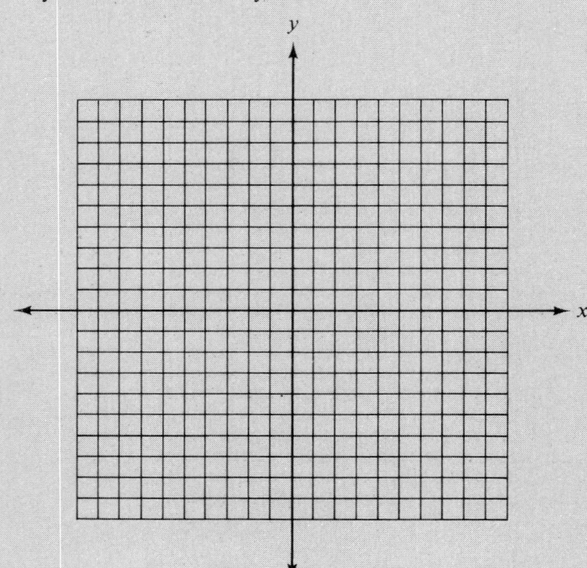

3. $3x + y = 5$ and $2x - y = {}^{-}5$.

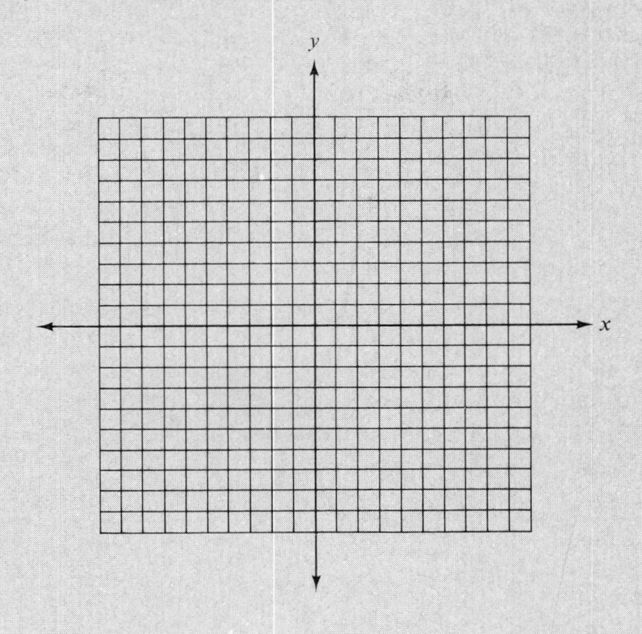

4. $x + 2y = 5$ and $3x + y = 0$.

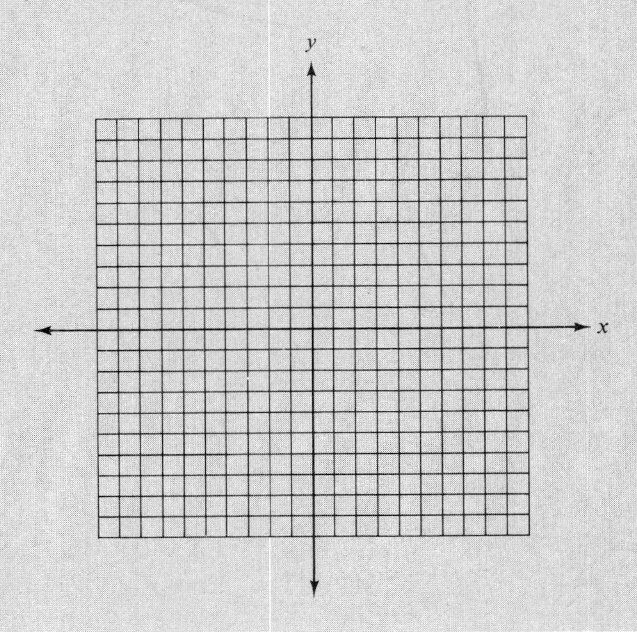

Chapter Summary

This chapter introduced the graphing of equations with two variables.

Each equation of this chapter was directly related to a straight line in the following ways. Every ordered-pair solution of the equation represents a point on the straight line. Every ordered pair that is not a solution of the equation represents a point that is not on the straight line.

Each equation has an unlimited (infinite) number of ordered-pair solutions. However, when two equations are graphed on one pair of axes, the intersection point of the two lines represents the only solution that is common to both equations.

Graphing is not the most accurate method of determining the common solution of two equations with two variables. Methods of determining the exact common solution of two equations with two variables are explained in the next chapter.

CHAPTER POST-TEST

The following questions test the objectives of this chapter. A score of 90% indicates sufficient mastery, and the student may proceed to the next chapter.

1. Is $(^{-}2, 5)$ a solution for $x + y = 3$?

2. Is $(8, ^{-}1)$ a solution for $2x - y = 17$?

3. Is $(^{-}5, ^{-}2)$ a solution for $2x - 3y = 6$?

4. Is $(3, 4)$ a solution for $5x - 2y = 23$?

5. (4, ________) is a solution for $3x - y = 9$.

6. (________, 7) is a solution for $2x - 3y = 9$.

7. (3, ________) is a solution for $x + 2y = 8$.

8. (________, $^{-}2$) is a solution for $5x - 3y = 13$.

9. Write ordered pairs (x, y) for each point shown on the graph by a capital letter.

A ________ .
B ________ .
C ________ .
D ________ .
E ________ .

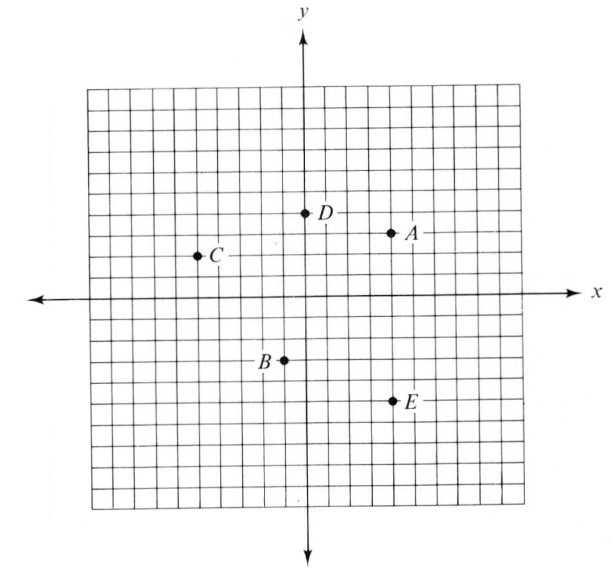

10. Mark the following points (by their capital letters) on the graph.

A $(3, {}^{-}2)$.
B $(5, 7)$.
C $({}^{-}2, 0)$.
D $({}^{-}6, {}^{-}2)$.
E $({}^{-}4, 3)$.

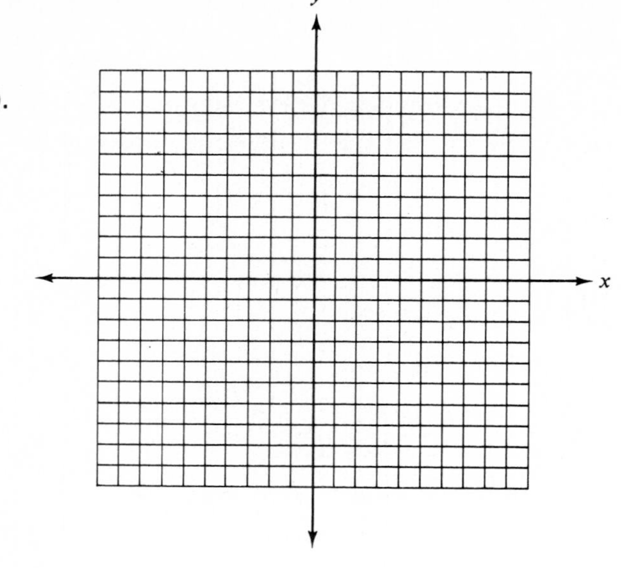

11. Graph $x + 3y = 8$.

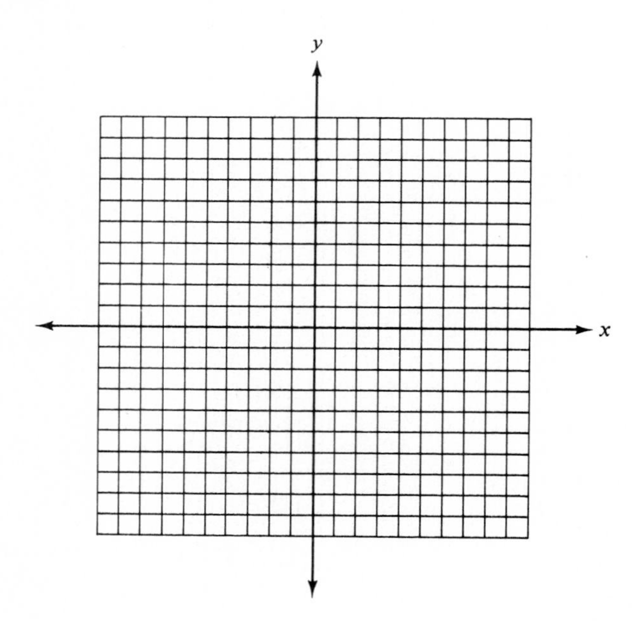

12. Graph $3x - 2y = 6$.

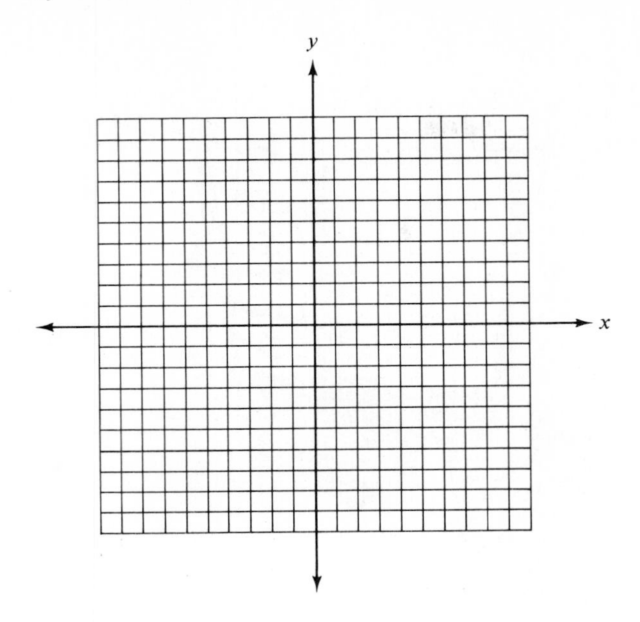

13. Graph both $2x - y = 8$ and $x + y = 1$ and find the point common to both lines.

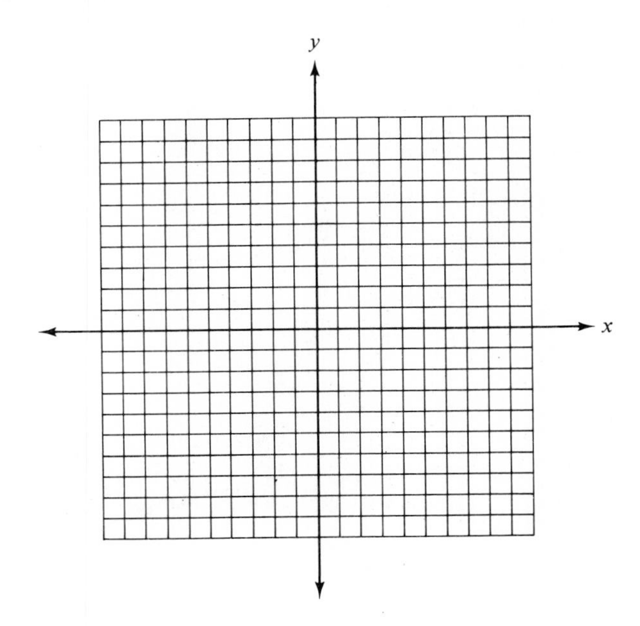

ALGEBRA

Solving Pairs of Equations

1
2
3
4
5
6
7
8
9
10

CHAPTER PRE-TEST

The following questions indicate the objectives of this chapter. A score of 90% indicates sufficient mastery, and the student may immediately take the Chapter Post-Test.

Find the common solution for each pair of equations or state that the equations have no common solution.

1. $x + y = 19$
$x - y = 31$

2. $3x + y = 17$
$x + y = 7$

3. $2x - 7y = 12$
$x - 3y = 5$

4. $3x - 6y = 13$
$x - 2y = 4$

5. $x = 4y$
$3x - y = 22$

6. $y = 3x - 1$
$x - 2y = {}^{-}8$

7. $2x - 3y = 13$
$3x + 4y = 11$

8. $5x - 2y = {}^{-}4$
$3x + 7y = 55$

9. $2x - 3y = 9$
$8x - 12y = 1$

10. $7x - 6y = 33$
$4x + 5y = 2$

There is no limit to the number of ordered-pair solutions for the equation $3x + y = 7$. There is also no limit to the ordered-pair solutions of $2x - y = 3$. However, there is only one ordered pair (2, 1) that is a solution for both equations:

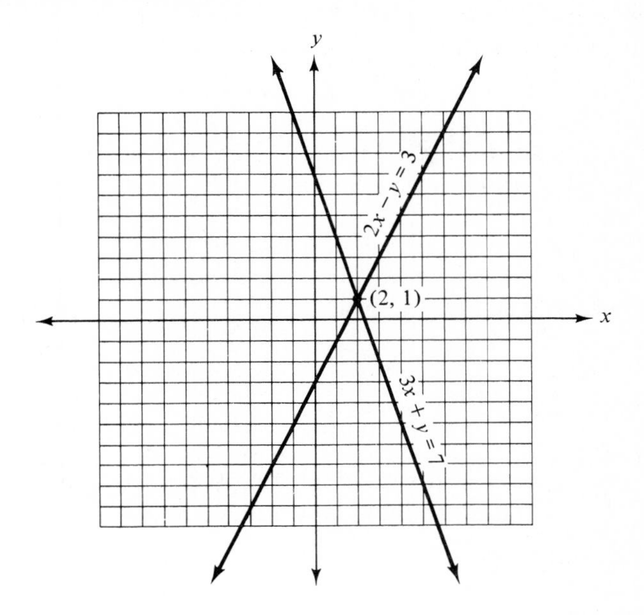

The solution (2, 1) is called the *common solution* for the pair of equations $3x + y = 7$ and $2x - y = 3$. In the following frames methods other than graphing for finding the common solution for two equations will be explained.

---

1 In solving equations such as $2x + 7 = 19$, $^-7$ can be added to each side to give

$$\begin{array}{rl} 2x + 7 &= 19 \\ ^-7 &= {}^-7 \\ \hline 2x \quad &= 12 \end{array}$$

Can $^-5$ be added to both sides of $3x + 5 = 14$?

_______ .

Yes

2 $^{-}4$ can be added to each side of $2x + 4 = 10$ to give

$$\begin{array}{l} 2x + \ \ 4 = 10 \\ \quad\quad {}^{-}4 = {}^{-}4 \\ \hline 2x \quad\quad = \ \ 6 \end{array}$$

6 can be added to each side of $2x - 6 = 12$ to give

$$\begin{array}{l} 2x - 6 = 12 \\ \quad\quad 6 = \ \ 6 \\ \hline 2x \quad\ = \_\_\_\_\_ \end{array}$$

18

3 $3x + y = 7$ and $2x - y = 3$ is a pair of equations. Since $2x - y$ is equal to 3, $2x - y$ can be added to $3x + y$ and 3 added to 7 to give

$$\begin{array}{l} 3x + y = 7 \\ 2x - y = 3 \\ \hline 5x \quad\ = 10 \end{array}$$

$2x + y = 5$ and $5x - y = 9$ is a pair of equations. Since $5x - y$ is equal to 9, $5x - y$ can be added to $2x + y$ and 9 can be added to 5 to give

$$\begin{array}{l} 2x + y = \ \ 5 \\ 5x - y = \ \ 9 \\ \hline 7x \quad\ = \_\_\_\_\_ \end{array}$$

14

4 $4x + 2y = 9$ and $3x - 2y = 4$ is a pair of equations. Since $3x - 2y$ is equal to 4, add $3x - 2y$ to $4x + 2y$ and add 4 to 9 to give

$$\begin{array}{l} 4x + 2y = 9 \\ 3x - 2y = 4 \\ \hline \quad\quad\quad\ = \end{array}$$

____

$7x = 13$

5 The pair of equations $8x - 3y = {}^{-}19$ and $x + 3y = 10$ can be added to give

$$\begin{array}{rcl} 8x - 3y &=& {}^{-}19 \\ x + 3y &=& 10 \\ \hline 9x \quad\quad &=& {}^{-}9 \end{array}$$

Add the pair of equations $4x - 7y = 12$ and $9x + 7y = 14$. ____________

$$\begin{array}{rcl} 4x - 7y &=& 12 \\ 9x + 7y &=& 14 \\ \hline 13x \quad\quad &=& 26 \end{array}$$

6 Add the pair of equations $2x + 7y = 6$ and $5x - 7y = {}^{-}8$. ____________

$$\begin{array}{rcl} 2x + 7y &=& 6 \\ 5x - 7y &=& {}^{-}8 \\ \hline 7x \quad\quad &=& {}^{-}2 \end{array}$$

7 Add the pair of equations $4x + y = 8$ and $2x - y = 11$. ____________.

$$\begin{array}{rcl} 4x + y &=& 8 \\ 2x - y &=& 11 \\ \hline 6x \quad\quad &=& 19 \end{array}$$

8 Add the pair of equations $6x - 2y = 9$ and $3x + 2y = 5$. ____________.

$$\begin{array}{rcl} 6x - 2y &=& 9 \\ 3x + 2y &=& 5 \\ \hline 9x \quad\quad &=& 14 \end{array}$$

9 Add the pair of equations $5x + 3y = 13$ and ${}^{-}5x - 7y = 9$. ____________.

$$\begin{array}{rcl} 5x + 3y &=& 13 \\ {}^{-}5x - 7y &=& 9 \\ \hline {}^{-}4y &=& 22 \end{array}$$

10 Add the pair of equations ${}^{-}x - 4y = {}^{-}5$ and $x + 9y = {}^{-}2$. ____________.

$$\begin{array}{rcl} {}^{-}x - 4y &=& {}^{-}5 \\ x + 9y &=& {}^{-}2 \\ \hline 5y &=& {}^{-}7 \end{array}$$

11 By addition, y can be eliminated from the following equations as shown below.

$$\begin{array}{rcr} 5x + 2y & = & 9 \\ 2x - 2y & = & 5 \\ \hline 7x \quad\quad & = & 14 \end{array}$$

Eliminate y for the following equations, by addition:

$$\begin{array}{rcl} x - 3y & = & 9 \\ 3x + 3y & = & 7 \\ \hline \end{array}$$

$4x = 16$

12 Eliminate y by addition for the following equations:

$$\begin{array}{rcl} 3x + 7y & = & 15 \\ 2x - 7y & = & 21 \\ \hline \end{array}$$

$5x = 36$

13 Eliminate the x by addition in the following equations:

$$\begin{array}{rcl} 4x + 3y & = & 7 \\ {}^{-}4x - \quad y & = & {}^{-}5 \\ \hline \end{array}$$

$2y = 2$

14 Eliminate x by addition:

$$\begin{array}{rcl} 2x + 5y & = & 9 \\ {}^{-}2x - 3y & = & 5 \\ \hline \end{array}$$

$2y = 14$

15 Eliminate x by addition:

$$\begin{array}{rcr} {}^{-}x + 4y & = & 7 \\ x + \quad y & = & {}^{-}12 \\ \hline \end{array}$$

$5y = {}^{-}5$

16 Eliminate y by addition:

$$\begin{array}{rcr} 4x + 3y & = & 13 \\ x - 3y & = & 2 \\ \hline \end{array}$$

$5x = 15$

17 Eliminate y by addition:

$$\begin{array}{r} 7x + 5y = 4 \\ 2x - 5y = {}^{-}3 \\ \hline \end{array}$$

$9x = 1$

18 Eliminate x by addition:

$$\begin{array}{r} 4x - y = 9 \\ {}^{-}4x + 3y = {}^{-}2 \\ \hline \end{array}$$

$2y = 7$

To complete the ordered-pair solution (3, ______) for $2x - y = 7$, the following steps are used:

$$\begin{aligned} (3, \_\_\_\_\_\_) \quad 2x - y &= 7 \\ 2 \cdot 3 - y &= 7 \\ 6 - y &= 7 \\ {}^{-}y &= 1 \\ y &= {}^{-}1 \end{aligned}$$

Hence $(3, {}^{-}1)$ is the solution for $2x - y = 7$ when $x = 3$.

---

19 To find the common solution for $3x + y = 7$ and $2x - y = 3$, first eliminate y by addition:

$$\begin{array}{l} 3x + y = 7 \\ 2x - y = 3 \\ \hline 5x \quad\;\; = 10 \quad \text{or} \quad x = 2 \end{array}$$

If x is replaced by 2 in $3x + y = 7$, the solution is (2, ______).

$(2, \underline{1})$

20 To find the common solution for $5x - y = 3$ and $x + y = 3$, first eliminate y by addition:

$$\begin{array}{rcl} 5x - y &=& 3 \\ x + y &=& 3 \\ \hline 6x \quad &=& 6 \quad \text{or} \quad x = 1 \end{array}$$

If x is replaced by 1 in $5x - y = 3$, the ordered-pair solution is (1, ______). (1, 2)

21 Find a common solution for the following pair of equations by completing the steps:

$$\begin{array}{rcl} 2x + y &=& 13 \\ x - y &=& 2 \\ \hline 3x \quad &=& 15 \quad \text{or} \quad x = 5 \end{array}$$

(5, ______) is a solution for $2x + y = 13$. (5, 3)

22 Complete the common solution:

$$\begin{array}{rcl} 4x - 3y &=& {}^{-}1 \\ x + 3y &=& 11 \\ \hline 5x \quad &=& 10 \quad \text{or} \quad x = 2 \end{array}$$

(2, ______) is a solution for $4x - 3y = {}^{-}1$. (2, 3)

23 Find a common solution for the following pair of equations by completing the steps:

$$\begin{array}{rcl} 2x + y &=& 6 \\ {}^{-}2x + 3y &=& 2 \\ \hline 4y &=& 8 \quad \text{or} \quad y = 2 \end{array}$$

(______, 2) is a common solution for $2x + y = 6$ and ${}^{-}2x + 3y = 2$. [*Note:* 2 is to replace y.] (2, 2)

24 Complete the common solution:

$$\begin{array}{rcl} x + 5y &=& 16 \\ {}^{-}x + y &=& 2 \\ \hline 6y &=& 18 \quad \text{or} \quad y = 3 \end{array}$$

(______, 3) is a solution for $x + 5y = 16$. (1, 3)

25 Find the common solution by addition:

$$3x + y = {}^-3$$
$${}^-2x - y = 1 \quad \underline{\qquad\qquad} .$$

$({}^-2, 3)$

26 Find a common solution by addition:

$$7x - 3y = {}^-1$$
$$x + 3y = 17 \quad \underline{\qquad\qquad} .$$

$(2, 5)$

27 Find a common solution by addition:

$$3x + y = 17$$
$${}^-3x - 5y = {}^-1 \quad \underline{\qquad\qquad} .$$

$(7, {}^-4)$

28 Find a common solution by addition:

$$5x + 3y = 10$$
$${}^-5x - 2y = {}^-10 \quad \underline{\qquad\qquad} .$$

$(2, 0)$

29 Find a common solution by addition:

$$x + 4y = 7$$
$$5x - 4y = {}^-13 \quad \underline{\qquad\qquad} .$$

$({}^-1, 2)$

30 Find a common solution by addition:

$$2x + 3y = 0$$
$${}^-2x + y = {}^-16. \quad \underline{\qquad\qquad} .$$

$(6, {}^-4)$

Self-Quiz # 1

The following questions test the objectives of the preceding section. 100% mastery is desired.

Find the common solution by addition:

1. $x + y = 5$
$3x - y = 3.$ ________ .

2. $3x + 2y = 8$
$4x - 2y = 20.$ ________ .

3. $2x + 3y = 7$
$^{-}2x + y = 5.$ ________ .

4. $x - 5y = 2$
$^{-}2x + 5y = 1.$ ________ .

5. $x - 3y = 11$
$4x + 3y = 14.$ ________ .

Neither x nor y can be eliminated by addition from the following pair of equations:

$$3x + 2y = 8$$
$$2x - y = 3$$

The following frames explain how to find the common solution for such equations.

---

31 $4x = 7$ and $\frac{1}{4} \cdot 4x = \frac{1}{4} \cdot 7$ have exactly the same truth set because $\frac{1}{4}$ has been multiplied by each side of $4x = 7$. If 2 is multiplied by both sides of $2x - y = 3$, the result is $2(2x - y) = 2 \cdot 3$ or
$4x - 2y =$ ________ .

6

32 If both sides of $3x - y = 7$ are multiplied by 4, the result is $4(3x - y) = 4 \cdot 7$ or $12x - 4y = 28$. Since 4 was multiplied by both sides of $3x - y = 7$, then the equations $3x - y = 7$ and ______________ have exactly the same solutions.

$12x - 4y = 28$

33 If both sides of $x - 3y = 4$ are multiplied by $^-1$, the equation $^-1(x - 3y) = {}^-1 \cdot 4$ or $^-x + 3y = {}^-4$ is obtained. Do $x - 3y = 4$ and $^-x + 3y = {}^-4$ have exactly the same solutions? ________ .

Yes

34 Do $x - 2y = 6$ and $^-3(x - 2y) = {}^-3 \cdot 6$ have exactly the same solutions? ________ .

Yes

35 Neither x nor y can be eliminated by addition from the following pair of equations:

$$3x - 2y = 6$$
$$x + y = 2$$

To find the common solution, multiply both sides of $x + y = 2$ by 2 to obtain $2x + 2y = 4$. Find the common solution:

$3x - 2y = 6$
$2x + 2y = 4$. ________ .

(2, 0)

36 Neither x nor y can be eliminated by addition from the following pair of equations:

$$5x + y = 12$$
$$2x + y = 3$$

To find the common solution, multiply both sides of $2x + y = 3$ by $^-1$ to obtain $^-2x - y = {}^-3$. Find the common solution:

$5x + y = 12$
$^-2x - y = {}^-3$. ________ .

$(3, {}^-3)$

37 Neither x nor y can be eliminated by addition from the following pair of equations:

$$4x - 3y = 9$$
$$6x + y = 19$$

To find the common solution, multiply both sides of $6x + y = 19$ by 3 to obtain $18x + 3y = 57$. Find the common solution:
$4x - 3y = 9$
$18x + 3y = 57$. ________ .

(3, 1)

38 Neither x nor y can be eliminated by addition from the following pair of equations:

$$3x + y = 5$$
$$2x - 3y = 7$$

To find the common solution, multiply both sides of $3x + y = 5$ by 3 to obtain $9x + 3y = 15$. Find the common solution:
$9x + 3y = 15$
$2x - 3y = 7$. ________ .

$(2, {}^{-}1)$

39 Neither x nor y can be eliminated by addition from the following pair of equations:

$$3x + 5y = 7$$
$$x - 4y = {}^{-}9$$

To find the common solution, multiply both sides of $x - 4y = {}^{-}9$ by ${}^{-}3$ to obtain ${}^{-}3x + 12y = 27$. Find the common solution:
$3x + 5y = 7$
${}^{-}3x + 12y = 27$. ________ .

$({}^{-}1, 2)$

40 Since $4x$ added to ${}^{-}4x$ is zero, what number could be multiplied by both sides of the second equation so that the x's can be eliminated by addition?
$4x + 3y = 17$
$x - 5y = 6$. ________ .

${}^{-}4$

41 ^-3y added to $3y$ is zero. What number can be multiplied by both sides of the first equation so that the y's can be eliminated by addition?

$2x - y = 4$
$5x + 3y = 19.$ ________ .

3

42 What number can be multiplied by both sides of the first equation so that the x's can be eliminated by addition?

$^-x + 3y = 7$
$4x - 2y = 5.$ ________ .

4

43 What number can be multiplied by both sides of the second equation so that the x's can be eliminated by addition?

$10x - 3y = 16$
$2x + 7y = 9.$ ________ .

$^-5$

44 What number can be multiplied by both sides of the first equation so that the y's can be eliminated by addition?

$x - 2y = 6$
$3x - 4y = 13.$ ________ .

$^-2$

45 What number can be multiplied by both sides of the second equation so that the y's can be eliminated by addition?

$3x - 5y = 9$
$2x + y = 4.$ ________ .

5

46 Multiply both sides of the second equation by 2 and find the common solution:

$3x + 2y = 6$
$x - y = 2.$ ________ .

(2, 0)

47 Multiply both sides of the first equation by $^-1$ and find the common solution:

$x + 4y = {}^-5$
$x - 3y = 9.$ ________ .

(3, $^-2$)

48 Multiply both sides of the second equation by $^{-}3$ and find the common solution:

$$2x + 3y = 7$$
$$x + y = 4. \_\_\_\_\_\_.$$

(5, $^{-}1$)

49 Multiply both sides of the second equation by 2 and find the common solution:

$$3x - 2y = 5$$
$$x + y = 5. \_\_\_\_\_\_.$$

(3, 2)

50 Multiply both sides of the first equation by 2 and find the common solution:

$$3x + y = 2$$
$$x - 2y = 10. \_\_\_\_\_\_.$$

(2, $^{-}4$)

51 Multiply both sides of the first equation by $^{-}4$ and find the common solution:

$$x - y = 5$$
$$5x - 4y = 28. \_\_\_\_\_\_.$$

(8, 3)

In finding the common solution of the following pair of equations, the three-step process shown should be used:

$$4x - 3y = 9$$
$$6x + y = 19$$

1. Multiply both sides of the second equation by 3 so that the y terms of the two equations are opposites.

$$\begin{aligned} 4x - 3y &= 9 \\ 6x + y &= 19 \\ \hline 4x - 3y &= 9 \\ 3(6x + y) &= 3(19) \\ \hline 4x - 3y &= 9 \\ 18x + 3y &= 57 \\ \hline \end{aligned}$$

($^{-}3y$ and $^{+}3y$ are opposites)

2. Add $4x - 3y = 9$ and $18x + 3y = 57$ to eliminate the y terms, and solve for x.

$$\begin{aligned} 4x - 3y &= 9 \\ 18x + 3y &= 57 \\ \hline 22x &= 66 \\ x &= 3 \end{aligned}$$

3. Use the fact, $x = 3$, in either of the two original equations to find y.

$$\begin{aligned} 4x - 3y &= 9 \qquad \text{and } x = 3 \\ 4(3) - 3y &= 9 \\ 12 - 3y &= 9 \\ {}^{-}12 &= {}^{-}12 \\ \hline {}^{-}3y &= {}^{-}3 \\ y &= 1 \end{aligned}$$

Therefore, the common solution of

$$\begin{aligned} 4x - 3y &= 9 \\ 6x + y &= 19 \end{aligned}$$

is (3, 1).

This solution should be checked in both of the original equations to show that true numerical statements are obtained in both cases.

$4x - 3y = 9$	(3, 1)	$6x + y = 19$
$4(3) - 3(1) = 9$		$6(3) + 1 = 19$
$12 - 3 = 9$		$18 + 1 = 19$
(true)		(true)

---

52 Find the common solution for
$x - 3y = 7$
${}^{-}2x + y = {}^{-}4$. ________ .

(1, ⁻2)

53 Find the common solution for
$5x + 2y = 3$
$8x - 4y = 12$. ________ .

(1, ⁻1)

54 Find the common solution for
$2x - 3y = 6$
$5x - y = 2.$ ________ . (0, $^-2$)

55 Find the common solution for
$x - 5y = {}^-2$
$2x + 5y = 26.$ ________ . (8, 2)

56 Find the common solution for
$3x - 4y = 7$
$x - 3y = 9.$ ________ . ($^-3$, $^-4$)

57 Find the common solution for
${}^-x + y = 9$
$3x + 5y = {}^-3.$ ________ . ($^-6$, 3)

58 Find the common solution for
$5x - y = 8$
$2x + y = 13.$ ________ . (3, 7)

59 Find the common solution for
$4x - y = 22$
$2x - y = 12.$ ________ . (5, $^-2$)

60 Find the common solution for
$x + 3y = 10$
$2x - 5y = 9.$ ________ . (7, 1)

61 Find the common solution for
$4x - 2y = 10$
$3x + 6y = 0.$ ________ . (2, $^-1$)

62 Find the common solution for
$3x + y = 7$
$2x - y = 3.$ ________ . (2, 1)

63 Find the common solution for
$2x + 3y = 3$
$x - 3y = {}^-21.$ ________ . ($^-6$, 5)

Self-Quiz # 2

The following questions test the objectives of the preceding section. 100% mastery is desired.

Find the common solution for the following:

1. $3x - y = 5$
$2x + 3y = 7$. ________ .

2. $5x - 2y = 19$
$3x + y = 7$. ________ .

3. $^{-}x + 3y = 10$
$5x + 2y = {}^{-}16$. ________ .

4. $3x - 4y = {}^{-}3$
$^{-}5x + 2y = 19$. ________ .

5. $3x + 7y = 12$
$^{-}x - 3y = {}^{-}4$. ________ .

To find a common solution for

$$5x - 3y = 8$$
$$3x + 2y = 6$$

both equations should be multiplied by a number. If the equation $5x - 3y = 8$ is multiplied by 2 and if $3x + 2y = 6$ is multiplied by 3, then the y's will be eliminated by addition in

$$10x - 6y = 16$$
$$9x + 6y = 18$$

---

64 Find the common solution for
$3x + 2y = 14$
$5x - 3y = 17$. ________ .

[*Note*: Multiply the first equation by 3 and the second equation by 2 so that the y's are eliminated by addition.]

(4, 1)

65 Find the common solution for

$2x + 5y = 6$

$3x - 2y = {}^{-}10.$ ________ . $({}^{-}2, 2)$

[*Note*: Multiply the first equation by 3 and the second equation by ${}^{-}2$ to eliminate the x's by addition.]

66 Find the common solution for

$2x - 7y = 1$

$5x - 6y = 14.$ ________ . $(4, 1)$

[*Note*: Multiply the first equation by ${}^{-}5$ and the second equation by 2 so that the x's are eliminated by addition.]

67 Find the common solution for

$5x - 2y = 0$

$2x + 3y = 19.$ ________ . $(2, 5)$

[*Note*: Multiply the first equation by 3 and the second equation by 2 so that the y's are eliminated by addition.]

In finding the common solution of the following pair of equations, the three-step process shown should be used:

$$3x + 2y = 7$$
$$5x + 3y = 10$$

1. Multiply both sides of the first equation by 3 and both sides of the second equation by ${}^{-}2$ so that the y terms of the two equations are opposites.

$$\begin{aligned} 3x + 2y &= 7 \\ 5x + 3y &= 10 \\ \hline 3(3x + 2y) &= 3(7) \\ {}^{-}2(5x + 3y) &= {}^{-}2(10) \\ \hline 9x + 6y &= 21 \\ {}^{-}10x - 6y &= {}^{-}20 \\ \hline \end{aligned}$$

(${}^{+}6y$ and ${}^{-}6y$ are opposites)

2. Add $9x + 6y = 21$ and $^{-}10x - 6y = {}^{-}20$ to eliminate the y terms, and solve for x.

$$\begin{array}{rcl} 9x + 6y &=& 21 \\ ^{-}10x - 6y &=& ^{-}20 \\ \hline ^{-}x &=& 1 \\ x &=& ^{-}1 \end{array}$$

3. Use the fact, $x = {}^{-}1$, in either of the two original equations to find y.

$$\begin{array}{rcl} 3x + 2y &=& 7 \qquad \text{and} \quad x = {}^{-}1 \\ 3({}^{-}1) + 2y &=& 7 \\ ^{-}3 + 2y &=& 7 \\ ^{+}3 \qquad && ^{+}3 \\ \hline 2y &=& 10 \\ y &=& 5 \end{array}$$

Therefore, the common solution of

$$\begin{array}{rcl} 3x + 2y &=& 7 \\ 5x + 3y &=& 10 \end{array}$$

is $({}^{-}1, 5)$.

This solution should be checked in both of the original equations to show that true numerical statements are obtained.

$$\begin{array}{rcc} 3x + 2y = 7 & ({}^{-}1, 5) & 5x + 3y = 10 \\ 3({}^{-}1) + 2(5) = 7 & & 5({}^{-}1) + 3(5) = 10 \\ ^{-}3 + 10 = 7 & & ^{-}5 + 15 = 10 \\ \text{(true)} & & \text{(true)} \end{array}$$

---

68 Find the common solution for
$5x - 4y = {}^{-}2$
$2x + 3y = 13$. ________ .

(2, 3)

69 Find the common solution for
$2x + 3y = 9$
$5x - 2y = 13$. ________ .

(3, 1)

70 Find the common solution for
$4x - 3y = 1$
$3x + 7y = 10.$ ________ . (1, 1)

71 Find the common solution for
$x - \frac{1}{3}y = 2$
$2x + y = 14.$ ________ . (4, 6)
[*Note*: Multiply both sides of the first equation by 3.]

72 Find the common solution for
$\frac{1}{2}x + 3y = {}^{-}1$
${}^{-}x + 2y = {}^{-}6.$ ________ . $(4, {}^{-}1)$
[*Note*: Multiply both sides of the first equation by 2.]

73 Find the common solution for
${}^{-}2x + y = {}^{-}7$
$\frac{1}{3}x - 6y = 7.$ ________ . $(3, {}^{-}1)$

74 Find the common solution for
$x + 3y = {}^{-}6$
$x - 2y = 14.$ ________ . $(6, {}^{-}4)$

75 Find the common solution for
$4x - 3y = 5$
${}^{-}3x + 2y = {}^{-}2.$ ________ . $({}^{-}4, {}^{-}7)$

76 Find the common solution for
$2x - 7y = 15$
$3x - 2y = 14.$ ________ . $(4, {}^{-}1)$

77 Find the common solution for
$2x - 8y = 2$
$x - 5y = 0.$ ________ . (5, 1)

78 Find the common solution for

$4x + 3y = 15$
$2x - 5y = {}^{-}25.$ ______ . (0, 5)

79 Find the common solution for

$2x + y = 6$
$3x - \frac{1}{4}y = 2.$ ______ . (1, 4)

80 Find the common solution for

$7x - 3y = 6$
$3x - 2y = {}^{-}1.$ ______ . (3, 5)

81 Find the common solution for

$3x - 5y = {}^{-}3$
$x + y = 7.$ ______ . (4, 3)

Self-Quiz # 3

The following questions test the objectives of the preceding section. 100% mastery is desired.

Find the common solution:

1. $3x + 2y = 11$
$4x + 5y = 3.$ ______ .

2. $x - \frac{1}{3}y = 1$
$2x + y = 7.$ ______ .

3. $2x + y = 1$
$x + y = 3.$ ______ .

4. $4x - 3y = 3$
$2x + y = 19.$ ______ .

5. $2x - 5y = 7$
$3x + 2y = 1.$ ______ .

The addition method for eliminating either x or y can always be used to find the common solution for two equations.

The substitution method, which will be explained in the following frames, can also be used to find the common solution for two equations.

---

82 The equation $x = 4y$ has been solved for x and shows that x is equal to $4y$. The equation $y = {}^{-}3x$ has been solved for y and shows that y is equal to ______ .

${}^{-}3x$

83 The equation $y = 10x$ has been solved for y and shows that y is equal to ______ .

$10x$

84 $3x + y$ is an open expression with two letters. If $x = 4y$, then $4y$ can be substituted for x in the open expression $3x + y$ as follows:

$$3x + y$$
$$3(4y) + y$$
$$12y + y$$
$$13y$$

For the open expression $2x + y$, if $x = 5y$, substitute $5y$ for x in $2x + y$ and simplify the result. ______ .

$11y$

85 If $x = 3y$, replace x by $3y$ in the open expression $5x - 7y$ as follows:

$$5x - 7y$$
$$5(3y) - 7y$$
$$15y - 7y$$
$$8y$$

If $x = 6y$, substitute $6y$ for x in $2x - 3y$ and simplify. ______ .

$9y$

86 If $y = 2x$, substitute $2x$ for y in $3x + 5y$ and simplify. ______ .

$13x$

87 If $y = {}^-3x$, substitute ${}^-3x$ for y in $5x + 2y$ and simplify. ________ .

${}^-x$

88 If $x = y + 3$, then to substitute $y + 3$ for x in $2x + 3y$ the following steps are used:

$$2x + 3y$$
$$2(y + 3) + 3y$$
$$2y + 6 + 3y$$
$$5y + 6$$

If $x = y + 7$, substitute $y + 7$ for x in $3x + 5y$ and simplify. ________ .

$8y + 21$

89 If $y = x - 4$, then to substitute $x - 4$ for y in $5x - 2y$, the following steps are used:

$$5x - 2y$$
$$5x - 2(x - 4)$$
$$5x - 2x + 8$$
$$3x + 8$$

If $y = x - 5$, substitute $x - 5$ for y in $8x - 3y$ and simplify. ________ .

$5x + 15$

90 If $x = y - 2$, substitute $y - 2$ for x in $3x - 4y$ and simplify. ________ .

${}^-y - 6$

91 To find the common solution for the pair of equations $2x - 3y = 5$ and $x = 4y$, the following steps are used. Since the second equation is solved for x, x can be replaced by $4y$ in the first equation to give

$$2x - 3y = 5$$
$$2(4y) - 3y = 5$$
$$8y - 3y = 5$$
$$5y = 5$$
$$y = 1$$

Since $y = 1$ and $x = 4y$, then $x =$ ________ .

4

92

$$3x - y = 10$$
$$x = 2y$$

The second equation has been solved for x. Using the substitution method, replace x by $2y$ in the first equation as follows:

$$3x - y = 10$$
$$3(2y) - y = 10$$
$$6y - y = 10$$
$$5y = 10$$
$$y = 2$$

Since $x = 2y$, (________, 2) is the common solution for the pair of equations.

(4, 2)

93

$$4x - y = 12$$
$$y = {}^{-}2x$$

The second equation has been solved for y. Using the substitution method, replace y by $^{-}2x$ in the first equation as follows:

$$4x - y = 12$$
$$4x - ({}^{-}2x) = 12$$
$$4x + 2x = 12$$
$$6x = 12$$
$$x = 2$$

(2, ________) is the common solution.

(2, $^{-}4$)

94

$$y = 3x$$
$$2x - 5y = {}^-13$$

The first equation has been solved for y. Using the substitution method, replace y by $3x$ in the second equation as follows:

$$2x - 5y = {}^-13$$
$$2x - 5(3x) = {}^-13$$
$$2x - 15x = {}^-13$$
$${}^-13x = {}^-13$$
$$x = 1$$

(1, ______) is the common solution. (1, 3)

95

$${}^-2y + x = 9$$
$$x = 5y$$

The second equation has been solved for x. Using the substitution method, replace x by $5y$ in the first equation as follows:

$${}^-2y + x = 9$$
$${}^-2y + 5y = 9$$
$$3y = 9$$
$$y = 3$$

(______, 3) is the common solution. (15, 3)

96

$$4x + 3y = 25$$
$$x = y + 1$$

The second equation has been solved for x. Replace x by $y + 1$ in the first equation as follows:

$$4x + 3y = 25$$
$$4(y + 1) + 3y = 25$$
$$4y + 4 + 3y = 25$$
$$7y + 4 = 25$$
$$7y = 21$$
$$y = 3$$

(________, 3) is the common solution. (<u>4</u>, 3)

97

$$y = 2x - 1$$
$$3x - y = 8$$

The first equation is solved for y. Using the substitution method, replace y by $2x - 1$ in the second equation as follows:

$$3x - y = 8$$
$$3x - (2x - 1) = 8$$
$$3x - 2x + 1 = 8$$
$$x + 1 = 8$$
$$x = 7$$

(7, ________) is the common solution. (7, <u>13</u>)

98 Find the common solution for
$2x + y = 10$
$y = 3x$. ________ . (2, 6)
[*Hint*: Replace y by $3x$ in the first equation.]

99 Find the common solution for
$4x + y = 21$
$x = 5y$. ________ . (5, 1)
[*Hint*: Replace x in the first equation by $5y$.]

To find the common solution for

$$2x + 5y = {}^{-}2$$
$$y = x + 8$$

by the substitution method, the following procedure must be followed:

1. Since the equation $y = x + 8$ shows that y is equal to $(x + 8)$, the y in the equation, $2x + 5y = {}^{-}2$, should be replaced by $(x + 8)$.

$$\begin{aligned} 2x + 5y &= {}^{-}2 \qquad y = x + 8 \\ 2x + 5(x + 8) &= {}^{-}2 \\ 2x + 5x + 40 &= {}^{-}2 \\ 7x + 40 &= {}^{-}2 \\ {}^{-}40 &\quad {}^{-}40 \\ \hline 7x &= {}^{-}42 \\ x &= {}^{-}6 \end{aligned}$$

2. Using ${}^{-}6$ as a replacement for x in either of the original equations makes it possible to determine the value of y.

$$\begin{aligned} y &= x + 8 \qquad x = {}^{-}6 \\ y &= {}^{-}6 + 8 \\ y &= 2 \end{aligned}$$

The common solution is $({}^{-}6, 2)$.

3. The common solution $({}^{-}6, 2)$ is checked in both of the original equations to show that true numerical statements are obtained.

$$\begin{aligned} 2x + 5y &= {}^{-}2 \qquad ({}^{-}6, 2) \qquad y = x + 8 \\ 2({}^{-}6) + 5(2) &= {}^{-}2 \qquad\qquad\qquad 2 = {}^{-}6 + 8 \\ {}^{-}12 + 10 &= {}^{-}2 \qquad\qquad\qquad \text{(true)} \\ &\text{(true)} \end{aligned}$$

---

100 Find the common solution for

$y = 7x$

$2x + y = 18.$ ________ .

(2, 14)

101 Find the common solution for

$x = 4y$

$x + y = 10.$ ________ .

(8, 2)

102 Find the common solution for

$x - 3y = 25$
$y = 2x$. ________ .

$(^-5, ^-10)$

103 Find the common solution for

$\frac{1}{2}x + 2y = 15$
$x = 6y$. ________ .

[*Hint*: Replace x by $6y$ in the first equation.]

$(18, 3)$

104 Find the common solution for

$2x - y = 9$
$x = y + 2$. ________ .

[*Hint*: Replace x in the first equation by $y + 2$.]

$(7, 5)$

105 Find the common solution for

$x = 2y + 3$
$2x + 3y = ^-1$. ________ .

[*Hint*: Replace x in the second equation by $2y + 3$.]

$(1, ^-1)$

106 Find the common solution for

$y = 3x$
$2x - y = 2$. ________ .

$(^-2, ^-6)$

107 Find the common solution for

$2x + y = 11$
$y = x + 2$. ________ .

$(3, 5)$

108 Find the common solution for

$2x - 3y = 4$
$x = 2y$. ________ .

$(8, 4)$

109 Find the common solution for

$x + 6y = 0$
$y = ^-2x$. ________ .

$(0, 0)$

110 Find the common solution for

$y = 5x$
$x + 2y = 22$. ________ .

$(2, 10)$

111 Find the common solution for

$3x - 2y = 8$

$x = y + 3.$ ______ . $(2, {}^{-}1)$

112 Find the common solution for

$y = 4x$

$5x - 2y = 9.$ ______ . $({}^{-}3, {}^{-}12)$

Self-Quiz # 4

The following questions test the objectives of the preceding section. 100% mastery is desired.

Find the common solution for each of the following pairs of equations:

1. $x = 3y$

$2x + y = 14.$ ______ .

2. $3x - 2y = {}^{-}5$

$y = 4x.$ ______ .

3. $x = y + 2$

$2x - y = 3.$ ______ .

4. $3x - 2y = {}^{-}8$

$x = 2y.$ ______ .

Two ways for finding common solutions have been shown in this chapter. Either method, addition to eliminate one of the letters or substitution, can be used with any pair of equations. Sometimes one method may be more convenient to use than the other, but either method may be used to arrive at the common solution.

In the following frames, find the common solution for each pair of equations. Use whichever method seems most convenient.

---

113 Find the common solution for
$2x - 5y = 3$
$x + 5y = 9.$ ________ . (4, 1)

114 Find the common solution for
$x + y = 1$
$2x + 5y = 14.$ ________ . ($^{-}3$, 4)

115 Find the common solution for
$y = 3x$
$x - 4y = 11.$ ________ . ($^{-}1$, $^{-}3$)

116 Find the common solution for
$4x + 2y = 16$
$x - 3y = 11.$ ________ . (5, $^{-}2$)

117 Find the common solution for
$x = 4y$
$x - 3y = 2.$ ________ . (8, 2)

118 Find the common solution for
$3x + y = 16$
$2x - 5y = 5.$ ________ . (5, 1)

Not every pair of equations has a common solution. The graphs of $3x - 2y = 6$ and $3x - 2y = {}^{-}2$ are shown below.

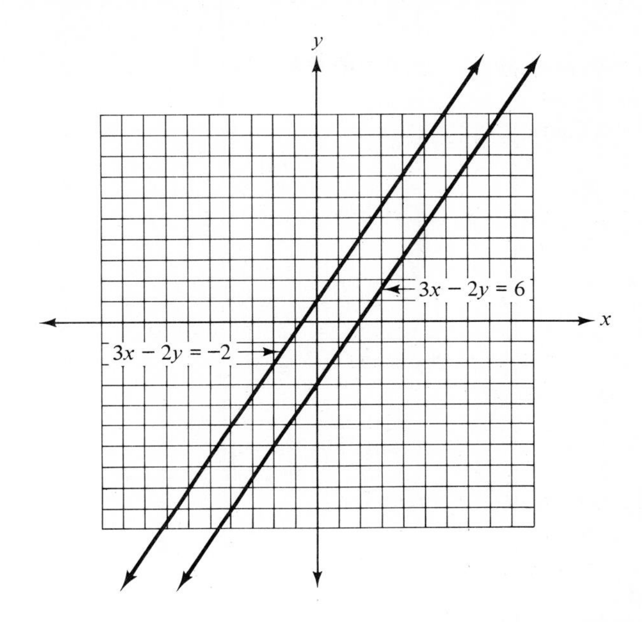

Notice that the lines are parallel and do not have a common point. The fact that there is no point common to both lines means that there is no ordered pair (x, y) that is a solution for both equations.

A method for determining when two equations do not have a common solution is shown in the following frames.

---

119 The graphs of $3x - 2y = 6$ and $3x - 2y = {}^{-}2$ are parallel lines. If both sides of the second equation are multiplied by $^{-}1$ and the equations are added, the result is

$$\begin{array}{r} 3x - 2y = 6 \\ {}^{-}3x + 2y = 2 \\ \hline 0 + 0 = \end{array}$$

______.

8

120

$$\begin{array}{r} 3x - 2y = 6 \\ ^{-}3x + 2y = 2 \\ \hline 0 = 8 \end{array}$$

$0 = 8$ is a false statement. This result shows that there is no common solution for the equations $3x - 2y = 6$ and $3x - 2y = {}^{-}2$.

Complete the addition of the following two equations:

$$\begin{array}{r} 2x + 5y = {}^{-}3 \\ ^{-}2x - 5y = {}^{-}8 \\ \hline \end{array}$$

$0 = {}^{-}11$

121

$$\begin{array}{r} 2x + 5y = {}^{-}3 \\ ^{-}2x - 5y = {}^{-}8 \\ \hline 0 = {}^{-}11 \end{array}$$

$0 = {}^{-}11$ is a false statement. Do the equations $2x + 5y = {}^{-}3$ and ${}^{-}2x - 5y = {}^{-}8$ have a common solution? ________ .

No

122 Whenever the elimination of either x or y results in a false statement like $0 = 8$ or $0 = {}^{-}11$, the two equations do not have a common solution. Does the following pair of equations have a common solution?

$3x + 8y = 6$

$3x + 8y = {}^{-}9$. ________________

No, because $0 = 15$ is false.

123 Add

$$\begin{array}{r} 7x - 2y = 6 \\ ^{-}7x + 2y = {}^{-}3 \\ \hline \end{array}$$

Do the equations have a common solution?

________________ .

No, because $0 = 3$ is false.

124 Does the following pair of equations have a common solution?

$2x + 4y = 9$

$2x + 4y = {}^{-}13$. ________ .

No

125 Not every pair of equations has a common solution. Find the common solution for the following pair of equations or show that there is no common solution.

$x + y = 7$
$x - y = 9.$ ______ . (8, ⁻1)

126 Not every pair of equations has a common solution. Find the common solution for the following pair of equations or show that there is no common solution.

$4x - 2y = 5$
$2x - y \ = 7.$ ____________ . No solution

For each of the following frames, 127–136, find the common solution or state that the equations have no common solution.

---

127 $5x + 3y = 1$
$2x - y \ = 7.$ ______ . (2, ⁻3)

128 $3x - 4y = 6$
$x + 2y = 12.$ ______ . (6, 3)

129 $5x + y = 7$
$5x + y = 1.$ ____________ . No solution

130 $2x - y \ = 4$
$x + 3y = 16.$ ______ . (4, 4)

131 $5x + 3y = {}^{-}5$
$x - 2y = {}^{-}1.$ ______ . (⁻1, 0)

132 $2x + 5y = 4$
$x + 3y = 3.$ ______ . (⁻3, 2)

133 $y = 2x$
$2x + y = 20.$ ______ . (5, 10)

134 $x = 2y$
$x - 2y = 7$. ________________ . No solution

135 $x + y = 1$
$2x - 3y = 17$. ________ . $(4, {}^{-}3)$

136 $x - y = 2$
$3x - 2y = 4$. ________ . $(0, {}^{-}2)$

137 $6x - 4y = 17$
${}^{-}3x + 2y = 5$. ________________ . No solution

138 $y = {}^{-}2x$
$2x - 3y = 24$. ________ . $(3, {}^{-}6)$

139 $3x - 4y = {}^{-}7$
$5x - 3y = 3$. ________ . $(3, 4)$

140 $5x - 2y = {}^{-}20$
${}^{-}3x + y = 11$. ________ . $({}^{-}2, 5)$

141 $3x + 5y = 18$
$8x - 3y = {}^{-}1$. ________ . $(1, 3)$

142 $y = 4x$
$4x - y = 9$. ________________ . No solution

143 $3x - 7y = 4$
$x + 5y = 16$. ________ . $(6, 2)$

144 $x = 3y$
$3x - 7y = 4.$ ________ . (6, 2)

145 $\frac{1}{3}x + 2y = 6$
$^{-}x + 5y = 4.$ ________ . (6, 2)

146 $3x - 6y = 10$
$x - 2y = {}^{-}7.$ ________________ . No solution

147 $7x - 4y = {}^{-}5$
$2x + 5y = 17.$ ________ . (1, 3)

148 $y = x + 2$
$3x - y = 8.$ ________ . (5, 7)

Self-Quiz # 5

The following questions test the objectives of the preceding section. 100% mastery is desired.

For each of the following pairs of equations find the common solution or state that there is no common solution:

1. $3x + y = 13$
$x - y = {}^{-}5.$ ________ .

2. $x = 3y$
$x - 2y = 4.$ ________ .

3. $3x - 2y = 7$
$6x - 4y = 1.$ ________ .

4. $2x - 5y = 3$
$3x - 2y = 10.$ ________ .

5. $x + y = 1$
$2x + 3y = 5.$ ________ .

Chapter Summary

The common solution for a pair of equations with two variables can be found in the following ways:

1. By eliminating one of the variables by addition.
2. By obtaining the value of one variable in terms of the other from one of the equations, and substituting in the other equation.
3. By locating the point of intersection of the two lines by graphing both equations on the same set of axes.

Not all pairs of equations with two variables have a solution. Example: parallel lines do not intersect.

The common solution of a pair of equations is checked by replacing x and y in both of the original equations to show that true numerical statements are obtained.

CHAPTER POST-TEST

The following questions test the objectives of this chapter. A score of 90% indicates sufficient mastery, and the student may proceed to the next chapter.

Find the common solution and check each pair of equations or state that the equations have no common solution.

1. $x + y = 17$
$x - y = 21$

2. $2x + y = 11$
$x + y = 7$

3. $3x - 5y = 5$
$x + 2y = 9$

4. $x + 2y = 7$
$2x + 4y = 3$

5. $x = 3y$
$x - 7y = 12$

6. $x = 2y - 1$
$3x - 5y = 0$

7. $3x - 2y = 11$
$2x + 5y = 39$

8. $2x + 4y = 14$
$7x + 2y = -11$

9. $x + y = 4$
$3x - y = 2$

10. $2x + y = 5$
$2x + 3y = 2$

(27 - 4x)
N
÷ 2Y =
1 x 3
5
÷
x
49
26
=

An Introduction to Polynomials

1
2
3
4
5
6
7
8
9
10

CHAPTER PRE-TEST

The following questions indicate the objectives of this chapter. A score of 90% indicates sufficient mastery, and the student may immediately take the Chapter Post-Test.

Multiply:

1. $(^{-}9x^3) \cdot (2x^2)$
2. $(^{-}x^5y^4) \cdot (^{-}4xy^3)$
3. $^{-}6(3x - 2)$
4. $7x(x^2 - 3)$
5. $^{-}2(x^2 - 5x - 3)$
6. $(x - 9)(x - 3)$
7. $(x + 7)(3x - 2)$
8. $(3x + 8)(2x - 5)$
9. $(3x - 1)^2$
10. $(x - 3)(x^2 - 5x + 8)$

Add:

11. $\dfrac{3}{x} - \dfrac{y}{5}$
12. $\dfrac{2}{x - 5} - \dfrac{x - 1}{7}$
13. $\dfrac{x - 5}{x + 2} + \dfrac{3x - 2}{x - 3}$

Divide:

14. $x - 2\overline{)x^2 - 9x + 14}$
15. $3x - 1\overline{)6x^2 + x - 5}$

The following frames are a review of exponents, the multiplication of monomial algebraic factors, and the Distributive Law of Multiplication over Addition.

---

1 x^5 means $xxxxx$
x^3 means ________ . xxx

2 x^4 means ________ . $xxxx$

3 xx means x^2.
$xxxx$ means ________ . x^4

4 $x^4 \cdot x^3 = xxxx \cdot xxx = xxxxxxx = x^7$.
$x^2 \cdot x^3 = xx \cdot xxx = xxxxx =$ ________ . x^5

5 $x \cdot x^2 = x \cdot xx = xxx = x^3$.
$x^4 \cdot x = xxxx \cdot x = xxxxx =$ ________ . x^5

6 $y^4 \cdot y =$ ________ . y^5

7 $x^7 \cdot x^2 =$ ________ . x^9

8 $z^2 \cdot z^4 =$ ________ . z^6

9 To multiply $(2x^2) \cdot (3x^5)$, the following steps are used:

$$(2x^2) \cdot (3x^5)$$
$$(2 \cdot 3) \cdot (x^2 \cdot x^5)$$
$$6 \cdot x^7$$
$$6x^7$$

Multiply $(3x^4) \cdot (5x^2)$. ________ . $15x^6$

10 $(4x^3) \cdot (^-2x) =$ ________ . $^-8x^4$

11 $(4x^3) \cdot (6x^5) =$ ________ . $24x^8$

12 $(^{-}7y^3) \cdot (^{-}3y^2) =$ ________ .

$21y^5$

13 To multiply $(^{-}5x^2y) \cdot (^{-}x^3y^2)$, the following steps are used:

$$(^{-}5x^2y) \cdot (^{-}x^3y^2)$$
$$(^{-}5x^2y) \cdot (^{-}1x^3y^2)$$
$$(^{-}5 \cdot {}^{-}1) \cdot (x^2x^3) \cdot (y \cdot y^2)$$
$$5x^5y^3$$

Multiply $(4xy^4) \cdot (^{-}x^2y^2)$. ________ .

$^{-}4x^3y^6$

14 $(^{-}2x^5y^2) \cdot (5x^3y) =$ ________ .

$^{-}10x^8y^3$

15 $(3x^2y^5) \cdot (^{-}5xy^4) =$ ________ .

$^{-}15x^3y^9$

16 $(^{-}4x^3y^3) \cdot (^{-}xy^2) =$ ________ .

$4x^4y^5$

17 $(7xy) \cdot (^{-}4xy) =$ ________ .

$^{-}28x^2y^2$

18 To remove the parentheses of $5(x - 2)$, the following steps are used:

$$5(x - 2)$$
$$5 \cdot x + 5 \cdot {}^{-}2$$
$$5x - 10$$

Remove the parentheses from $7(x - 8)$.

________ .

$7x - 56$

19 Remove the parentheses from $2(3x - 7)$.

________ .

$6x - 14$

20 Remove the parentheses from $6(2x - 8)$.

________________ .

$12x - 48$

21 Remove the parentheses from $3(2x - 5)$.

________________ .

$6x - 15$

$4x + 2x$ can be simplified as follows:

$$4x + 2x$$
$$(4 + 2)x$$
$$6x$$

This simplification is possible only because $4 + 2 = 6$. $5x^2 + 3x$ *cannot* be simplified to $8x$ or $8x^2$ because

$$5x^2 + 3x = (5x + 3)x$$

and $5x + 3$ *cannot* be simplified. Therefore, $5x^2 + 3x$ *cannot* be simplified.

---

22 To remove the parentheses from $3x(2x - 7)$, the following steps are used:

$$3x(2x - 7)$$
$$3x \cdot 2x + 3x \cdot {}^{-}7$$
$$6x^2 - 21x$$

Remove the parentheses from $5x(3x - 8)$.

________________ .

$15x^2 - 40x$

23 Remove the parentheses from $2x(4x - 3)$.

________________ .

$8x^2 - 6x$

24 Remove the parentheses from ${}^{-}3x(4x - 7)$.

________________ .

${}^{-}12x^2 + 21x$

25 To remove the parentheses from $6x^2y(3x - 4y^5)$, the following steps are used:

$$6x^2y(3x - 4y^5)$$
$$6x^2y \cdot 3x + 6x^2y \cdot {}^{-}4y^5$$
$$18x^3y - 24x^2y^6$$

Remove the parentheses from $5x^3y(2x^2y^3 - 7xy^8)$.

________________ .

$10x^5y^4 - 35x^4y^9$

26 Remove the parentheses from $2x^3y(3x^5 - 5x^4y^3)$.

________________ . $6x^8y - 10x^7y^4$

27 Remove the parentheses: $5x(x - 3)$.

________________ . $5x^2 - 15x$

28 Remove the parentheses: $^-3x(5 - 6x^2)$.

________________ . $^-15x + 18x^3$

29 Remove the parentheses: $-(2x - 5)$.

________________ . $^-2x + 5$

[*Hint:* $-(2x - 5)$ is equivalent to $^-1(2x - 5)$.]

30 Remove the parentheses: $-(5 - 7x)$.

________________ . $^-5 + 7x$

31 Remove the parentheses: $4x(2x^3 - 4x)$.

________________ . $8x^4 - 16x^2$

32 To remove the parentheses from $(4x + 3) \cdot 2x$, the following steps are used:

$$(4x + 3) \cdot 2x$$
$$4x \cdot 2x + 3 \cdot 2x$$
$$8x^2 + 6x$$

Remove the parentheses: $(5x + 6) \cdot 3x$.

________________ . $15x^2 + 18x$

33 Remove the parentheses: $(5x + 3) \cdot 6x$.

________________ . $30x^2 + 18x$

34 Remove the parentheses: $(3 + 5x) \cdot {}^-3x$.

________________ . $^-9x - 15x^2$

35 To remove the parentheses from $(4x - 7y^2) \cdot 5x^2y^2$, the following steps are used:

$$(4x - 7y^2) \cdot 5x^2y^2$$
$$4x \cdot 5x^2y^2 + {}^{-}7y^2 \cdot 5x^2y^2$$
$$20x^3y^2 - 35x^2y^4$$

Remove the parentheses: $(3x - 10y^3) \cdot 2xy$.

______________ .

$6x^2y - 20xy^4$

36 Remove the parentheses: $(x + 7) \cdot 7x$.

______________ .

$7x^2 + 49x$

37 Remove the parentheses: $(2x - 5) \cdot {}^{-}3x$.

______________ .

${}^{-}6x^2 + 15x$

38 Remove the parentheses: $2x^2y(3x - 5y)$.

______________ .

$6x^3y - 10x^2y^2$

39 Remove the parentheses: ${}^{-}4(3x - 2)$.

______________ .

${}^{-}12x + 8$

40 Remove the parentheses: $(4 - 5x) \cdot x$.

______________ .

$4x - 5x^2$

41 Remove the parentheses: $(2x - 5) \cdot {}^{-}6$.

______________ .

${}^{-}12x + 30$

42 Remove the parentheses: $4x(7 - x)$.

______________ .

$28x - 4x^2$

43 Remove the parentheses: ${}^{-}(4x^2 - 3y)$.

______________ .

${}^{-}4x^2 + 3y$

44 Remove the parentheses: $(7x - 2) \cdot 5x$.

______________ .

$35x^2 - 10x$

45 Remove the parentheses: $(2 - x) \cdot 5$.

______________ .

$10 - 5x$

46 Remove the parentheses: $^{-}5x(4 - 10x^3)$.

________________ .

$^{-}20x + 50x^4$

47 To remove the parentheses from $5(x + 3y - 7)$, the following steps are used:

$$5(x + 3y - 7)$$
$$5 \cdot x + 5 \cdot 3y + 5 \cdot {}^{-}7$$
$$5x + 15y + {}^{-}35$$
$$5x + 15y - 35$$

Remove the parentheses: $2(3x - 5y + 4)$.

________________ .

$6x - 10y + 8$

48 Remove the parentheses: $4(3a^2 - 2a + 5)$.

________________ .

$12a^2 - 8a + 20$

49 Remove the parentheses: $^{-}2(3x^2 + x - 7)$.

________________ .

$^{-}6x^2 - 2x + 14$

50 Remove the parentheses: $x(x^2 + 5x + 2)$.

________________ .

$x^3 + 5x^2 + 2x$

51 Remove the parentheses: $(2x^2 - 3x + 5) \cdot 3$.

________________ .

$6x^2 - 9x + 15$

52 Remove the parentheses: $5a(2a^2 + a - 3)$.

________________ .

$10a^3 + 5a^2 - 15a$

53 Remove the parentheses: $^{-}(4x^2 - x + 5)$.

________________ .

$^{-}4x^2 + x - 5$

54 Remove the parentheses: $c(c^2 - 5c - 7)$.

________________ .

$c^3 - 5c^2 - 7c$

55 Remove the parentheses: $(4a - 3b + 2c) \cdot 4$.

________________ .

$16a - 12b + 8c$

56 Remove the parentheses: $4x^2(3x^2 - x + 3)$.

____________________ .

$12x^4 - 4x^3 + 12x^2$

57 Remove the parentheses: $^-4(2x^2 - 3x - 1)$.

____________________ .

$^-8x^2 + 12x + 4$

58 Remove the parentheses: $3a(2a^2 - 5a + 4)$.

____________________ .

$6a^3 - 15a^2 + 12a$

59 Remove the parentheses: $2(x^2 - 3x + 2)$.

____________________ .

$2x^2 - 6x + 4$

60 Remove the parentheses: $^-2(x^2 - 5x + 6)$.

____________________ .

$^-2x^2 + 10x - 12$

61 Remove the parentheses: $(5x^2 - 3xy + 2) \cdot 3y$.

____________________ .

$15x^2y - 9xy^2 + 6y$

62 Remove the parentheses: $5(x^2 - 2x + 7)$.

____________________ .

$5x^2 - 10x + 35$

63 Remove the parentheses: $^-(^-2x^2 + 5x - 1)$.

____________________ .

$2x^2 - 5x + 1$

64 Remove the parentheses: $^-3(b^2 - 7b + 8)$.

____________________ .

$-3b^2 + 21b - 24$

65 Remove the parentheses: $2x(x^2 - 3x - 7)$.

____________________ .

$2x^3 - 6x^2 - 14x$

Self-Quiz # 1

The following questions test the objectives of the preceding section. 100% mastery is desired.

1. $(4x^5) \cdot (^-2x) =$ ________ .
2. $(5x^2y) \cdot (^-x^4y^7) =$ ________ .
3. $8(x - 4) =$ ________ .
4. $^-(3x - 5) =$ ________ .
5. $2x(x^2 + 7) =$ ________ .
6. $3(5x - 4) =$ ________ .
7. $^-5(x^2 + 3x - 2) =$ ________ .
8. $x^2y(x^2 - 2y^2 + xy) =$ ________ .

66 $3x$ is an expression that consists of *one* term because there is no addition indicated in the expression. $2y$ is an expression that consists of ________ term because there is no addition involved.

one

67 $4x^2y$ is an expression of *one* term because there is no addition involved. $3x^5y^2$ is an expression consisting of ________ term.

one

68 The expression $7x^4y^5z^8w^3$ is one term because there is no addition involved. How many terms are there in the expression $14x^5y^2z$? ________ .

One

69 How many terms are there in the expression $7x$?

________ .

One

70 How many terms are there in the expression $^-10x^2y$?

________ .

One

$2x^2y$ is an expression consisting of one term. An expression consisting of one term is called a *monomial*. Expressions involving more than one term are called *polynomials* and will be introduced in the following frames.

---

71 $2x + y$ is an expression involving two terms because $2x$ is added to y. How many terms are in the expression $4x + 5y$? ________ .

Two

72 How many terms are in the expression $5x + 3$?
________ .

Two

73 How many terms are in the expression $x + 6$?
________ .

Two

74 How many terms are in the expression $5x - 7y$?
________ . [*Hint:* $5x - 7y$ means $5x + {}^{-}7y$.]

Two

75 How many terms are in the expression $x^2 - 5x$?
________ .

Two

76 $4x + 3y + z$ is a polynomial consisting of three terms. How many terms are in the polynomial expression $2a + 3b - 2c$? ________ .

Three

77 How many terms are in the expression $2x^2 + 5x - 7$?
________ .

Three

78 $3x^2 + 4x + 5$ is a polynomial expression consisting of three terms because $3x^2$, $4x$, and 5 are monomial expressions that are being added. How many terms are in the expression $2x^2 + 5x + 9$? ________ .

Three

79 How many terms are in the expression $x - 5$?
________ .

Two

80 How many terms are in the expression $4x^2 - 7x + 6$? ______ . Three

81 How many terms are in the expression $4x^2y^3$? ______ . One

82 How many terms are in the expression $3x + 5y - 7z + 6$? ______ . Four

83 Any expression in which the operation addition appears one or more times is called a *polynomial.* $2x + 3y$ and $4x^2 + 3x + 7$ are expressions that have addition indicated at least one time and are, therefore, *polynomials.* Is $5x + 2y$ a polynomial? ______ . Yes

84 $2x - 5y + 4$ is a polynomial consisting of three terms. Is $4x^2 - 3x + 1$ a polynomial? ______ . Yes

85 $2x + 3y$ is a polynomial consisting of two terms. Is $5x + 7$ a polynomial? ______ . Yes

86 $6x^2y + x^2 - 5y - 7$ is a polynomial consisting of four terms. Is $3x^3 - x^2 + 5x + 4$ a polynomial? ______ . Yes

87 A polynomial that consists of two terms is called a *binomial.* $5a + 3b$ is a *binomial.* Is $4a + 7$ a binomial? ______ . Yes

88 $2x^2 + 5x$ is a binomial because it consists of two terms. Is ${}^{-}3x^2 + 4$ a binomial? ______ . Yes

89 Is $4a - 3b$ a binomial expression? ______ . Yes

90 Is $x^2 - y^2$ a binomial expression? ______ . Yes

91 Is $2x + 10$ a binomial expression? ______ . Yes

92 A polynomial that consists of three terms is called a *trinomial.* $2x^2 - x + 5$ is a *trinomial* because it is a polynomial with three terms. Is $2x^2 + x - 9$ a trinomial expression? ________ .

Yes

93 $4x - 3y + 2$ is a *trinomial* because it consists of three terms. Is $3x^2 - x + 7$ a trinomial expression? ________ .

Yes

94 Is $3a - 2b + c$ a trinomial expression? ________ .

Yes

95 Is $x^2 - 7x + 12$ a trinomial expression? ________ .

Yes

The expression $2x^2y$ is a *monomial* because it consists of just *one* term.

The expression $2x - 5y$ is a *binomial* because it consists of *two* terms.

The expression $4x - 5y + z$ is a *trinomial* because it consists of *three* terms.

An expression that consists of *more than one* term is called a *polynomial.* Binomial and trinomial expressions are polynomials. Binomials and trinomials are referred to by specific names because of their importance. Any expression that has four or more terms will be referred to as a polynomial with no specific name.

---

96 $5y$ is a monomial because it has just one term. Is $4xy$ a monomial? ________ .

Yes

97 $2x + 5y$ is a binomial because it has two terms. Is $4a - 3b$ a binomial? ________ .

Yes

98 $3x + 5y - 7w$ is a trinomial because it has three terms? Is $4a - 4b + 3c$ a trinomial? ________ .

Yes

Each of the following ten frames is to be answered by monomial, binomial, or trinomial.

---

99 The expression $2x - 3y$ is a ________________ . binomial

100 The expression $x^2 + 7x - 23$ is a ________________ . trinomial

101 The expression $^-4x^3$ is a ________________ . monomial

102 The expression $4x^2 + 9$ is a ________________ . binomial

103 The expression $3x^2yz$ is a ________________ . monomial

104 The expression $2x^2 - 5x$ is a ________________ . binomial

105 The expression $3x^2y^5z^8$ is a ________________ . monomial

106 The expression $2x^2 + x - 1$ is a ________________ . trinomial

107 The expression $x^2 - y^2$ is a ________________ . binomial

108 The expression $x^2 - 49$ is a ________________ . binomial

109 The binomial $2x + 4x$ can be simplified to the monomial $6x$ as follows:

$$2x + 4x$$
$$(2 + 4)x$$
$$6x$$

Simplify the binomial $8x + 3x$. ________ . $11x$

110 Simplify the binomial $7x - 2x$. ________ . $5x$

111 Simplify the binomial $^-8x + 2x$. ________ . ^-6x

112 The binomial $8x^2 + 5x^2$ can be simplified to the monomial $13x^2$ as follows:

$$8x^2 + 5x^2$$
$$(8 + 5)x^2$$
$$13x^2$$

Simplify the binomial $12x^2 + 3x^2$. ______ . $15x^2$

113 Simplify the binomial $4a^2 + 6a^2$. ______ . $10a^2$

114 Simplify the binomial $6a^2 - 4a^2$. ______ . $2a^2$

115 Simplify the binomial $17x^2 - 12x^2$. ______ . $5x^2$

116 Simplify the binomial expression $4x^2 - x^2$ by writing a monomial. ______ . $3x^2$

117 Simplify the binomial expression $4x - 7x$ by writing a monomial. ______ . ^-3x

118 Simplify $3xy + 13xy$. ______ . $16xy$

119 Simplify the trinomial expression $9y - 4y + 2y$ by writing a monomial. ______ . $7y$

120 Simplify $3a - 5a + 4a$. ______ . $2a$

121 Simplify $4x^2 - x^2 + 5x^2$. ______ . $8x^2$

122 Simplify $5x^2 - x^2$. ______ . $4x^2$

123 Simplify the binomial expression $5x^2y + 8x^2y$ by writing a monomial. ______ . $13x^2y$

124 The binomial $8x^2 + 3x$ *can*not be simplified to a monomial.

$$8x^2 + 3x = (8x + 3)x$$

$8x + 3$ *cannot* be simplified; therefore, $8x^2 + 3x$ *cannot* be written as a monomial. Can $9x^2+4x$ be simplified to a monomial?________ .

No

125 $7x + 4x$ can be simplified to the monomial $11x$. $8x^2 + 7x$ *cannot* be simplified to a monomial. Can $5x + 9x$ be simplified to a monomial? ________ .

Yes, $14x$

126 $12x^3 - 2x^2$ *cannot* be simplified to a monomial. $5x^3 - 2x^3$ can be simplified to the monomial $3x^3$. Can $12x^2 - 5x$ be simplified to a monomial? ________ .

No

127 Can $6x - 4x$ be simplified to a monomial? ________ .

Yes, $2x$

128 Can $4x^2 + 3x$ be simplified to a monomial? ________ .

No

129 Can $16x^3 - x^3$ be simplified to a monomial? ________ .

Yes, $15x^3$

130 Can $^-3x^2 + 7x^2$ be simplified to a monomial? ________ .

Yes, $4x^2$

131 Can $5x^3 + 2x^2$ be simplified to a monomial? ________ .

No

132 Can $8x + 5x$ be simplified to a monomial? ________ .

Yes, $13x$

133 Can $12x^2 - 9x^2$ be simplified to a monomial? ________ .

Yes, $3x^2$

134 Can $4x^5 + 3x^5$ be simplified to a monomial? ________ .

Yes, $7x^5$

Self-Quiz # 2

The following questions test the objectives of the preceding section. 100% mastery is desired.

1. $^{-}3x^2y$ is a monomial. True or false? ________ .
2. $x^2 - 3x + 4$ is a trinomial. True or false? ________ .
3. $a^2 - 4$ is a binomial. True or false? ________ .
4. $x^2 - 7$ is a monomial. True or false? ________ .
5. $x^2 + y^2$ is a binomial. True or false? ________ .
6. Can $9x + 2x$ be simplified to a monomial? (Yes or no.) ________ .
7. Can $3x^2 + 7$ be simplified to a monomial? (Yes or no.) ________ .
8. Can $5x^2 - 3x^2$ be simplified to a monomial? (Yes or no.) ________ .
9. Can $3y^2 - y$ be simplified to a monomial? (Yes or no.) ________ .

135 The first step in multiplying $8(x + 7)$ is $8 \cdot x + 8 \cdot 7$. The first step in multiplying $6(x + 4)$ is $6 \cdot x +$ ________________ .

$6 \cdot 4$

136 The first step in multiplying $2x^2(x + 5)$ is $2x^2 \cdot x + 2x^2 \cdot 5$. The first step in multiplying $3x^2(x + 8)$ is $3x^2 \cdot x +$ ________ .

$3x^2 \cdot 8$

137 The first step in multiplying $5xy^2z(2x + 3y)$ is $5xy^2z \cdot 2x + 5xy^2z \cdot 3y$. The first step in multiplying $2x^2yz^3(4x + 7y)$ is $2x^2yz^3 \cdot 4x +$ ________________ .

$2x^2yz^3 \cdot 7y$

138 Multiply $3x(4x + 3z)$. ________________ .

$12x^2 + 9xz$

139 Multiply $2x^2y(3x + 2y)$. ________________ .

$6x^3y + 4x^2y^2$

140 Multiply $5x(2x - 3)$. ________________ .

$10x^2 - 15x$

141 Multiply $10a^2b^3(3a + 7)$. ______________ . $30a^3b^3 + 70a^2b^3$

142 Multiply $^-3(2x + 6)$. ______ . $^-6x - 18$

143 Multiply $4x(3x^2 - 5x)$. ____________ . $12x^3 - 20x^2$

144 Multiply $7(4x + 5y)$. ____________ . $28x + 35y$

145 Multiply $10(5x - 8y)$. ____________ . $50x - 80y$

146 Multiply $^-5(x - 8)$. ______ . $^-5x + 40$

To remove the parentheses from $5(x + 8)$, each term of the binomial $x + 8$ is multiplied by 5.

To remove the parentheses from $(x + 4)(y + 6)$, each term of the binomial $y + 6$ is multiplied by the binomial $x + 4$.

Below are shown two problems: $5(x + 8)$ and $(x + 4)(y + 6)$. Compare the two problems. Note how $\underline{5}$ and $\underline{x + 4}$ are used.

$$\underline{5}(x + 8) \qquad (\underline{x + 4})(y + 6)$$

$$\underline{5} \cdot x + \underline{5} \cdot 8 \qquad (\underline{x + 4})(y) + (\underline{x + 4})(6)$$

$$5x + 40 \qquad xy + 4y + 6x + 24$$

---

147 To remove the parentheses of $\underline{5}(x + 8)$, each term of the binomial $x + 8$ is multiplied by $\underline{5}$. To remove the parentheses of $(\underline{x + 4})(y + 6)$, each term of the binomial $y + 6$ is multiplied by the binomial ______ . $x + 4$

148 The first step in multiplying $(x + 3)(y + 7)$ is $(x + 3)y + (x + 3) \cdot 7$. The first step in multiplying $(x + 5)(y + 3)$ is $(x + 5)y +$ ____________ . $(x + 5) \cdot 3$

149 The first step in multiplying $(x + 7)(y + 4)$ is $(x + 7)y + (x + 7) \cdot 4$. The first step in multiplying $(x + 3)(y + 11)$ is $(x + 3)y +$ ____________ . $(x + 3) \cdot 11$

150 The first step in multiplying $(x + 4)(y + 9)$ is $(x + 4)y + (x + 4) \cdot 9$. The first step in multiplying $(x + 3)(y + 10)$ is $(x + 3)y +$ ____________ .

$(x + 3) \cdot 10$

151 To multiply $(x + 4)(y + 6)$, the following steps are used:

$$(x + 4)(y + 6)$$
$$(x + 4)y + (x + 4)6$$
$$xy + 4y + x \cdot 6 + 4 \cdot 6$$
$$xy + 4y + 6x + 24$$

Multiply $(x + 8)(y + 2)$. ____________ .

$xy + 8y + 2x + 16$

152 To multiply $(x + 9)(y - 3)$, the following steps are used:

$$(x + 9)(y - 3) = (x + 9)(y + {}^{-}3)$$
$$(x + 9)y + (x + 9) \cdot {}^{-}3$$
$$xy + 9y - 3x - 27$$

Multiply $(x + 7)(y - 4)$. ____________ .

$xy + 7y - 4x - 28$

153 Multiply $(x + 6)(y - 5)$. ____________ .

$xy + 6y - 5x - 30$

154 Multiply $(x - 6)(y - 7)$. ____________ .

$xy - 6y - 7x + 42$

155 Multiply $(2x + 3)(y + 4)$. ____________ .

$2xy + 3y + 8x + 12$

156 Multiply $(x + 2)(3y + 4)$. ____________ .

$3xy + 6y + 4x + 8$

157 Multiply $(4x - 3)(y + 4)$. ____________ .

$4xy - 3y + 16x - 12$

158 $x^2 + 7x + 3x + 21$ can be simplified to $x^2 + 10x + 21$. Simplify $x^2 + \underline{4x + 2x} + 8$.

____________ .

$x^2 + 6x + 8$

159 Multiply the binomials and simplify as follows:

$$(x + 4)(x + 3)$$
$$(x + 4)x + (x + 4)3$$
$$xx + 4x + x \cdot 3 + 4 \cdot 3$$
$$x^2 + \underline{4x + 3x} + 12$$
$$x^2 + 7x + 12$$

Multiply and simplify:

$(x + 6)(x + 2)$. ______________ .

$x^2 + 8x + 12$

160 Complete the multiplication and simplification:

$$(x + 5)(x + 8)$$
$$(x + 5)x + (x + 5)8$$
$$xx + 5x + x \cdot 8 + 5 \cdot 8$$
$$x^2 + \underline{5x + 8x} + 40$$
$$x^2 + \underline{\qquad\qquad} + 40$$

$x^2 + \underline{13x} + 40$

The first step in multiplying $(x + 2)(x + 3)$ is:

$$(x + 2)x + (x + 2)3 \quad \text{or} \quad x(x + 3) + 2(x + 3)$$

Either first step is acceptable. Choose the first step that seems easier to apply correctly.

$(x + 2)(x + 3)$	$(x + 2)(x + 3)$
$(x + 2)x + (x + 2)3$	$x(x + 3) + 2(x + 3)$
$x^2 + 2x + 3x + 6$	$x^2 + 3x + 2x + 6$
$x^2 + 5x + 6$	$x^2 + 5x + 6$

---

161 Multiply and simplify the result:
$(x + 5)(x + 4)$. ______________ .

$x^2 + 9x + 20$

162 Multiply $(x + 7)(x + 4)$. ______________ .
[*Note:* Always simplify your result whenever possible.]

$x^2 + 11x + 28$

163 Multiply $(x + 5)(x + 7)$. ______________ .

$x^2 + 12x + 35$

164 Multiply $(x - 10)(x - 3)$. ________________ .

$x^2 - 13x + 30$

165 Multiply $(x + 2)(x - 2)$. ________________ .
[*Note:* $^-2x + 2x$ is 0.]

$x^2 - 4$

166 Multiply $(x + 6)(x - 6)$. ________________ .

$x^2 - 36$

167 Multiply $(x - 8)(x + 2)$. ________________ .

$x^2 - 6x - 16$

168 Multiply $(x + 7)(x + 3)$. ________________ .

$x^2 + 10x + 21$

169 Multiply $(x - 5)(x + 5)$. ________________ .

$x^2 - 25$

A shortcut method for multiplying binomials is called the FOIL *method.* Each capital letter stands for one step in the FOIL method of multiplying binomials.

---

170 In the binomial $2x - 3$ the first term (F) is $2x$, and the last term (L) is $^-3$. In the binomial $4x - 7$ the first term (F) is $4x$, and the last term (L) is ________ .

$^-7$

171 In the binomial $5x + 4$ the first term (F) is ________, and the last term (L) is ________ .

$5x$, 4

172 In the multiplication expression

$$(5x - 6)(4x + 9)$$

$5x$ and 9 are the outside (O) terms, and $^-6$ and $4x$ are the inside (I) terms. What are the outside terms (O) of $(2x + 5)(3x - 2)$? ________________ .

$2x$ and $^-2$

173 In the multiplication expression

$$(3x + 5)(x - 2)$$

the outside (O) terms are $3x$ and $^{-}2$. The inside terms (I) are 5 and x. What are the inside terms (I) of $(4x - 7)(9x - 1)$? ________________ .

$^{-}7$ and $9x$

174 To multiply

$$(3x - 2)(x + 7)$$

by the FOIL method, it is necessary to find:
(a) F terms, $3x$ and x.
(b) O terms, $3x$ and 7.
(c) I terms, $^{-}2$ and x.
(d) L terms, ________ and ________ .

$^{-}2$ and 7

175 To multiply $(2x + 5)(3x - 7)$ by the FOIL method:
(a) Multiply the F terms

$$2x \cdot 3x = 6x^2$$

(b) Multiply the O terms

$$2x \cdot {^{-}7} = {^{-}14x}$$

(c) Multiply the I terms

$$5 \cdot 3x = 15x$$

(d) Multiply the L terms
$5 \cdot {^{-}7} =$ ________ .

$^{-}35$

176

$$(2x + 5)(3x - 7)$$

(F, O, I, L brackets: F joins $2x$ and $3x$; O joins $2x$ and $^{-}7$; I joins 5 and $3x$; L joins 5 and $^{-}7$)

$$\text{F} + \text{O} + \text{I} + \text{L}$$

$$6x^2 + {^{-}14x} + 15x + {^{-}35}$$

Simplify $6x^2 - 14x + 15x - 35$ by combining $^{-}14x$ and $15x$. ________________ .

$6x^2 + x - 35$

177 To multiply $(x - 7)(3x + 2)$ using the FOIL method:

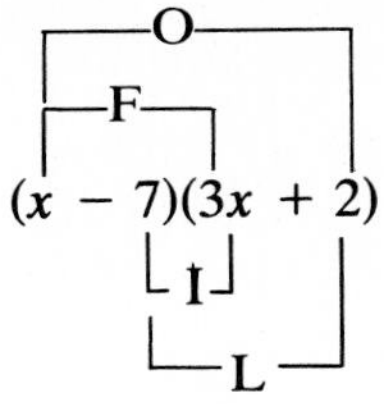

$$(x - 7)(3x + 2)$$

O, F, I, L

$$\begin{array}{cccc} F & + O + & I & + L \\ 3x^2 & + 2x + & ^-21x & + ^-14 \end{array}$$

Simplify the sum. ______________ .

$3x^2 - 19x - 14$

178 To multiply $(x + 3)(x - 8)$ by the FOIL method:

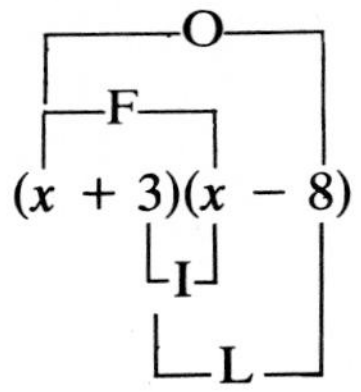

$$(x + 3)(x - 8)$$

O, F, I, L

$$\begin{array}{cccc} F & + O & + I & + L \\ x^2 & + ^-8x & + 3x & + ^-24 \end{array}$$

Simplify the sum. ______________ .

$x^2 - 5x - 24$

179 Multiply $(x + 8)(3x + 1)$ by using the FOIL method:

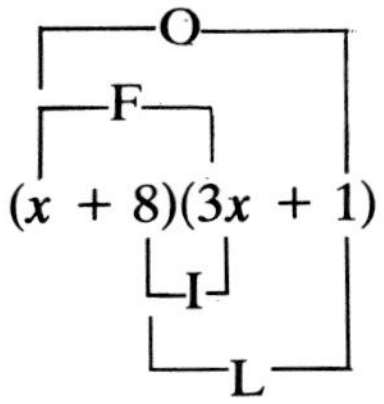

$$(x + 8)(3x + 1)$$

O, F, I, L

______________ .

$3x^2 + 25x + 8$

180 Multiply $(x - 6)(x + 4)$ by using the FOIL method:

$$(x - 6)(x + 4)$$

________________ . 	$x^2 - 2x - 24$

181 Multiply $(x + 3)(x - 1)$. ________________ . 	$x^2 + 2x - 3$

182 $(3x - 5)(x + 4) =$ ________________ . 	$3x^2 + 7x - 20$

183 $(x - 7)^2 =$ ________________ . 	$x^2 - 14x + 49$
[*Note:* $(x - 7)^2$ means $(x - 7)(x - 7)$.]

184 $(5x - 2)(3x - 1) =$ ________________ . 	$15x^2 - 11x + 2$

185 $(2x + 3)(x + 7) =$ ________________ . 	$2x^2 + 17x + 21$

186 $(x + 7)^2 =$ ________________ . 	$x^2 + 14x + 49$

187 $(x + 7)(x - 1) =$ ________________ . 	$x^2 + 6x - 7$

188 $(x + 8)(x - 8) =$ ________________ . 	$x^2 - 64$

189 $(x + 4)(x + 5) =$ ________________ . 	$x^2 + 9x + 20$

190 $(x - 6)(x - 5) =$ ________________ . 	$x^2 - 11x + 30$

The binomials $(x + 5)$ and $(x - 5)$ are called the *sum and difference* of x and 5.

The product of $(x + 5)(x - 5)$ is a binomial because the x terms obtained in the multiplication always have a sum of *zero*, as shown in the following example:

$$(x + 5)(x - 5)$$
$$x^2 - 5x + 5x - 25$$
$$x^2 - 25$$

The binomial product $x^2 - 25$ is called the *difference of two squares* because $x^2 = x \cdot x$, and $25 = 5 \cdot 5$.

---

191 $(x - 1)(x + 1) =$ ________________ . $x^2 - 1$

192 $(x - 3)(x - 9) =$ ________________ . $x^2 - 12x + 27$

193 $(x + 4)(x - 6) =$ ________________ . $x^2 - 2x - 24$

194 $(x + 12)(x + 5) =$ ________________ . $x^2 + 17x + 60$

195 $(x + 3)(x - 10) =$ ________________ . $x^2 - 7x - 30$

196 $(x - 5)(x - 7) =$ ________________ . $x^2 - 12x + 35$

197 $(x - 2)(x + 3) =$ ________________ . $x^2 + x - 6$

198 $(x - 5)^2 =$ ________________ . $x^2 - 10x + 25$
[*Note:* $(x - 5)^2$ means $(x - 5)(x - 5)$.]

199 $(x - 3)(x + 3) =$ ________________ . $x^2 - 9$

200 $(x - 5)(x + 8) =$ ________________ . $x^2 + 3x - 40$

201 $(x + 7)(x + 3) =$ ________________ . $x^2 + 10x + 21$

202 $(x - 1)(x + 3) =$ ________________ . $x^2 + 2x - 3$

203 $(x + 10)(x + 3) =$ ________________ .

$x^2 + 13x + 30$

204 $(x + 9)(x - 4) =$ ________________ .

$x^2 + 5x - 36$

205 $(x + 4)^2 =$ ________________ .

$x^2 + 8x + 16$

206 To remove the parentheses of $\underset{\sim}{5}(x^2 + 3x - 2)$, each term of the trinomial is multiplied by $\underset{\sim}{5}$. To remove the parentheses of $\underline{(x + 4)}(x^2 + 3x - 2)$, each term of the trinomial is multiplied by ________________ .

$x + 4$

207 To remove the parentheses of $(x + 4)(x^2 + 3x - 2)$, the following steps are used:

$$\underline{(x + 4)}(x^2 + 3x - 2)$$
$$\underline{(x + 4)}x^2 + \underline{(x + 4)}3x + \underline{(x + 4)} \cdot {}^{-}2$$
$$x \cdot x^2 + 4x^2 + x \cdot 3x + 4 \cdot 3x + x \cdot {}^{-}2 + 4 \cdot {}^{-}2$$
$$x^3 + \underline{4x^2 + 3x^2} + \underline{12x - 2x} - 8$$
$$x^3 + 7x^2 + 10x - 8$$

Multiply and simplify $(x + 3)(x^2 + 5x - 4)$.

________________ .

$x^3 + 8x^2 + 11x - 12$

208 Multiply $(x + 5)(x^2 + 2x + 3)$. ________________ .
[*Note:* Simplify your result.]

$x^3 + 7x^2 + 13x + 15$

209 Multiply (always simplify your result)
$(x + 6)(x^2 - 3x - 1)$ ________________ .

$x^3 + 3x^2 - 19x - 6$

210 Multiply $(x - 3)(x^2 + 5x - 7)$. ________________ .

$x^3 + 2x^2 - 22x + 21$

211 Use the FOIL method to multiply $(x + 5)(2x + 3)$.

________________ .

$2x^2 + 13x + 15$

212 Multiply $(3x - 1)(x + 2)$. ________________ .

$3x^2 + 5x - 2$

213 $(2x + 3)(3x + 1) =$ ________________ .

$6x^2 + 11x + 3$

214 $(6x - 1)(6x + 1) =$ ________________ . $36x^2 - 1$

215 $(2x - 7)(2x + 7) =$ ________________ . $4x^2 - 49$

216 $(2x - 1)(5x + 1) =$ ________________ . $10x^2 - 3x - 1$

217 $(4x + 3)(x - 3) =$ ________________ . $4x^2 - 9x - 9$

218 $(2x - 3)^2 =$ ________________ . $4x^2 - 12x + 9$
[*Note:* $(2x - 3)^2$ means $(2x - 3)(2x - 3)$.]

219 $(3x - 1)^2 =$ ________________ . $9x^2 - 6x + 1$

220 $(x + 9)(x - 3) =$ ________________ . $x^2 + 6x - 27$

221 $(2x - 1)(x + 5) =$ ________________ . $2x^2 + 9x - 5$

222 $(x - 9)(x + 9) =$ ________________ . $x^2 - 81$

223 $(x + 10)^2 =$ ________________ . $x^2 + 20x + 100$

224 $(2x + 3)(3x - 7) =$ ________________ . $6x^2 - 5x - 21$

225 $(x + 10)(x - 10) =$ ________________ . $x^2 - 100$

226 $(3x - 2)^2 =$ ________________ . $9x^2 - 12x + 4$

227 $(x + 4)(x^2 - 3x + 4) =$ ________________ . $x^3 + x^2 - 8x + 16$

Self-Quiz # 3

The following questions test the objectives of the preceding section. 100% mastery is desired.

1. $3(2x - 3) =$ _______ .
2. $^{-}5(^{-}3x + 2y - 7) =$ _______ .
3. $(x + 3)(x + 5) =$ _______ .
4. $(x - 7)(x - 4) =$ _______ .
5. $(x + 6)(x - 2) =$ _______ .
6. $(x + 3)(x - 8) =$ _______ .
7. $(2x - 3)(x + 6) =$ _______ .
8. $(5x - 3)(2x + 7) =$ _______ .
9. $(x + 5)(x - 5) =$ _______ .
10. $(3x - 2)(3x + 2) =$ _______ .
11. $(x + 3)^2 =$ _______ .
12. $(2x - 3)^2 =$ _______ .
13. $(x + 3)(x^2 - 2x + 2) =$ _______ .
14. $(x - 2)(x^2 + 4x - 3) =$ _______ .

Polynomial fractions are added by using the same procedure that is used in the addition of rational numbers:

1. The least common multiple (LCM) of the denominators is found.
2. Each fraction is written with the LCM as its new denominator.
3. The numerators are added; their sum is placed over the LCM as the denominator.

In the following section the addition of polynomial fractions is introduced.

---

228 The least common multiple (LCM) of 4 and 5 is 20. Complete the following addition:

$$\frac{3}{4} + \frac{1}{5}$$

$$\frac{3}{4} \cdot \frac{5}{5} + \frac{1}{5} \cdot \frac{4}{4}$$

$$\frac{15}{20} + \frac{4}{20}$$

$$\frac{15 + 4}{20}$$

________________ .

$\frac{19}{20}$

229 Add $\frac{2}{3} + \frac{1}{4}$. ________ .

$\frac{11}{12}$

230 Add $\frac{2}{3} + \frac{1}{7}$. ________ .

$\frac{17}{21}$

231 Add $\frac{7}{9} + \frac{{}^{-}1}{4}$. ________ .

$\frac{19}{36}$

232 The least common multiple (LCM) of y and 3 is $3y$. Complete the following addition:

$$\frac{2x}{y} + \frac{1}{3}$$

$$\frac{2x}{y} \cdot \frac{3}{3} + \frac{1}{3} \cdot \frac{y}{y}$$

$$\frac{6x}{3y} + \frac{y}{3y}$$

________________ .

$\frac{6x + y}{3y}$

233 Add $\frac{3x}{5} + \frac{1}{y}$. ________________ .

$\frac{3xy + 5}{5y}$

234 The least common multiple (LCM) of x and y is xy. Complete the following addition:

$$\frac{3}{x} + \frac{5}{y}$$

$$\frac{3}{x} \cdot \frac{y}{y} + \frac{5}{y} \cdot \frac{x}{x}$$

________________ .

$$\frac{3y + 5x}{xy}$$

235 Add $\frac{7}{x} + \frac{2}{y}$. ________________ .

$$\frac{7y + 2x}{xy}$$

236 Add $\frac{5x}{2} + \frac{3}{y}$. ________________ .

$$\frac{5xy + 6}{2y}$$

237 Add $\frac{6}{5x} + \frac{2}{y}$. ________________ .

$$\frac{6y + 10x}{5xy}$$

238 Add $\frac{5}{x} + \frac{{}^{-}2x}{y}$. ________________ .

$$\frac{5y - 2x^2}{xy}$$

239 Add $\frac{6}{5} - \frac{2}{x}$. ________________ .

$$\frac{6x - 10}{5x}$$

240 The LCM of 5 and 7 is 35. Complete the following addition:

$$\frac{x - 3}{5} + \frac{2}{7}$$

$$\frac{x - 3}{5} \cdot \frac{7}{7} + \frac{2}{7} \cdot \frac{5}{5}$$

$$\frac{7(x - 3)}{35} + \frac{2 \cdot 5}{35}$$

________________ .

$$\frac{7x - 11}{35}$$

241 Add $\frac{x - 3}{2} + \frac{1}{5}$. ________________ .

$$\frac{5x - 13}{10}$$

242. Add $\frac{2x + 1}{3} - \frac{1}{2}$. ________________ .

$\frac{4x - 1}{6}$

243 The LCM of 3 and $x - 3$ is $3(x - 3)$. Complete the following addition:

$$\frac{x + 1}{3} + \frac{1}{x - 3}$$

$$\frac{x + 1}{3} \cdot \frac{x - 3}{x - 3} + \frac{1}{x - 3} \cdot \frac{3}{3}$$

$$\frac{(x + 1)(x - 3)}{3(x - 3)} + \frac{1 \cdot 3}{3(x - 3)}$$

$$\frac{x^2 - 2x - 3 + 3}{3x - 9}$$

________________ .

$\frac{x^2 - 2x}{3x - 9}$

244 Add $\frac{x + 3}{x - 1} + \frac{2}{3}$. ________________ .

$\frac{5x + 7}{3x - 3}$

245 The LCM of $x - 1$ and $x + 3$ is $(x - 1)(x + 3)$. Complete the following addition:

$$\frac{x + 4}{x - 1} + \frac{x - 2}{x + 3}$$

$$\frac{x + 4}{x - 1} \cdot \frac{x + 3}{x + 3} + \frac{x - 2}{x + 3} \cdot \frac{x - 1}{x - 1}$$

$$\frac{(x + 4)(x + 3)}{(x - 1)(x + 3)} + \frac{(x - 2)(x - 1)}{(x + 3)(x - 1)}$$

$$\frac{x^2 + 7x + 12 + x^2 - 3x + 2}{x^2 + 2x - 3}$$

________________ .

$\frac{2x^2 + 4x + 14}{x^2 + 2x - 3}$

246 Add $\frac{x - 6}{x + 3} + \frac{x - 2}{x - 1}$. ________________ .

$\frac{2x^2 - 6x}{x^2 + 2x - 3}$

247 Add $\frac{x + 8}{x + 4} + \frac{x - 2}{x + 2}$. ________________ .

$\frac{2x^2 + 12x + 8}{x^2 + 6x + 8}$

248 Add $\dfrac{x-6}{x+5}+\dfrac{x-5}{x-6}$. ______________ .

$\dfrac{2x^2-12x+11}{x^2-x-30}$

249 Add $\dfrac{x-7}{x+2}+\dfrac{x+3}{x-4}$. ______________ .

$\dfrac{2x^2-6x+34}{x^2-2x-8}$

250 To add $\dfrac{x+2}{x-3}-\dfrac{x-7}{x-1}$, the following steps are used:

$$\frac{x+2}{x-3}-\frac{x-7}{x-1}$$

$$\frac{(x+2)(x-1)-(x-7)(x-3)}{(x-3)(x-1)}$$

$$\frac{x^2+x-2-(x^2-10x+21)}{x^2-4x+3}$$

$$\frac{x^2+x-2-x^2+10x-21}{x^2-4x+3}$$

$$\frac{11x-23}{x^2-4x+3}$$

Add $\dfrac{x-1}{x+2}-\dfrac{x+3}{x-7}$. ______________ .

$\dfrac{{}^-13x+1}{x^2-5x-14}$

251 Add $\dfrac{x-1}{x+3}-\dfrac{x+5}{x-2}$. ______________ .

$\dfrac{{}^-11x-13}{x^2+x-6}$

252 $\dfrac{x-6}{x+3}-\dfrac{x+3}{x-2}=$ ______________ .

$\dfrac{{}^-14x+3}{x^2+x-6}$

253 $\dfrac{5x}{2}+\dfrac{3x}{5}=$ ______________ .

$\dfrac{31x}{10}$

254 $\dfrac{2x+1}{5}-\dfrac{5x+2}{3}=$ ______________ .

$\dfrac{{}^-19x-7}{15}$

255 $\dfrac{2x-1}{x+4}+\dfrac{x-7}{3x+1}=$ ______________ .

$\dfrac{7x^2-4x-29}{3x^2+13x+4}$

256 $\frac{x}{5} - \frac{y}{6} = $ ________________ .

$\frac{6x - 5y}{30}$

257 $\frac{4}{3x} - \frac{5}{4x} = $ ________________ .

$\frac{1}{12x}$

Self-Quiz # 4

The following questions test the objectives of the preceding section. 100% mastery is desired.

1. $\frac{5x}{2} + \frac{4x}{3} = $ ________ .

2. $\frac{2x}{5} - \frac{4y}{3} = $ ________ .

3. $\frac{x + 3}{2} + \frac{x - 4}{5} = $ ________ .

4. $\frac{x - 5}{x + 2} + \frac{x + 3}{x - 4} = $ ________ .

5. $\frac{2x - 7}{5} - \frac{3x + 1}{6} = $ ________ .

6. $\frac{a}{b} + \frac{c}{d} = $ ________ .

7. $\frac{2x + 1}{x + 4} + \frac{3x + 2}{3x - 1} = $ ________ .

8. $\frac{3x}{4} - \frac{2y}{7} = $ ________ .

258 5 subtracted from 7 is shown by $7 - 5$.

$$7 - 5 = 7 + {}^{-}5 = 2$$

8 subtracted from 12 is ________ .

4

259 $^-3$ subtracted from 5 is shown by $5 - {}^-3$.

$$5 - {}^-3 = 5 + 3 = 8$$

$^-4$ subtracted from 10 is ______ . 14

260 $7 - 3 =$ ______ . 4

261 $9 - {}^-4 =$ ______ . 13

262 Subtraction means the addition of the opposite of the second number (the number following the "–" sign).
$7 - 5$ means $7 + {}^-5$.
$3 - {}^-6$ means $3 + 6$.
$^-4 - 5$ means $^-4 + {}^-5$.
$^-2 - {}^-8$ means ______ . $^-2 + 8$

263 $8 - 3 = 8 + {}^-3 =$ ______ . 5

264 $^-7 - 3 = {}^-7 + {}^-3 =$ ______ . $^-10$

265 $9 - {}^-2 = 9 + 2 =$ ______ . 11

266 The opposite of $5x + 6$ is $^-5x - 6$. Write the opposite of $7x + 9$. ______ . $^-7x - 9$

267 The opposite of $3x - 8$ is $^-3x + 8$. Write the opposite of $8x - 5$. ______ . $^-8x + 5$

268 The opposite of $^-3x + 2$ is $3x - 2$. Write the opposite of $^-7x + 3$. ______ . $7x - 3$

269 To subtract $3x + 7$ from $5x - 2$, the opposite of $3x + 7$ is added to $5x - 2$. Write the opposite of $3x + 7$. ______ . $^-3x - 7$

270 Write the opposite of $4x - 9$. ______ . $^-4x + 9$

271 To subtract

$$\begin{array}{r} 5x - 2 \\ \underline{3x + 7} \end{array}$$

first write the opposite of $3x + 7$ below $5x - 2$ as follows:

$$\begin{array}{r} 5x - 2 \\ \underline{{}^{-}3x - 7} \end{array}$$

Complete the subtraction by adding the binomials. ______ .

$2x - 9$

272 To subtract

$$\begin{array}{r} 4x + 7 \\ \underline{2x - 3} \end{array}$$

first write the opposite of $2x - 3$ below $4x + 7$ as follows:

$$\begin{array}{r} 4x + 7 \\ \underline{{}^{-}2x + 3} \end{array}$$

Then add the binomials. ______ .

$2x + 10$

273 Subtract:

$$\begin{array}{r} 3x - 7 \\ \underline{5x + 2} \end{array}$$

${}^{-}2x - 9$

[*Note*: Change $5x + 2$ to its opposite, ${}^{-}5x - 2$; then add to $3x - 7$.]

274 Subtract:

$$\begin{array}{r} x - 3 \\ \underline{3x - 4} \end{array}$$

${}^{-}2x + 1$

275 Subtract:

$$\begin{array}{r} x^2 + 4x \\ \underline{2x^2 - 3x} \end{array}$$

$^{-}x^2 + 7x$

276 Subtract:

$$\begin{array}{r} 2x^2 - 3x \\ \underline{2x^2 - 3x} \end{array}$$

0

277 Subtract:

$$\begin{array}{r} x^2 - 10x \\ \underline{x^2 + \ \ 2x} \end{array}$$

$^{-}12x$

278 Subtract:

$$\begin{array}{r} x^2 + 3x \\ \underline{x^2 - 2x} \end{array}$$

$5x$

279 Subtract:

$$\begin{array}{r} 4x^2 - 3x \\ \underline{2x^2 - 3x} \end{array}$$

$2x^2$

280 Subtract:

$$\begin{array}{r} x^2 + 2x \\ \underline{x^2 + 5x} \end{array}$$

$^{-}3x$

The following frames will be devoted to division of polynomials, which follows the same pattern as long division in arithmetic. For example,

$$\begin{array}{r} 41 \\ 24\overline{)984} \\ 96 \\ \hline 24 \\ 24 \\ \hline 0 \end{array}$$

24 is the divisor
984 is the dividend
41 is the quotient
0 is the remainder

281 In the division problem

$$x - 3\overline{)x^2 + 7x + 13}$$

$(x - 3)$ is the divisor and $(x^2 + 7x + 13)$ is the dividend. In the division problem

$$x - 5\overline{)x^2 - x + 30}$$

$(x - 5)$ is the divisor and $(x^2 - x + 30)$ is the ________________ .

dividend

282 In the problem

$$x + 5\overline{)x^2 + 7x + 10}$$

$(x^2 + 7x + 10)$ is the ________________ .

dividend

283 In the problem

$$x - 7\overline{)x^2 + 6x + 3}$$

$(x^2 + 6x + 3)$ is the dividend and $x - 7$ is the divisor. In

$$x + 5\overline{)x^2 + 3x + 2}$$

$x + 5$ is the ________________ .

divisor

284 In the division problem

$$x + 10\overline{)x^2 - 3x + 17}$$

$x + 10$ is the ________________ .

divisor

285 In the problem

$$\begin{array}{r} x + 5 \\ x - 3\overline{)x^2 + 2x - 15} \\ \underline{x^2 - 3x} \\ 5x - 15 \\ \underline{5x - 15} \\ 0 \end{array}$$

$x - 3$ is the divisor, $x^2 + 2x - 15$ is the dividend, and $x + 5$ is the quotient. What is the quotient in the following example? ________ .

$x + 4$

$$\begin{array}{r} x + 4 \\ x + 3\overline{)x^2 + 7x + 12} \\ \underline{x^2 + 3x} \\ 4x + 12 \\ \underline{4x + 12} \\ 0 \end{array}$$

286 In the problem

$$\begin{array}{r} x - 3 \\ x - 7\overline{)x^2 - 10x + 21} \\ \underline{x^2 - 7x} \\ {}^-3x + 21 \\ \underline{{}^-3x + 21} \\ 0 \end{array}$$

$x - 7$ is the divisor, $x^2 - 10x + 21$ is the dividend, and ________ is the quotient.

$x - 3$

287 In the following problem

$$
\begin{array}{r}
x - 4 \\
x + 1 \overline{) x^2 - 3x - 4} \\
\underline{x^2 + x } \\
{}^-4x - 4 \\
\underline{{}^-4x - 4} \\
0
\end{array}
$$

$x + 1$ is the divisor, $x^2 - 3x - 4$ is the dividend, $x - 4$ is the quotient, and 0 is the remainder. In the following problem

$$
\begin{array}{r}
x + 2 \\
x + 5 \overline{) x^2 + 7x + 10} \\
\underline{x^2 + 5x } \\
2x + 10 \\
\underline{2x + 10} \\
0
\end{array}
$$

$x + 2$ is the quotient, and the remainder is ________ .

0

288 In the problem

$$
\begin{array}{r}
x + 5 \\
x + 3 \overline{) x^2 + 8x + 17} \\
\underline{x^2 + 3x } \\
5x + 17 \\
\underline{5x + 15} \\
2
\end{array}
$$

the quotient is $x + 5$, and the remainder is ________ .

2

289 In the problem

$$
\begin{array}{r}
x - 2 \\
x - 5 \overline{) x^2 - 7x + 11} \\
\underline{x^2 - 5x } \\
{}^-2x + 11 \\
\underline{{}^-2x + 10} \\
1
\end{array}
$$

$x - 5$ is the ________________ ,
$x^2 - 7x + 11$ is the ________________ ,
$x - 2$ is the ________________ , and
1 is the ________________ .

divisor
dividend
quotient
remainder

In the following frames the division of polynomials will be explained. Division of polynomials is similar in method to long division in arithmetic.

---

290 To divide $x + 3\overline{)x^2 + 5x}$, *first* divide x^2 by x:

$$\begin{array}{r} x \\ x + 3\overline{)x^2 + 5x} \end{array}$$

Second, multiply $x(x + 3)$, and place the product below the dividend:

$$\begin{array}{r} x \\ x + 3\overline{)x^2 + 5x} \\ x^2 + 3x \end{array}$$

Third, subtract $x^2 + 3x$ from $x^2 + 5x$:

$$\begin{array}{r} x \\ x + 3\overline{)x^2 + 5x} \\ \underline{x^2 + 3x} \\ 2x \end{array}$$

Using the same three steps as shown above, divide $x + 7\overline{)x^2 + 10x}$.

$$\begin{array}{r} x \\ x + 7\overline{)x^2 + 10x} \\ \underline{x^2 + 7x} \\ 3x \end{array}$$

291 Divide $x + 6\overline{)x^2 + 2x}$.

$$\begin{array}{r} x \\ x + 6\overline{)x^2 + 2x} \\ \underline{x^2 + 6x} \\ {}^{-}4x \end{array}$$

292 Divide $x - 2\overline{)x^2 + 6x}$.

$$\begin{array}{r} x \\ x - 2\overline{)x^2 + 6x} \\ \underline{x^2 - 2x} \\ 8x \end{array}$$

293 Divide $x + 5\overline{)x^2 + 5x}$.

$$\begin{array}{r} x \\ x + 5\overline{)x^2 + 5x} \\ \underline{x^2 + 5x} \\ 0 \end{array}$$

294 Divide $x - 8\overline{)x^2 - 3x}$.

$$\begin{array}{r} x \\ x - 8\overline{)x^2 - 3x} \\ \underline{x^2 - 8x} \\ 5x \end{array}$$

295 Divide $x + 4\overline{)x^2 - 6x}$.

$$\begin{array}{r} x \\ x + 4\overline{)x^2 - 6x} \\ \underline{x^2 + 4x} \\ ^{-}10x \end{array}$$

296 To divide $x - 1\overline{)2x + 9}$, *first*, divide $2x$ by x:

$$\begin{array}{r} 2 \\ x - 1\overline{)2x + 9} \end{array}$$

Second, multiply $2(x - 1)$ and place the product below the dividend:

$$\begin{array}{r} 2 \\ x - 1\overline{)2x + 9} \\ 2x - 2 \end{array}$$

Third, subtract $2x - 2$ from $2x + 9$:

$$\begin{array}{r} 2 \\ x - 1\overline{)2x + 9} \\ \underline{2x - 2} \\ 11 \end{array}$$

Using the same three steps as shown above, divide $x - 3\overline{)3x + 4}$.

$$\begin{array}{r} 3 \\ x - 3\overline{)3x + 4} \\ \underline{3x - 9} \\ 13 \end{array}$$

297 Divide $x + 8\overline{)2x - 3}$.

$$\begin{array}{r} 2 \\ x + 8\overline{)2x - 3} \\ \underline{2x + 16} \\ ^{-}19 \end{array}$$

298 Divide $x + 1\overline{)4x - 5}$.

$$\begin{array}{r} 4 \\ x + 1\overline{)4x - 5} \\ \underline{4x + 4} \\ ^{-}9 \end{array}$$

299 Divide $x - 3\overline{)3x + 1}$.

$$\begin{array}{r} 3 \\ x - 3\overline{)3x + 1} \\ \underline{3x - 9} \\ 10 \end{array}$$

300 Divide $x + 4\overline{)5x + 20}$.

$$\begin{array}{r} 5 \\ x + 4\overline{)5x + 20} \\ \underline{5x + 20} \\ 0 \end{array}$$

301 Divide $x - 2\overline{)3x + 6}$.

$$\begin{array}{r} 3 \\ x - 2\overline{)3x + 6} \\ \underline{3x - 6} \\ 12 \end{array}$$

302 To find the first term of $x + 3\overline{)x^2 + 7x + 12}$, the first term of the dividend, x^2, is divided by the first term of the divisor, x.

$\frac{x^2}{x} =$ ________ .

x

303 The next step is to multiply the first term of the quotient, x, by the divisor, $x + 3$.

$x(x + 3) =$ ________ .

$x^2 + 3x$

304

$$\begin{array}{r} x \\ x + 3\overline{)x^2 + 7x + 12} \\ x^2 + 3x \end{array}$$

The next step is to subtract $x^2 + 3x$ from the first two terms of the dividend, $x^2 + 7x$.

$$\begin{array}{r} x \\ x + 3\overline{)x^2 + 7x + 12} \\ \underline{x^2 + 3x} \\ \underline{} \end{array}$$

$4x$

305 Combine the next term of the dividend, 12, with $4x$ as shown below:

$$\begin{array}{r} x \\ x + 3\overline{)x^2 + 7x + 12} \\ \underline{x^2 + 3x} \\ 4x + 12 \end{array}$$

To obtain the next term of the quotient, divide $4x$ by the first term of the divisor, x.

$\frac{4x}{x} =$ ________ .

4

306 Multiply the second term of the quotient, 4, by the entire divisor, $x + 3$.

$4(x + 3) =$ ________ .

$4x + 12$

$$\begin{array}{r} x + 4 \\ x + 3\overline{)x^2 + 7x + 12} \\ \underline{x^2 + 3x} \\ 4x + 12 \end{array}$$

307 Complete the following division problem by completing the subtraction shown below:

$$\begin{array}{r} x + 4 \\ x + 3\overline{)x^2 + 7x + 12} \\ \underline{x^2 + 3x} \quad\quad \\ 4x + 12 \\ \underline{4x + 12} \\ \_\_\_\_ . \end{array}$$

0

308 To divide, the following steps are used:

$$\begin{array}{r} x + 5 \\ x + 2\overline{)x^2 + 7x + 10} \\ \underline{x^2 + 2x} \quad\quad \\ 5x + 10 \\ \underline{5x + 10} \\ 0 \end{array}$$

Divide $x + 5\overline{)x^2 + 8x + 15}$.

$x + 3$

309 Complete the following division problem:

$$\begin{array}{r} x \quad\quad\quad\quad \\ x + 7\overline{)x^2 + 9x + 14} \\ \underline{x^2 + 7x} \quad\quad \end{array}$$

$$\begin{array}{r} x + 2 \\ x + 7\overline{)x^2 + 9x + 14} \\ \underline{x^2 + 7x} \quad\quad \\ 2x + 14 \\ \underline{2x + 14} \\ 0 \end{array}$$

[*Note:* The next step is to subtract $x^2 + 7x$ from $x^2 + 9x$, and bring down the 14.]

310 Complete the following division problem:

$$\begin{array}{r} x \\ x - 2\overline{)x^2 + 4x - 12} \\ \underline{x^2 - 2x} \\ 6x \end{array}$$

$$\begin{array}{r} x + 6 \\ x - 2\overline{)x^2 + 4x - 12} \\ \underline{x^2 - 2x} \\ 6x - 12 \\ \underline{6x - 12} \\ 0 \end{array}$$

311 Complete the following division problem:

$$\begin{array}{r} x \\ x - 3\overline{)x^2 - 7x + 12.} \end{array}$$

$$\begin{array}{r} x - 4 \\ x - 3\overline{)x^2 - 7x + 12} \\ \underline{x^2 - 3x} \\ {}^{-}4x + 12 \\ \underline{{}^{-}4x + 12} \\ 0 \end{array}$$

[*Note*: First multiply $x(x - 3)$.]

312 Divide $x + 8\overline{)x^2 + 10x + 16}$.

$$\begin{array}{r} x + 2 \\ x + 8\overline{)x^2 + 10x + 16} \\ \underline{x^2 + 8x} \\ 2x + 16 \\ \underline{2x + 16} \\ 0 \end{array}$$

[*Note:* First divide x^2 by x.]

313 Divide $x - 4\overline{)x^2 - x - 12}$.

$$\begin{array}{r} x + 3 \\ x - 4\overline{)x^2 - x - 12} \\ \underline{x^2 - 4x} \\ 3x - 12 \\ \underline{3x - 12} \\ 0 \end{array}$$

314 Divide $x + 6\overline{)x^2 + 14x + 48}$.

$$\begin{array}{r} x + 8 \\ x + 6\overline{)x^2 + 14x + 48} \\ \underline{x^2 + 6x} \phantom{{}+48} \\ 8x + 48 \\ \underline{8x + 48} \\ 0 \end{array}$$

315 Divide $x - 2\overline{)x^2 - 4x + 4}$.

$x - 2$

316 Divide $x - 5\overline{)x^2 - 2x - 15}$.

$x + 3$

317 Divide $x + 3\overline{)x^2 - 3x - 18}$.

$x - 6$

318 Divide $x - 7\overline{)x^2 - 6x - 7}$.

$x + 1$

319 To divide $x + 3\overline{)x^2 - 9}$, first write $x^2 - 9$ as $x^2 + 0x - 9$, as follows:

$$\begin{array}{r} x - 3 \phantom{{}-9} \\ x + 3\overline{)x^2 + 0x - 9} \\ \underline{x^2 + 3x} \phantom{{}-9} \\ {}^{-}3x - 9 \\ \underline{{}^{-}3x - 9} \\ 0 \end{array}$$

Divide $x + 5\overline{)x^2 - 25}$ by first writing $x^2 - 25$ as $x^2 + 0x - 25$. ________ .

$x - 5$

320 To divide $x - 8\overline{)x^2 - 64}$, the following steps are used:

$$\begin{array}{r} x + 8 \\ x - 8\overline{)x^2 + 0x - 64} \\ \underline{x^2 - 8x} \\ 8x - 64 \\ \underline{8x - 64} \\ 0 \end{array}$$

Divide $x - 7\overline{)x^2 - 49}$.

$x + 7$

321 Divide $x + 4\overline{)x^2 - 16}$.

$x - 4$

322 Divide $x - 6\overline{)x^2 - 2x - 24}$.

$x + 4$

323 Divide $x - 8\overline{)x^2 - 11x + 24}$.

$x - 3$

324 To check any division problem, the product of the quotient and divisor must be equal to the dividend, as shown in the following example:

$$\begin{array}{r} x + 5 \\ x + 2\overline{)x^2 + 7x + 10} \\ \underline{x^2 + 2x} \\ 5x + 10 \\ \underline{5x + 10} \\ 0 \end{array}$$

Check: $(x + 5)(x + 2) = \underline{x^2 + 7x + 10}$.

Divide and check:

$x + 3\overline{)x^2 + 10x + 21}$.

$x + 7$, because $(x + 7)(x + 3) = x^2 + 10x + 21$

325 Divide $x - 3\overline{)x^2 + x - 12}$.

$x + 4$

326 To check the division problem in the previous frame, $(x + 4)(x - 3) =$ ________________ .

$x^2 + x - 12$

327 Divide $x + 6\overline{)x^2 + 10x + 24}$.

$x + 4$

328 To check the division problem in the previous frame, $(x + 4)(x + 6) =$ ________________ .

$x^2 + 10x + 24$

329 Divide $x - 7\overline{)x^2 - 49}$.

$x + 7$

330 To check the division problem in the previous frame, $(x + 7)(x - 7) =$ ________________ .

$x^2 - 49$

331 Divide $x - 3\overline{)x^2 - 8x + 15}$.

$x - 5$

332 To check the division problem in the previous frame, $(x - 5)(x - 3) =$ ________________ .

$x^2 - 8x + 15$

333 Divide $x + 6\overline{)x^2 + 3x - 18}$.

$x - 3$

334 To check the division problem in the previous frame, $(x - 3)(x + 6) =$ ________________ .

$x^2 + 3x - 18$

335 Divide $x + 1\overline{)2x^2 + 5x + 3}$.

[*Note:* $2x^2$ divided by x is $2x$.]

$2x + 3$

336 Divide $x - 5\overline{)5x^2 - 29x + 20}$.

$5x - 4$

337 Divide $2x - 3\overline{)4x^2 - 9}$. $2x + 3$

[*Note:* $4x^2 - 9$ should be written as $4x^2 + 0x - 9$.]

338 Divide $5x + 4\overline{)25x^2 - 16}$. $5x - 4$

339 Divide $x - 3\overline{)x^2 + 5x - 24}$. $x + 8$

340 Divide $x + 2\overline{)3x^2 + 7x + 2}$. $3x + 1$

341 To check the division problem in the previous frame, $(3x + 1)(x + 2) =$ ______________ . $3x^2 + 7x + 2$

342 Divide $x - 2\overline{)3x^2 - 5x - 2}$. $3x + 1$

343 To check the division problem in the previous frame, $(3x + 1)(x - 2) =$ ______________ . $3x^2 - 5x - 2$

344 Divide $x - 7\overline{)2x^2 - 13x - 7}$. $2x + 1$

345 Divide $x + 4\overline{)x^2 + 14x + 40}$. $x + 10$

346 Divide $x - 6\overline{)x^2 - 7x + 6}$. $x - 1$

347 Divide $2x + 1\overline{)6x^2 + 7x + 2}$. $3x + 2$

[*Note:* First divide $6x^2$ by $2x$.]

348 Divide $3x - 1\overline{)12x^2 + 5x - 3}$. $4x + 3$

349 Divide $6x - 5\overline{)36x^2 - 25}$. $6x + 5$

350 To check the division problem in the previous frame, $(6x + 5)(6x - 5) =$ ______________ . $36x^2 - 25$

351 Divide $5x + 1\overline{)10x^2 - 13x - 3}$. $2x - 3$

352 Divide $3x + 4\overline{)9x^2 - 9x - 28}$. $3x - 7$

353 The following problem is an example of a division problem that has a remainder:

$$\begin{array}{r} x + 2 \\ x + 3\overline{)x^2 + 5x - 3} \\ \underline{x^2 + 3x} \\ 2x - 3 \\ \underline{2x + 6} \\ {}^{-}9 \end{array}$$

The answer is written, $x + 2$ R $^{-}9$. Divide the following problem, which has a remainder:

$x + 2\overline{)x^2 + 5x + 3}$. $x + 3$ R $^{-}3$

354 Divide $x - 3\overline{)x^2 - 7x - 15}$. $x - 4$ R $^{-}27$

355 Divide $x + 3\overline{)2x^2 + 11x + 20}$. $2x + 5$ R 5

356 Divide $x + 2\overline{)x^2 + 7x + 13}$. $x + 5$ R 3

357 The answer to the division problem in the previous frame is $x + 5$ R 3. To check the answer, which has a remainder, the following steps are used:

$$[(x + 5)(x + 2)] + 3$$
$$[x^2 + 7x + 10] + 3$$
$$\underline{x^2 + 7x + 13}$$

Is this the dividend in the previous frame? ________ . — Yes

358 Divide $x + 4\overline{)x^2 + 9x + 18}$. — $x + 5$ R $^-2$

359 To check the division problem in the previous frame, $[(x + 5)(x + 4)] + {}^-2 =$ ____________ . — $x^2 + 9x + 18$

360 Divide $x + 3\overline{)x^2 - 2x + 6}$. — $x - 5$ R 21

361 To check the division problem in the previous frame, $[(x - 5)(x + 3)] + 21 =$ ____________ . — $x^2 - 2x + 6$

362 Divide $x - 4\overline{)x^2 - 7x + 3}$. — $x - 3$ R $^-9$

363 To check the division problem in the previous frame, $[(x - 3)(x - 4)] + {}^-9 =$ ____________ . — $x^2 - 7x + 3$

Self-Quiz # 5

The following questions test the objectives of the preceding section. 100% mastery is desired.

1. $x - 3\overline{)x^2 - 7x + 12}$

2. $x + 5\overline{)x^2 - x - 30}$

3. $x + 2\overline{)x^2 + 9x + 16}$

4. $x + 3\overline{)5x^2 + 14x - 3}$

5. $2x + 1\overline{)6x^2 - x + 3}$

6. $x - 6\overline{)x^2 - 36}$

Chapter Summary

In this chapter some basic skills and language for an open expression called a polynomial were introduced.

The multiplication of monomials, binomials, and trinomials was explained. The FOIL method was used to simplify the multiplication of two binomials.

Fractions involving polynomials in numerators or denominators were added. Each problem had the product of the denominators as the least common multiple (LCM).

The final skill shown in the chapter was the long division process as it can be applied to the division of polynomials by binomials.

CHAPTER POST-TEST

The following questions test the objectives of this chapter. A score of 90% indicates sufficient mastery, and the student may proceed to the next chapter.

Multiply:

1. $(5x^4) \cdot ({}^{-}x^2)$

2. $({}^{-}2x^2y^3) \cdot ({}^{-}3xy)$

3. ${}^{-}5(2x - 7)$

4. $3x(x^2 + 4)$

5. $6(x^2 + 7x - 2)$

6. $(x + 6)(x - 7)$

7. $(2x + 7)(x - 3)$

8. $(3x - 1)(2x + 5)$

9. $(2x + 3)^2$

10. $(x + 4)(x^2 - 3x + 5)$

Add:

11. $\frac{6}{x} - \frac{y}{7}$

12. $\frac{3}{x - 7} - \frac{x + 4}{5}$

13. $\frac{x - 1}{x + 9} + \frac{2x - 5}{x - 3}$

Divide:

14. $x + 4\overline{)x^2 - 2x - 24}$

15. $2x + 3\overline{)6x^2 - x - 1}$

ALGEBRA

Factoring Polynomials

4

1 2 3 5 6 7 8 9 10

CHAPTER PRE-TEST

The following questions indicate the objectives of this chapter. A score of 90% indicates sufficient mastery, and the student may immediately take the Chapter Post-Test.

I Factor:

1. $6x - 24$
2. $8x^2 - 9x$
3. $^-8x - 12$
4. $5x + 5$

II Factor:

1. $x^2 + 8x + 15$
2. $x^2 - 4x - 21$
3. $x^2 - 17x + 70$
4. $x^2 + 6x - 7$

III Factor:

1. $2x^2 + 7x + 5$
2. $3x^2 - 4x - 4$
3. $6x^2 - 5x + 1$
4. $2x^2 - x - 36$

IV Factor:

1. $x^2 + 7x - 30$
2. $4x^3 - 9x^2 - x$
3. $x^2 - 25$
4. $3x^2 - 11x + 6$
5. $16x^2 - 1$
6. $5x^2 + 20x + 15$
7. $6x^2 - 3x - 18$
8. $x^2 + 11x + 24$

Three methods of factoring are explained in this chapter: (1) the common factor method, (2) factoring trinomials when the coefficient of x^2 is 1, and (3) factoring trinomials when the coefficient of x^2 is not 1.

---

1 $6 \cdot 3$ is 18. 6 and 3 are *factors* of 18. $2 \cdot 9$ is also 18. Therefore, 2 and 9 are ________ of 18.

factors

2 $3x^2 \cdot 2x$ is $6x^3$. $3x^2$ and $2x$ are called *factors* of $6x^3$. $4xy \cdot 3xy^2$ is $12x^2y^3$. $4xy$ and $3xy^2$ are called ________ of $12x^2y^3$.

factors

3 $2x$ is a factor of $6x$ because $2x \cdot 3 = 6x$. $5x$ is a factor of $20x$ because $5x \cdot$ ________ $= 20x$.

4

4 $3x^2$ is a factor of $3x^3$ because $3x^2 \cdot x = 3x^3$. $7x$ is a factor of $28x^3$ because $7x \cdot$ ________ $= 28x^3$.

$4x^2$

5 Every monomial is a factor of itself. $3x^2y$ is a factor of $3x^2y$ because $3x^2y \cdot 1 = 3x^2y$. $5xy^4$ is a factor of $5xy^4$ because $5xy^4 \cdot$ ________ $= 5xy^4$.

1

6 1 is a factor of 7 because $1 \cdot$ ________ $= 7$.

7

7 $^-1$ is a factor of 8 because $^-1 \cdot$ ________ $= 8$.

$^-8$

8 2 is a factor of $10x$ because $2 \cdot$ ________ $= 10x$.

$5x$

9 5 is a factor of $^-20$ because $5 \cdot$ ________ $= {}^-20$.

$^-4$

10 Is $2x$ a factor of $6x$? ______________ .

Yes, because $2x \cdot 3 = 6x$

11 Is $^-5$ a factor of 35? ______________ .

Yes, because $^-5 \cdot {}^-7 = 35$

12 Is ^-3x a factor of $6x^2$? ______________ .

Yes, because $^-3x \cdot {}^-2x = 6x^2$

13 Which of the following monomials is a factor of $6x^2y$? $2x$ or $5x$? ________ . 　　$2x$

14 Which of the following monomials is a factor of $12xy^3$? $4xy^2$ or $7xy$? ________ . 　　$4xy^2$

15 Which of the following monomials is a factor of $8x^2$? $2x$ or $5x^5$? ________ . 　　$2x$

16 Which of the following monomials is a factor of $12x^2y$? $7xy^4$ or $12x^2y$? ________ . 　　$12x^2y$

17 Which of the following monomials is a factor of $16x$? 10 or 8? ________ . 　　8

---

1 and $^-1$ are always factors of any monomial. For example, 1 is a factor of $7x^2$ because $1 \cdot 7x^2 = 7x^2$. $^-1$ is also a factor of $7x^2$ because $^-1 \cdot {^-7x^2} = 7x^2$.

---

18 1 is a factor of 8 because $1 \cdot$ ________ $= 8$. 　　8

19 $^-1$ is a factor of 9 because $^-1 \cdot$ ________ $= 9$. 　　$^-9$

20 6 is a factor of $^-6$ because $6 \cdot$ ________ $= {^-6}$. 　　$^-1$

21 $^-2$ is a factor of 2 because $^-2 \cdot$ ________ $= 2$. 　　$^-1$

22 7 is a factor of 7 because $7 \cdot$ ________ $= 7$. 　　1

23 11 is a factor of $^-11$ because $11 \cdot$ ________ $= {^-11}$. 　　$^-1$

24 Is $^-12$ a factor of 12? ________ . 　　Yes

25 Is 2 a factor of 2? ________ . 　　Yes

26 Is $^-5$ a factor of 5? ________ . 　　Yes

27 Is 4 a factor of $^-4$? ________ . Yes

28 2 is a factor of both 8 and 14 because $2 \cdot 4 = 8$ and $2 \cdot 7 = 14$. Is 5 a factor of both 10 and 15? ________ . Yes

29 4 is a factor of both $8x$ and 12 because $4 \cdot 2x = 8x$ and $4 \cdot 3 = 12$. Is 3 a factor of both $6x$ and 21? ________ . Yes

30 $^-3$ is a factor of both $6x$ and $^-9$ because $^-3 \cdot {}^-2x = 6x$ and $^-3 \cdot 3 = {}^-9$. Is $^-7$ a factor of both $21x$ and $^-14$? ________ . Yes

31 6 is a factor of both 6 and 12 because $6 \cdot 1 = 6$ and $6 \cdot 2 = 12$. Is 5 a factor of both 15 and 5? ________ . Yes

32 x is a factor of both $3x$ and $5x$ because $x \cdot 3 = 3x$ and $x \cdot 5 = 5x$. Is x a factor of both $7x$ and $9x$? ________ . Yes

33 Is 7 a factor of both 14 and 49? ________ . Yes

34 Is $^-8$ a factor of both 24 and $^-48$? ________ . Yes

35 $2x$ is a factor of both $4x^2$ and $6x$ because $2x \cdot 2x = 4x^2$ and $2x \cdot 3 = 6x$. Is $3x$ a factor of both $3x^2$ and $12x$? ________ . Yes

36 Is $9x$ a factor of both $18x^3$ and $27x$? ________ . Yes

---

5 is a *common factor* for 10 and 15 because $5 \cdot 2 = 10$ and $5 \cdot 3 = 15$.

$2x$ is a *common factor* for $2x^2$ and $6x$ because $2x \cdot x = 2x^2$ and $2x \cdot 3 = 6x$.

1 and $^-1$ are always *common factors* for any two monomials. In the following frames find the *common factors*, other than 1 or $^-1$, for the two monomials given.

---

37 The common factors for 4 and 10 are 2 and $^-2$. What are the common factors for 8 and 6? ______ .

2, $^-2$

38 The common factors for 9 and 12 are 3 and $^-3$. What are the common factors for 6 and 21? ______ .

3, $^-3$

39 The common factors of $10x$ and $13x^2$ are x and ^-x. What are the common factors of $5x^2$ and $17x$? ______ .

x, ^-x

40 The common factors of $6x$ and 15 are 3 and $^-3$. What are the common factors of $25x$ and 30? ______ .

5, $^-5$

41 The common factors of 14 and 35 are ______ and ______ .

7, $^-7$

42 The common factors of 15 and 40 are ______ and ______ .

5, $^-5$

43 The common factors of $7x$ and $2x$ are ______ and ______ .

x, ^-x

44 The common factors of 9 and 24 are ______ and ______ .

3, $^-3$

45 The common factors of 8 and 18 are ______ and ______ .

2, $^-2$

46 The common factors of $12x$ and 14 are ______ and ______ .

2, $^-2$

47 The common factors of $19x$ and $21x^2$ are ______ and ______ .

x, ^-x

48 Every monomial is a factor of itself. The common factors of 3 and 6 are ______ and ______ .

3, $^-3$

49 The common factors of 7 and 63 are ______ and ______ .

7, $^-7$

50 The common factors of 11 and 33 are ______ and ______ .

11, $^{-}11$

51 Two monomials may have more than two common factors besides 1 and $^{-}1$. The common factors of 4 and 8 are 2, $^{-}2$, 4, and $^{-}4$. What are the common factors of 9 and 18? ______________ .

3, $^{-}3$, 9, $^{-}9$

52 What are the common factors for 6 and 12? (There are six common factors besides 1 and $^{-}1$.) ________________ .

2, $^{-}2$, 3, $^{-}3$, 6, $^{-}6$

53 $4x^2$ and $2x$ have six common factors besides 1 and $^{-}1$. The common factors of $4x^2$ and $2x$ are: 2, $^{-}2$, x, ^{-}x, $2x$, and $^{-}2x$. Find six common factors for $5x$ and $10x$. ________________ .

5, $^{-}5$, x, ^{-}x, $5x$, $^{-}5x$

54 The common factors for 12 and $33x$ are ______ and ______ .

3, $^{-}3$

55 The common factors for $5x$ and 10 are ______ and ______ .

5, $^{-}5$

56 The common factors for $3x$ and 12 are ______ and ______ .

3, $^{-}3$

To factor the binomial $5x + 10$ means to write a multiplication expression that has $5x + 10$ as its product.

$5(x + 2) = 5x + 10$, so 5 and $(x + 2)$ are factors of $5x + 10$. Notice that 5 is a common factor for the two terms of the binomial $5x$ and 10.

---

57 To factor $3x + 12$, the first step is to find a common factor for $3x$ and 12. Since 3 is a common factor, the binomial $3x + 12$ may be written as $3 \cdot x + 3 \cdot 4$. Is $3 \cdot x + 3 \cdot 4$ equal to $3(x + 4)$? ______ .

Yes

58 To factor $7x - 21$, the following steps are used:

$$7x - 21$$

$$7 \cdot x + 7 \cdot {}^{-}3$$

$$7(x - 3)$$

Complete the factoring of $3x - 15$.

$3x - 15$

$3 \cdot x + 3 \cdot {}^{-}5$

__________ $3(x - 5)$

59 4 is a common factor for $8x$ and ${}^{-}12$. To factor $8x - 12$, the following steps are used:

$$8x - 12$$

$$4 \cdot 2x + 4 \cdot {}^{-}3$$

$$4(2x - 3)$$

7 is a common factor for $14x$ and ${}^{-}35$. Factor $14x - 35$. __________ . $7(2x - 5)$

60 Factor $5x + 25$ by first finding a common factor for $5x$ and 25. __________ . $5(x + 5)$

61 Factor $2x - 16$ by first finding a common factor for $2x + {}^{-}16$. __________ . $2(x - 8)$

62 Factor $11x + 33$ by first finding a common factor for $11x$ and 33. __________ . $11(x + 3)$

63 Factor $4x - 6$ by first finding a common factor for $4x$ and ${}^{-}6$. __________ . $2(2x - 3)$

64 Factor $6x + 15$ by first finding a common factor for $6x$ and 15. __________ . $3(2x + 5)$

65 7 is a common factor of $14x$ and 7. To factor $14x + 7$, the following steps are used:

$$14x + 7$$
$$7 \cdot 2x + 7 \cdot 1$$
$$7(2x + 1)$$

3 is a common factor of $9x$ and 3. Factor $9x + 3$.

_______________ . $3(3x + 1)$

66 Factor $12x + 3$. _______________ . $3(4x + 1)$

67 Factor $2x + 2$. _______________ . $2(x + 1)$

68 Factor $42x - 7$. _______________ . $7(6x - 1)$

69 Factor $5x - 5$. _______________ . $5(x - 1)$

70 Factor $26x - 13$. _______________ . $13(2x - 1)$

71 In factoring $^-6x + 8$ either 2 or $^-2$ could be used as a common factor to give $2(^-3x + 4)$ or $^-2(3x - 4)$. $^-2(3x - 4)$ is preferred. Factor $^-6x + 15$ by using $^-3$ as the common factor. _______________ . $^-3(2x - 5)$

72 The factors of $^-3x + 6$ are $3(^-x + 2)$ or $^-3(x - 2)$. $^-3(x - 2)$ is the preferred factorization. The factors of $^-4x + 8$ are $4(^-x + 2)$ or $^-4(x - 2)$. Which factorization is preferred? _______________ . $^-4(x - 2)$

73 $^-5(x - 3)$ or $5(^-x + 3)$ are factorizations of $^-5x + 15$. Which factorization is preferred? _______________ . $^-5(x - 3)$

74 $^-2(4x - 3)$ or $2(^-4x + 3)$ are factorizations of $^-8x + 6$. Which factorization is preferred?

_______________ . $^-2(4x - 3)$

75 $5(^-2x - 1)$ or $^-5(2x + 1)$ are factorizations of $^-10x - 5$. Which factorization is preferred?

_______________ . $^-5(2x + 1)$

76 $7(^-2x - 3)$ or $^-7(2x + 3)$ are factorizations of $^-14x - 21$. Which factorization is preferred?
______________ .

$^-7(2x + 3)$

77 In factoring $^-4x - 8$, either 4 or $^-4$ can be used as a common factor to give $4(^-x - 2)$ or $^-4(x + 2)$. $^-4(x + 2)$ is preferred. Factor $^-6x - 18$.
______________ .

$^-6(x + 3)$

78 Factor $^-5x + 10$. ______________ .

$^-5(x - 2)$

79 Factor $^-15x - 3$. ______________ .

$^-3(5x + 1)$

80 $^-3x + 5$ can be factored as $^-1(3x - 5)$ or, more simply, $^-(3x - 5)$. Factor $^-2x + 7$. ______________ .

$^-(2x - 7)$

81 Factor $^-2x + 17$. ______________ .

$^-(2x - 17)$

82 Factor $5x + 10$. ______________ .

$5(x + 2)$

83 Factor $^-4x + 12$. ______________ .

$^-4(x - 3)$

84 Factor $7x - 21$. ______________ .

$7(x - 3)$

85 Factor $^-6x - 3$. ______________ .

$^-3(2x + 1)$

86 $2x$ is a common factor of $2x^2$ and $4x$. To factor $2x^2 + 4x$, the following steps are used:

$$2x^2 + 4x$$

$$2x \cdot x + 2x \cdot 2$$

$$2x(x + 2)$$

$5x$ is a common factor of $5x^2$ and $20x$. Factor $5x^2 + 20x$. ______________ .

$5x(x + 4)$

87 Factor $3x^2 + 15x$. ______________ .

$3x(x + 5)$

88 Factor $13x^2 + 26x$. ______________ .

$13x(x + 2)$

89 Factor $^{-}2x^2 + 8x$. ________________ . $^{-}2x(x - 4)$

90 Factor $14x^2 + 7x$. ________________ . $7x(2x + 1)$

$4x + 8$ could be factored as $2(2x + 4)$, but the factoring is incomplete because $2x$ and 4 still have a common factor of 2.

To completely factor $4x + 8$, it is necessary to use the common factor 4 to give $4(x + 2)$. The only common factors of x and 2 are 1 and $^{-}1$. A polynomial is completely factored only when the common factors are 1 and $^{-}1$.

---

91 Factor $4x + 16$. ________________ . $4(x + 4)$

92 Factor $6x - 18$. ________________ . $6(x - 3)$

93 Factor $^{-}10x + 20$. ________________ . $^{-}10(x - 2)$

94 Factor $8x + 20$. ________________ . $4(2x + 5)$

95 Factor $2x^2 - 8x$. ________________ . $2x(x - 4)$

96 Factor $^{-}8x^2 - 24x$. ________________ . $^{-}8x(x + 3)$

97 Factor $9x^2 + 3x$. ________________ . $3x(3x + 1)$

98 Factor $5x^3 - 15x^2$. ________________ . $5x^2(x - 3)$

99 Factor $^{-}12x^2 + 30x$. ________________ . $^{-}6x(2x - 5)$

Polynomials with three or more terms can be factored in a manner similar to the factorization of $5x - 10$.

To factor $2x^2 - 4x + 6$, notice that 2 is a common factor for $2x^2$, ^-4x, and 6. The factorization of $2x^2 - 4x + 6$ is done as follows:

$$2x^2 - 4x + 6$$
$$2 \cdot x^2 + 2 \cdot {}^-2x + 2 \cdot 3$$
$$2(x^2 - 2x + 3)$$

100 Complete the factorization of $3x^2 - 12x + 15$.
$3x^2 - 12x + 15$
$3 \cdot x^2 + 3 \cdot {}^-4x + 3 \cdot 5$
________________ .

$3(x^2 - 4x + 5)$

101 Factor $4x^2 + 2x - 8$. ________________ .

$2(2x^2 + x - 4)$

102 Factor $^-6x^2 + 12x - 9$. ________________ .

$^-3(2x^2 - 4x + 3)$

103 Factor $10x^2 - 15x + 25$. ________________ .

$5(2x^2 - 3x + 5)$

104 Factor $8x^2 + 16x - 8$. ________________ .

$8(x^2 + 2x - 1)$

105 Factor $x^3 + 5x^2 + 3x$. ________________ .

$x(x^2 + 5x + 3)$

106 Factor $6x^3 - 3x^2 + 27x$. ________________ .

$3x(2x^2 - x + 9)$

107 Factor $^-5x^3 + 10x^2 + 25x$. ________________ .

$^-5x(x^2 - 2x - 5)$

108 Factor $4x^3 - 10x^2 - 16x$. ________________ .

$2x(2x^2 - 5x - 8)$

4 is a common factor for $4 \cdot x + 4 \cdot 7$. Therefore, $4 \cdot x + 4 \cdot 7$ can be factored as $4(x + 7)$.

$(x + 3)$ is a common factor for $x(x + 3) + 5(x + 3)$. Therefore, $x(x + 3) + 5(x + 3)$ can be factored as $(x + 3)(x + 5)$.

109 To factor $x(x - 5) + 7(x - 5)$, notice that $(x - 5)$ is a common factor.

$$x(x - 5) + 7(x - 5)$$
$$(x - 5)(x + 7)$$

Factor $x(x + 8) + 3(x + 8)$. ______________ .

$(x + 8)(x + 3)$

110 $x(x - 5) - 3(x - 5)$ is factored as $(x - 5)(x - 3)$ because $(x - 5)$ is a common factor. Factor $x(x + 7) - 2(x + 7)$. ______________ .

$(x + 7)(x - 2)$

111 Factor $x(x + 9) + 2(x + 9)$. ______________ .

$(x + 9)(x + 2)$

112 Factor $3x(x + 6) - 5(x + 6)$. ______________ .

$(x + 6)(3x - 5)$

113 Factor $x(2x + 1) - 5(2x + 1)$. ______________ .

$(2x + 1)(x - 5)$

114 Factor $2x(x + 3) + (x + 3)$. ______________ .
[*Hint*: $(x + 3)$ is equivalent to $1(x + 3)$.]

$(x + 3)(2x + 1)$

115 Factor $x(x - 7) - (x - 7)$. ______________ .
[*Hint*: $^{-}(x - 7)$ is equivalent to $^{-}1(x - 7)$.]

$(x - 7)(x - 1)$

116 Factor $x(x + 4) - 5(x + 4)$. ______________ .

$(x + 4)(x - 5)$

117 Factor $x(x + 8) + (x + 8)$. ______________ .

$(x + 8)(x + 1)$

118 Factor $x(x - 5) - (x - 5)$. ______________ .

$(x - 5)(x - 1)$

119 Factor $x(x - 5) - 3(x - 5)$. ______________ .

$(x - 5)(x - 3)$

120 Factor $x(x - 8) + 7(x - 8)$. ______________ .

$(x - 8)(x + 7)$

121 Factor $x(x + 11) - (x + 11)$. ______________ .

$(x + 11)(x - 1)$

Self-Quiz # 1

The following questions test the objectives of the preceding section. 100% mastery is desired.

Factor each of the following:

1. $3x + 12$. ______ .

2. $5x - 35$. ______ .

3. $2x^2 - 6x$. ______ .

4. $x^2 + 6x$. ______ .

5. $^{-}10x^2 - 15x$. ______ .

6. $4x^2 + 8x - 16$. ______ .

7. $7x^2 - 21x$. ______ .

8. $5x^3 - 6x^2 + 7x$. ______ .

9. $4x^2 + 16x$. ______ .

10. $6x^3 - 8x^2 - 10x$. ______ .

11. $x(x + 2) - 3(x + 2)$. ______ .

12. $3x(x - 5) - 2(x - 5)$. ______ .

13. $x(x - 7) + (x - 7)$. ______ .

14. $5x(x + 3) - (x + 3)$. ______ .

15. $7x(2x - 3) + 4(2x - 3)$. ______ .

In this section polynomials such as $x^2 + 3x - 40$ are factored. The factorization of $x^2 + 3x - 40$ is $(x + 8)(x - 5)$. The factors $(x + 8)$ and $(x - 5)$ are binomials.

---

122 $4 \cdot 5 = 20$ and $4 + 5 = 9$. The integers 4 and 5 have a product of 20 and a sum of 9. What two integers have a product of 18 and a sum of 11? ______ .

9 and 2

123 $6 \cdot 8 = 48$ and $6 + 8 = 14$. The factors of 48 that have a sum of 14 are 6 and 8. What two factors of 30 have a sum of 11? ________ .

6 and 5

124 What factors of 24 have a sum of 10? ________ .

6 and 4

125 What factors of 20 have a sum of 9? ________ .

4 and 5

126 What factors of 15 have a sum of 8? ________ .

5 and 3

127 What factors of 12 have a sum of 13? ________ .

12 and 1

128 What factors of 35 have a sum of 12? ________ .

7 and 5

129 What factors of 25 have a sum of 10? ________ .

5 and 5

130 $^{-}8 \cdot {}^{-}3 = 24$ and $^{-}8 + {}^{-}3 = {}^{-}11$. The integers $^{-}8$ and $^{-}3$ are factors of 24 that have a sum of $^{-}11$. What factors of 15 have a sum of $^{-}8$? ________ .

$^{-}5$ and $^{-}3$

131 What factors of 18 have a sum of $^{-}9$? ________ .

$^{-}6$ and $^{-}3$

132 What factors of 24 have a sum of $^{-}11$? ________ .

$^{-}8$ and $^{-}3$

133 What factors of 6 have a sum of $^{-}5$? ________ .

$^{-}3$ and $^{-}2$

134 What factors of 9 have a sum of $^{-}10$? ________ .

$^{-}9$ and $^{-}1$

135 What factors of 28 have a sum of $^{-}16$? ________ .

$^{-}14$ and $^{-}2$

136 $4 \cdot {}^{-}6 = {}^{-}24$ and $4 + {}^{-}6 = {}^{-}2$. The integers 4 and $^{-}6$ are factors of $^{-}24$ that have a sum of $^{-}2$. What factors of $^{-}20$ have a sum of $^{-}8$? ________ .

$^{-}10$ and 2

137 What factors of $^{-}12$ have a sum of $^{-}4$? ________ .

$^{-}6$ and 2

138 What factors of $^{-}15$ have a sum of 2? ________ .

5 and $^{-}3$

139 What factors of $^{-}24$ have a sum of $^{-}5$? ________ . $^{-}8$ and 3

140 What factors of $^{-}16$ have a sum of 6? ________ . 8 and $^{-}2$

141 What factors of $^{-}25$ have a sum of 0? ________ . 5 and $^{-}5$

142 What factors of $^{-}20$ have a sum of 8? ________ . 10 and $^{-}2$

143 What factors of $^{-}9$ have a sum of 0? ________ . 3 and $^{-}3$

144 What factors of $^{-}35$ have a sum of 2? ________ . 7 and $^{-}5$

145 What factors of $^{-}30$ have a sum of $^{-}1$? ________ . $^{-}6$ and 5

146 1 and $^{-}1$ are factors of every number. What factors of 7 have a sum of 8? ________ . 1 and 7

147 What factors of $^{-}13$ have a sum of 12? ________ . 13 and $^{-}1$

148 What factors of $^{-}6$ have a sum of $^{-}5$? ________ . $^{-}6$ and 1

149 What factors of $^{-}1$ have a sum of 0? ________ . 1 and $^{-}1$

150 What factors of 1 have a sum of 2? ________ . 1 and 1

151 What factors of 1 have a sum of $^{-}2$? ________ . $^{-}1$ and $^{-}1$

To multiply $(x - 5)(x + 8)$ using the FOIL method, the following steps are used:

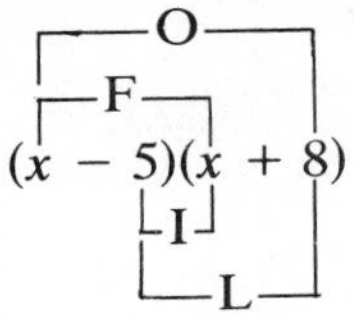

$$(x - 5)(x + 8)$$

(F, O, I, L)

$$x^2 + 8x - 5x - 40$$

$$x^2 + 3x - 40$$

In this section the FOIL procedure will be reversed to factor polynomials such as $x^2 + 3x - 40$.

$$x^2 + 3x - 40$$

$$x^2 + 8x - 5x - 40$$

$$(x + 8)(x - 5)$$

Notice that the factors of $x^2 + 3x - 40$ are the two binomials $(x + 8)$ and $(x - 5)$.

---

152 To multiply $(x + 8)(x - 5)$, the FOIL method is used.

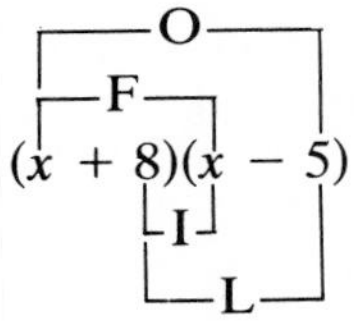

$$(x + 8)(x - 5)$$

(F, O, I, L)

$$x^2 - 5x + 8x - 40$$

$$x^2 + 3x - 40$$

Using the FOIL method, multiply $(x - 6)(x - 5)$.

________________ .

$x^2 - 5x - 6x + 30$
$x^2 - 11x + 30$

153 Using the FOIL method, multiply $(x - 6)(x + 9)$.

________________ .

$x^2 + 9x - 6x - 54$
$x^2 + 3x - 54$

154 To factor $x^2 + 7x + 10$, the FOIL procedure is reversed.

$$\begin{array}{c} x^2 + 7x + 10 \\ \swarrow \quad \swarrow\searrow \quad \downarrow \\ F + O + I + L \end{array}$$

The F term is x^2, the L term is 10, and the sum of the O and I terms is ______ .

$7x$

155 To factor $x^2 - 8x - 9$, the FOIL multiplication is reversed.

$$\begin{array}{c} x^2 - 8x - 9 \\ \swarrow \quad \swarrow\searrow \quad \searrow \\ F + O + I + L \end{array}$$

The F term is x^2, the L term is $^-9$, and the sum of the O and I terms is ______ .

^-8x

156 To factor $x^2 - 5x + 6$, the FOIL multiplication is reversed.

$$\begin{array}{c} x^2 - 5x + 6 \\ \swarrow \quad \swarrow\searrow \quad \searrow \\ F + O + I + L \end{array}$$

The F term is x^2, the L term is ______ , and the sum of the O and I terms is ______ .

6
^-5x

157 To factor $x^2 + 5x - 14$, the FOIL multiplication is reversed.

$$\begin{array}{c} x^2 + 5x - 14 \\ \swarrow \quad \swarrow\searrow \quad \searrow \\ F + O + I + L \end{array}$$

The F term is x^2, the L term is ______ , and the sum of the O and I terms is ______ .

$^-14$
$5x$

158 To factor $x^2 + 10x + 21$, the four-term FOIL expression is written as follows:

$$\begin{array}{c} x^2 + 10x + 21 \\ \swarrow \quad \swarrow\searrow \quad \searrow \\ x^2 + O + I + 21 \end{array}$$

Find the O and I terms by selecting factors of 21 with a sum of 10. ______ , ______ .

$3x$, $7x$

159 To factor $x^2 - 8x - 9$, the four-term FOIL expression is written. Write the four-term FOIL expression for $x^2 - 8x - 9$. Use factors of $^-9$ with a sum of $^-8$ to find the O and I terms. ________________ ,

________________ .

$x^2 - 9x + 1x - 9$
or
$x^2 + 1x - 9x - 9$

160 Write the four-term FOIL expression for $x^2 - 5x + 6$ by finding factors of 6 with a sum of $^-5$ as the O and I terms. ________________ .

$x^2 - 3x - 2x + 6$
or
$x^2 - 2x - 3x + 6$

161 Write the FOIL expression for $x^2 + 5x - 14$ by finding factors of $^-14$ with a sum of 5.

________________ .

$x^2 + 7x - 2x - 14$
or
$x^2 - 2x + 7x - 14$

162 To factor $x^2 + 8x + 12$, the FOIL expression is used.

$$x^2 + 8x + 12 \qquad 6 \cdot 2 = 12$$

$$\swarrow \quad \swarrow \searrow \quad \searrow \qquad \text{and}$$

$$x^2 + 6x + 2x + 12 \qquad 6 + 2 = 8$$

$$(x + 2)(x + 6)$$

The O term is $6x$, and the I term is ________ .

$2x$

163 The factorization of $x^2 + 13x + 30$ is completed as follows:

$$x^2 + 13x + 30 \qquad 3 \cdot 10 = 30$$

$$\swarrow \quad \swarrow \searrow \quad \searrow \qquad \text{and}$$

$$x^2 + 3x + 10x + 30 \qquad 3 + 10 = 13$$

$$(x + 10)(x + 3)$$

Since the O term is $3x$ and the I term is $10x$, the factors of $x^2 + 13x + 30$ are $(x + 10)$ and ________ .

$(x + 3)$

164 The factorization of $x^2 + 7x + 6$ is completed as follows:

$$x^2 + 7x + 6 \qquad 6 \cdot 1 = 6$$

$$\swarrow \quad \swarrow \searrow \quad \searrow \qquad \text{and}$$

$$x^2 + 6x + 1x + 6 \qquad 6 + 1 = 7$$

$$(x + 1)(x + 6)$$

The O term is $6x$ and the I term is $1x$. The factors of $x^2 + 7x + 6$ are ________ and ________ .

$(x + 1), (x + 6)$

165 To factor $x^2 + 9x + 18$, the FOIL expression is used.

$$x^2 + 9x + 18$$

$$\swarrow \quad \swarrow \searrow \quad \searrow$$

$$x^2 + 6x + 3x + 18$$

The O and the I terms are $6x$ and $3x$ because $6 \cdot 3 = 18$ and ________ .

$6 + 3 = 9$

166 The factorization of $x^2 + 9x + 18$ is shown below:

$$x^2 + 9x + 18$$

$$x^2 + 6x + 3x + 18$$

$$(x + 3)(x + 6)$$

Factor $x^2 + 9x + 14$ by first writing the FOIL expression. ________________ .

$(x + 7)(x + 2)$
or
$(x + 2)(x + 7)$

167 The factorization of $x^2 + 11x + 30$ is shown below:

$$x^2 + 11x + 30$$

$$x^2 + 5x + 6x + 30$$

$$(x + 6)(x + 5)$$

Factor $x^2 + 8x + 15$ by first writing the FOIL expression. ________________ .

$(x + 3)(x + 5)$
or
$(x + 5)(x + 3)$

168 Factor $x^2 + 8x + 7$ by first writing the FOIL expression. ________________ .

$x^2 + 1x + 7x + 7$
$(x + 7)(x + 1)$
or
$(x + 1)(x + 7)$

169 Factor $x^2 + 15x + 56$ by first writing the FOIL expression. ________________ .

$x^2 + 7x + 8x + 56$
$(x + 7)(x + 8)$

170 Factor $x^2 + 9x + 20$. ________________ .

$(x + 5)(x + 4)$

171 Factor $x^2 + 12x + 27$. ________________ .

$(x + 9)(x + 3)$

172 Factor $x^2 + 6x + 9$. ________________ .

$(x + 3)(x + 3)$

173 Factor $x^2 + 6x + 5$. ________________ .

$(x + 5)(x + 1)$

174 Factor $x^2 + 10x + 25$. ________________ .

$(x + 5)(x + 5)$

175 To factor $x^2 - 7x + 10$, the FOIL expression is used.

$$x^2 - 7x + 10$$
$$x^2 - 5x - 2x + 10$$

Why are $^{-}5x$ and $^{-}2x$ used as the O and I terms? ________________ and ________________ .

$^{-}5 \cdot {}^{-}2 = 10$
and
$^{-}5 + {}^{-}2 = {}^{-}7$

176 To factor $x^2 - 7x + 10$, the following steps are used:

$$x^2 - 7x + 10$$
$$x^2 - 5x - 2x + 10$$
$$(x - 2)(x - 5)$$

Factor $x^2 - 5x + 6$ by first writing the FOIL expression. ________________ .

$x^2 - 2x - 3x + 6$
$(x - 3)(x - 2)$

177 Factor $x^2 - 8x + 15$. (Find factors of 15 that have a sum of $^-8$.) ______________ .

$(x - 5)(x - 3)$

178 Factor $x^2 - 3x + 2$. (Find factors of 2 that have a sum of $^-3$.) ______________ .

$(x - 1)(x - 2)$

179 Factor $x^2 - 10x + 9$. ______________ .

$(x - 9)(x - 1)$

180 Factor $x^2 - 8x + 16$. ______________ .

$(x - 4)(x - 4)$

181 Factor $x^2 - 12x + 36$. ______________ .

$(x - 6)(x - 6)$

182 Factor $x^2 - 15x + 50$. ______________ .

$(x - 5)(x - 10)$

183 To factor $x^2 - 5x - 24$, the FOIL expression is used.

$$x^2 - 5x - 24$$
$$x^2 - 8x + 3x - 24$$

Why are ^-8x and $3x$ used as the O and I terms? ______________ and ______________ .

$^-8 \cdot 3 = {}^-24$
and
$^-8 + 3 = {}^-5$

184 The factorization of $x^2 - 5x - 24$ is shown below:

$$x^2 - 5x - 24$$
$$x^2 - 8x + 3x - 24$$
$$(x + 3)(x - 8)$$

Factor $x^2 - 9x - 22$ by first writing the FOIL expression. ______________ .

$(x - 11)(x + 2)$
or
$(x + 2)(x - 11)$

185 Factor $x^2 + 4x - 12$. (Find factors of $^-12$ that have a sum of 4.) ______________ .

$(x + 6)(x - 2)$

186 Factor $x^2 - 2x - 24$. (Find factors of $^-24$ that have a sum of $^-2$.) ______________ .

$(x - 6)(x + 4)$

187 Factor $x^2 - 4x - 5$. ________________ . $(x - 5)(x + 1)$

188 Factor $x^2 + 9x - 22$. ________________ . $(x + 11)(x - 2)$

189 Factor $x^2 - 4x - 21$. ________________ . $(x - 7)(x + 3)$

190 Factor $x^2 - x - 2$. ________________ . $(x - 2)(x + 1)$

191 Factor $x^2 + 3x - 18$. ________________ . $(x + 6)(x - 3)$

192 To factor any trinomial of the form $x^2 + bx + c$, it is necessary to find the factors of c that have a sum of b. The signs (positive or negative) of b and c are very important. Factor $x^2 + 7x + 6$. ________________ . $(x + 6)(x + 1)$

193 Factor $x^2 - 9x + 18$. ________________ . $(x - 6)(x - 3)$

194 Factor $x^2 - 5x - 24$. ________________ . $(x - 8)(x + 3)$

195 Factor $x^2 + 2x + 1$. ________________ . $(x + 1)(x + 1)$

196 Factor $x^2 - 8x + 12$. ________________ . $(x - 6)(x - 2)$

197 Factor $x^2 - x - 20$. ________________ . $(x - 5)(x + 4)$

198 Factor $x^2 - 18x + 81$. ________________ . $(x - 9)(x - 9)$

199 Factor $x^2 - 12x + 20$. ________________ . $(x - 10)(x - 2)$

200 Factor $x^2 + 8x + 7$. ________________ . $(x + 7)(x + 1)$

201 Factor $x^2 + 13x - 30$. ________________ . $(x + 15)(x - 2)$

202 Factor $x^2 - 10x + 21$. ________________ . $(x - 7)(x - 3)$

203 Factor $x^2 - 2x + 1$. ________________ . $(x - 1)(x - 1)$

204 Factor $x^2 - 10x + 24$. ________________ . $(x - 6)(x - 4)$

The binomial $x^2 - 9$ is called the difference of two squares because $x^2 = x \cdot x$ and $9 = 3 \cdot 3$.

Any binomial that is the difference of two squares may be factored by reversing the FOIL multiplication.

---

205 To factor $x^2 - 9$, the FOIL expression is used.

$$x^2 - 9$$

$$x^2 + 0x - 9$$

$$x^2 + 3x - 3x - 9$$

Why are $3x$ and $^{-}3x$ used as the O and I terms? ________ and ________ .

$3 \cdot {}^{-}3 = {}^{-}9$
and
$3 + {}^{-}3 = 0$

206 $x^2 - 9$ is the difference of two squares. Complete the factorization of $x^2 - 9$.

$$x^2 - 9$$

$$x^2 + 0x - 9$$

$$x^2 + 3x - 3x - 9$$

________________ .

$(x - 3)(x + 3)$

207 $x^2 - 49$ is the difference of two squares because $x^2 = x \cdot x$ and $49 = 7 \cdot 7$. Complete the factorization of $x^2 - 49$.

$$x^2 - 49$$

$$x^2 + 0x - 49$$

________________ .

$x^2 + 7x - 7x - 49$
$(x - 7)(x + 7)$

208 Factor $x^2 - 1$ as the difference of two squares, because $x^2 = x \cdot x$, and $1 \cdot 1 = 1$. ________________ .

$(x + 1)(x - 1)$

209 $x^2 - 16$ can be written as $x^2 + 0x - 16$ and factored as the difference of two squares. Factor $x^2 - 16$.
______________ . $(x + 4)(x - 4)$

210 Factor $x^2 - 25$. ______________ . $(x + 5)(x - 5)$

211 Factor $x^2 - 4$. ______________ . $(x + 2)(x - 2)$

212 Factor $x^2 - 64$. ______________ . $(x + 8)(x - 8)$

---

7 is a prime number because its only positive factors are 1 and itself.

Some polynomials are also prime. $5x - 3$ is a prime polynomial because the only common factors of $5x$ and $^-3$ are 1 and $^-1$.

$x^2 + 7x + 9$ is a prime polynomial because there are no factors of 9 with a sum of 7.

---

213 $4x + 6$ is factorable because $4x$ and 6 have 2 as a common factor.

$$4x + 6 = 2(2x + 3)$$

Is $9x - 13$ factorable? ________ . No

214 $9x - 13$ is prime because the only common factors of $9x$ and $^-13$ are 1 and $^-1$. Factor $5x + 12$ or state that it is prime. ________ . Prime

215 $x^3 + 5x^2 + 3x$ is factorable because each term has a factor of x; therefore, x is a common factor.

$$x^3 + 5x^2 + 3x = x(x^2 + 5x + 3)$$

Factor $x^3 - 7x^2 + 5x$. ______________ . $x(x^2 - 7x + 5)$

216 When factoring any polynomial always look for a common factor first; then attempt to factor by using the reverse of FOIL multiplication. Factor $x^2 + 11x + 24$.
______________ . $(x + 8)(x + 3)$

217 $x^2 - 9x + 7$ has no common factor. The next step in factoring $x^2 - 9x + 7$ involves the reverse of FOIL multiplication. Can $x^2 - 9x + 7$ be written as a FOIL expression? ________ .

No

218 $x^2 - 9x + 7$ is a prime polynomial because:
(a) It has no common factor.
(b) There are no factors of 7 with a sum of $^-9$.
Factor $x^2 + 7x + 13$ or state that it is prime.
________ .

Prime

219 Factor $x^2 + 12x + 36$ or state that it is prime.
________________ .

$(x + 6)(x + 6)$

220 Factor $2x + 5$ or state that it is prime.
________ .

Prime

221 Factor $3x^2 - 12x$ or state that it is prime.
________________ .

$3x(x - 4)$

222 Factor $x^2 + 19x + 18$ or state that it is prime.
________________ .

$(x + 18)(x + 1)$

223 Factor $x^2 - 4x - 21$ or state that it is prime.
________________ .

$(x - 7)(x + 3)$

224 Factor $x^2 - 8x + 25$ or state that it is prime.
________ .

Prime

225 Factor $x^2 - 14x + 49$ or state that it is prime.
______________ .

$(x - 7)(x - 7)$

226 Factor $x^3 - 3x^2 + 8x$ or state that it is prime.
________________ .

$x(x^2 - 3x + 8)$

227 Factor $x^2 + 16x + 64$ or state that it is prime.
________________ .

$(x + 8)(x + 8)$

Self-Quiz # 2

The following questions test the objectives of the preceding section. 100% mastery is desired.

Factor each of the following or state that it is prime:

1. $x^2 + 8x + 15$. _______ .

2. $x^2 + 6x + 8$. _______ .

3. $x^2 - 4x - 21$. _______ .

4. $x^2 + 2x - 35$. _______ .

5. $x^2 - 3x - 10$. _______ .

6. $x^2 + 9x + 18$. _______ .

7. $x^2 + 3x - 10$. _______ .

8. $x^2 + 2x + 5$. _______ .

9. $x^2 - 6x + 9$. _______ .

10. $x^2 + 9x - 22$. _______ .

The polynomial $15x - 10$ is factored by using 5 as the common factor.

$$15x - 10 = 5(3x - 2)$$

The polynomial $x^2 + 9x - 22$ is factored by finding two numbers that have a *product* of $^{-}22$ and a *sum* of $^{+}9$.

$$x^2 + 9x - 22 = (x + 11)(x - 2)$$

First attempt to find a common factor when factoring any polynomial. After any common factor has been removed, the remaining polynomial should then be factored whenever possible.

---

228 To factor $x^3 + 3x^2 - 4x$, first find the common factor x as shown in the following steps.

$$x^3 + 3x^2 - 4x$$
$$x(x^2 + 3x - 4)$$
$$x(x + 4)(x - 1)$$

Complete the factoring of $x^3 + 8x^2 - 9x$ by factoring $x^2 + 8x - 9$.

$x^3 + 8x^2 - 9x$

$x(x^2 + 8x - 9)$

________________ .

$x(x + 9)(x - 1)$

229 To factor $3x^2 - 12$ the following steps are used.

$$3x^2 - 12$$
$$3(x^2 - 4)$$
$$3(x + 2)(x - 2)$$

Complete the factoring of $5x^2 - 45$ by factoring $x^2 - 9$.

$5x^2 - 45$

$5(x^2 - 9)$

________________ .

$5(x + 3)(x - 3)$

230 Factor $2x^2 + 8x - 24$. ________________ .
[Use two steps. First find the common factor; then factor the remaining polynomial.]

$2(x + 6)(x - 2)$

231 The first step in factoring $x^4 - 36x^2$ is to find the common factor x^2. Factor $x^4 - 36x^2$.

________________ .

$x^2(x + 6)(x - 6)$

232 Factor $5x^2 + 10x + 5$. ________________ .
[Use 5 as the common factor; then factor the remaining polynomial.]

$5(x + 1)(x + 1)$

233 Factor $3x^3 + 6x^2 - 45x$. ________________ .
[Use $3x$ as the common factor; then factor the remaining polynomial.]

$3x(x + 5)(x - 3)$

234 Factor $7x^2 - 7$. ________________ . [First find the common factor; then factor the remaining polynomial.]

$7(x + 1)(x - 1)$

235 Factor $4x^2 - 12x - 40$. ________________ .

$4(x - 5)(x + 2)$

236 Factor $2x^2 + 8x + 8$. ________________ .

$2(x + 2)(x + 2)$

237 Factor $x^3 - 81x$. ________________ .

$x(x + 9)(x - 9)$

238 Factor $2x^2 - 8$. ________________ .

$2(x + 2)(x - 2)$

239 Factor $10x^2 + 50x + 60$. ________________ .

$10(x + 3)(x + 2)$

240 Factor $3x^2 - 9x - 30$. ________________ .

$3(x - 5)(x + 2)$

241 Factor $6x^2 + 24x + 18$. ________________ .

$6(x + 1)(x + 3)$

242 Factor $4x^3 - 4x^2 - 80x$. ________________ .

$4x(x + 4)(x - 5)$

243 Factor $x^3 - 5x^2 + 6x$. ________________ .

$x(x - 2)(x - 3)$

Whenever factoring is being done, the first step should always be an attempt to find a common factor.

The following fifteen frames require factoring. Most of the polynomials have a common factor, which should be factored out first. Some of the polynomials are not factorable and should be called prime.

---

244 $5x - 15$. ________________ .

$5(x - 3)$

245 $2x^2 + 10x + 12$. ________________ .

$2(x + 2)(x + 3)$

246 $5x^2 - 45$. ________________ .

$5(x + 3)(x - 3)$

247 $6x^2 - 12x + 6$. ________________ .

$6(x - 1)(x - 1)$

248 $2x^2 - 10x + 12$. ________________ . $2(x-3)(x-2)$

249 $x^2 + 17x + 1$. ________________ . Prime

250 $2x^2 - 3x$. ________________ . $x(2x - 3)$

251 $2x^2 - 2x - 24$. ________________ . $2(x - 4)(x + 3)$

252 $4x^2 - 16$. ________________ . $4(x + 2)(x - 2)$

253 $5x^2 - 35x - 40$. ________________ . $5(x - 8)(x + 1)$

254 $60x - 70$. ________________ . $10(6x - 7)$

255 $x^3 - 5x^2 + 4x$. ________________ . $x(x - 4)(x - 1)$

Self-Quiz # 3

The following questions test the objectives of the preceding section. 100% mastery is desired.

Factor:

1. $x^3 - 5x^2 + 6x$. ________ .
2. $3x^2 + 6x - 24$. ________ .
3. $5x^2 - 20$. ________ .
4. $2x^2 + 4x - 30$. ________ .
5. $x^3 + 3x^2 - 21x$. ________ .
6. $x^2 - 7x + 7$. ________ .

Two methods of factoring have been explained in this chapter. The common factor method is used to factor $7x^2 - 14x$.

$$7x^2 - 14x = 7x(x - 2)$$

The trinomial $x^2 - 6x - 16$ is factored by finding two numbers with a product of $^-16$ and a sum of $^-6$.

$$x^2 - 6x - 16 = (x - 8)(x + 2)$$

In this section a third method of factoring applies to trinomials, such as $2x^2 + 7x + 6$, in which the number of x^2's (the coefficient of x^2) is not equal to 1.

---

256 In the trinomial $4x^2 + 3x - 1$ the multiplier or coefficient of x^2 is 4. In the trinomial $3x^2 + 5x - 2$ the multiplier or coefficient of x^2 is ________ .

3

257 In the trinomial $x^2 + 9x + 18$ the coefficient of x^2 is 1, and the trinomial is factored as:

$$x^2 + 9x + 18 = (x + 6)(x + 3)$$

In the trinomial $2x^2 + 13x + 18$ the coefficient of x^2 is ________ .

2

258 The trinomial $2x^2 + 13x + 18$ has 2 as the coefficient of x^2. Consequently, it *cannot* be factored by finding numbers with a product of 18 and a sum of 13. Can $4x^2 + 3x - 1$ be factored by finding numbers with a product of $^-1$ and a sum of 3? ________________ .

No ($4x^2$ has a coefficient of 4)

259 x^2 does not have a coefficient of 1 in $2x^2 + 7x + 6$. To determine whether or not $2x^2 + 7x + 6$ is factorable, multiply 2 by 6 to give 12. Are there any factors of 12 that have a sum of 7? ________________ .

Yes (3 and 4)

260 $3x^2 + 5x - 2$ does not have 1 as a coefficient of x^2. To determine whether or not $3x^2 + 5x - 2$ is factorable, multiply 3 by $^-2$ to give $^-6$. Are there any factors of $^-6$ that have a sum of 5? ________________ .

Yes (6 and $^-1$)

261 $4x^2 + 3x - 1$ does not have 1 as a coefficient of x^2. To determine whether or not $4x^2 + 3x - 1$ is factorable, multiply 4 by $^-1$ to give $^-4$. Are there any factors of $^-4$ that have a sum of 3? ______________ .

Yes (4 and $^-1$)

262 To determine whether or not $2x^2 + 7x + 6$ is factorable:

(a) Multiply 2 by 6 to give 12.
(b) Are there any factors of 12 that have a sum of 7?
(c) If the answer to (b) is "Yes," then $2x^2 + 7x + 6$ is factorable.

Is $2x^2 + 7x + 6$ factorable? ________ .

Yes

263 To determine whether or not $3x^2 - 20x - 4$ is factorable:

(a) Multiply 3 by $^-4$ to give $^-12$.
(b) Are there any factors of $^-12$ that have a sum of $^-20$?
(c) If the answer to (b) is "No," then $3x^2 - 20x - 4$ is not factorable.

Is $3x^2 - 20x - 4$ factorable? ________ .

No

264 $5x^2 - 12x + 4$ is factorable because $5 \cdot 4 = 20$ and $^-10$ and $^-2$ are factors of 20 that have a sum of $^-12$. Is $3x^2 - 8x + 4$ factorable? ________ .

Yes

265 $2x^2 - 8x + 9$ is not factorable because $2 \cdot 9 = 18$ and there are no factors of 18 that have a sum of $^-8$. Is $3x^2 - 2x + 7$ factorable? ________ .

No

266 A polynomial of the form $ax^2 + bc + c$ is factorable only when the product ac has factors that have a sum of b. Is $2x^2 + 7x + 5$ factorable? ________ .

Yes

267 Is $4x^2 + 11x + 6$ factorable? ______________ . (Are there factors of $4 \cdot 6$ or 24 that have a sum of 11?)

Yes, 8 and 3

268 Is $3x^2 + 5x + 6$ factorable? ________ . (Are there factors of $3 \cdot 6$ or 18 that have a sum of 5?)

No

269 Is $6x^2 - 7x + 2$ factorable? ______________ . (Are there factors of $6 \cdot 2$ or 12 that have a sum of $^-7$?)

Yes, $^-4$ and $^-3$

270 Is $9x^2 - 3x - 2$ factorable? ______________ . (Are there factors of $9 \cdot {}^-2$ or $^-18$ that have a sum of $^-3$?)

Yes, $^-6$ and 3

271 Is $5x^2 - 13x - 6$ factorable? ______________ . (Are there factors of $5 \cdot {}^-6$ or $^-30$ that have a sum of $^-13$?)

Yes, $^-15$ and 2

272 $2x^2 + 7x + 6$ is factorable. To factor, find the factors of $2 \cdot 6$ or 12 that have a sum of 7. These factors are 3 and 4.

$$2x^2 + \underline{7x} + 6$$

is rewritten as the following FOIL expression

$$2x^2 + 4x + 3x + 6$$

Factor the F + O expression $2x^2 + 4x$ by using the common factor method. ______________ .

$2x(x + 2)$

273 The factorization of $2x^2 + 7x + 6$ proceeds as follows:

$$2x^2 + 7x + 6$$
$$2x^2 + 4x + 3x + 6$$
$$2x(x + 2) + 3(x + 2)$$

Factor the I and L terms, $3x + 6$, by using the common factor method. ______________ .

$3(x + 2)$

274 The factorization of $2x^2 + 7x + 6$ proceeds as follows:

$$2x^2 + 7x + 6$$
$$2x^2 + 4x + 3x + 6$$
$$2x(x + 2) + 3(x + 2)$$

The binomial $(x + 2)$ is a common factor, and $2x(x + 2) + 3(x + 2)$ is factored as $(x + 2)(2x + 3)$. Factor $2x(2x + 3) + 1(2x + 3)$ by using the binomial $(2x + 3)$ as the common factor.

______________ .

$(2x + 3)(2x + 1)$

275 The factorization of $4x^2 + 8x + 3$ begins by writing the FOIL expression.

$$4x^2 + 8x + 3$$

$$4x^2 + 6x + 2x + 3$$

Why are $6x$ and $2x$ used as the O and I terms? __________, __________, and ________.

$4 \cdot 3 = 12$,
$6 \cdot 2 = 12$,
$6 + 2 = 8$

276 The factorization of $4x^2 + 8x + 3$ proceeds as follows:

$$4x^2 + 8x + 3$$

$$4x^2 + 6x + 2x + 3$$

$$2x(2x + 3) + 1(2x + 3)$$

What factor is common to the expressions $2x(2x + 3)$ and $1(2x + 3)$? __________.

$(2x + 3)$

277 Factor $4x^2 + 8x + 3$. ____________________.

$(2x + 3)(2x + 1)$

278 The factorization of $5x^2 + 12x + 4$ is completed in three steps.

(a) The FOIL expression is written by using the facts that $5 \cdot 4 = 20$, $10 \cdot 2 = 20$, and $10 + 2 = 12$.

$$5x^2 + 10x + 2x + 4$$

(b) In the common factor method the F and O terms and I and L terms are factored separately.

$$5x(x + 2) + 2(x + 2)$$

(c) The common binomial factor $(x + 2)$ is removed from each term of the expression in step c.

$$(x + 2)(5x + 2)$$

Using the three steps shown above the factor $4x^2 + 11x + 6$. __________________.

$4x^2 + 3x + 8x + 6$
$x(4x + 3) + 2(4x + 3)$
$(4x + 3)(x + 2)$

279 $2x^2 - 7x + 5$ is factored in three steps:

(a) $2x^2 - 2x - 5x + 5$ is a FOIL expression because $2 \cdot 5 = 10$, $^{-}2 \cdot {}^{-}5 = 10$, and $^{-}2 + {}^{-}5 = {}^{-}7$.

(b) $2x(x - 1) - 5(x - 1)$, using $2x$ as the common factor of $2x^2 - 2x$, and $^{-}5$ as the common factor of $^{-}5x + 5$.

(c) $(x - 1)(2x - 5)$ are the factors of $2x^2 - 7x + 5$ because $(x - 1)$ is a common factor in each of the terms in step b.

Using the same three-step procedure, factor $4x^2 - 16x + 7$. ________________ .

$4x^2 - 14x - 2x + 7$
$2x(2x - 7) - 1(2x - 7)$
$(2x - 7)(2x - 1)$

280 $4x^2 - 8x - 5$ is factored in three steps:

(a) $4x^2 - 10x + 2x - 5$.

(b) $2x(2x - 5) + 1(2x - 5)$.

(c) $(2x - 5)(2x + 1)$.

Factor $3x^2 - 8x - 3$. ________________ .

$3x^2 - 9x + x - 3$
$3x(x - 3) + 1(x - 3)$
$(x - 3)(3x + 1)$

281 $4x^2 + 12x - 7$ is factored in three steps:

(a) $4x^2 + 14x - 2x - 7$.

(b) $2x(2x + 7) - 1(2x + 7)$.

(c) $(2x + 7)(2x - 1)$.

Factor $2x^2 - 5x - 12$. ________________ .

$2x^2 - 8x + 3x - 12$
$2x(x - 4) + 3(x - 4)$
$(x - 4)(2x + 3)$

282 Factor $6x^2 + 17x + 5$ by rewriting $17x$ as two terms.

________________ .

[*Hint*: What factors of $6 \cdot 5$ or 30 have a sum of 17?]

$(2x + 5)(3x + 1)$

283 Factor $5x^2 + 9x - 2$ by rewriting $9x$ as two terms.

________________ .

[*Hint*: What factors of $^{-}10$ have a sum of 9?]

$(x + 2)(5x - 1)$

284 Factor $3x^2 - 8x + 5$ by rewriting $^{-}8x$ as two terms.

________________ .

[*Hint*: What factors of 15 have a sum of $^{-}8$?]

$(3x - 5)(x - 1)$

285 Factor $2x^2 + 9x + 7$. ______________ . $(2x + 7)(x + 1)$

286 Factor $3x^2 - 14x + 8$. ______________ . $(x - 4)(3x - 2)$

287 Factor $8x^2 - 17x + 2$. ______________ . $(x - 2)(8x - 1)$

288 Factor $6x^2 + 5x - 4$. ______________ . $(3x + 4)(2x - 1)$

289 Factor $4x^2 - 12x + 9$. ________ . $(2x - 3)^2$

290 $4x^2 - 9$ is the difference of two squares.
$4x^2 = 2x \cdot 2x$, and $9 = 3 \cdot 3$. Therefore, $4x^2 - 9$ is factored as $(2x - 3)(2x + 3)$.
Factor $16x^2 - 25$ as the difference of two squares.
______________ . $(4x + 5)(4x - 5)$

291 $4x^2 - 25$ is the difference of two squares because $4x^2 = 2x \cdot 2x$, and $25 = 5 \cdot 5$.
Factor $4x^2 - 25$. ______________ . $(2x + 5)(2x - 5)$

292 Factor $5x^2 - 11x + 2$. ______________ . $(x - 2)(5x - 1)$

293 Factor $3x^2 + 10x + 8$. ______________ . $(x + 2)(3x + 4)$

294 Factor $4x^2 - 49$. ______________ . $(2x + 7)(2x - 7)$

Self-Quiz #4

The following questions test the objectives of the preceding section. 100% mastery is desired.

Factor:

1. $2x^2 + 5x - 3$. ________ .
2. $3x^2 + 11x + 6$. ________ .
3. $4x^2 - 11x - 3$. ________ .
4. $5x^2 + 13x + 6$. ________ .
5. $6x^2 - 11x + 3$. ________ .
6. $8x^2 - 6x - 9$. ________ .
7. $9x^2 - 25$. ________ .

Three methods of factoring have been shown in this chapter. The first method of factoring that should be attempted for any polynomial is the common factor method.

---

295 Each term of $6x - 10$ has 2 as a factor. 2 is a common factor of $6x$ and $^-10$, and the factoring of $6x - 10$ is completed as follows:

$$6x - 10 = 2(3x - 5)$$

Factor $12x - 28$. ________ .

$4(3x - 7)$

296 Each term of $5x^3 - 30x^2 + 50x$ has $5x$ as a common factor. $5x$ is the common factor for $5x^3 - 30x^2 + 50x$, and the trinomial is factored as follows:

$$5x^3 - 30x^2 + 50x = 5x(x^2 - 6x + 10)$$

Factor $3x^3 - 9x^2 + 27x$. ________________ .

$3x(x^2 - 3x + 9)$

297 Using the common factor method, factor $x^2 - 5x$. ________ .

$x(x - 5)$

298 Factor $6x^3 - 8x^2 + 12x$. _______________ .

$2x(3x^2 - 4x + 6)$

299 $x^2 + 10x + 24$ does not have a common factor. To factor $x^2 + 10x + 24$, first note that x^2 has a coefficient of 1. Consequently, it is necessary to find numbers with a product of 24 and a sum of 10.

$$x^2 + 10x + 24 = (x + 6)(x + 4)$$

Factor $x^2 - 19x + 90$. _______________ .

$(x - 9)(x - 10)$

300 $x^2 - 49$ has no common factor. To factor $x^2 - 49$, first write the expression as $x^2 + 0x - 49$. Since x^2 has a coefficient of 1, it is necessary to find numbers with a product of $^-49$ and a sum of 0.

$$x^2 - 49 = x^2 + 0x - 49 = (x + 7)(x - 7)$$

Factor $x^2 - 81$. _______________ .

$(x + 9)(x - 9)$

301 Trinomials such as $x^2 - 9x - 10$ and $x^2 + 7x - 18$ have 1 as the coefficients of their x^2 terms. To factor $x^2 - 9x - 10$, find the factors of $^-10$ that have a sum of $^-9$.

$$x^2 - 9x - 10 = (x - 10)(x + 1)$$

Factor $x^2 + 7x - 18$. _______________ .

$(x + 9)(x - 2)$

302 Factor $x^2 + 10x + 21$. _______________ .

$(x + 7)(x + 3)$

303 Factor $x^2 - 13x + 30$. _______________ .

$(x - 10)(x - 3)$

304 Factor $x^2 + 5x - 6$. _______________ .

$(x + 6)(x - 1)$

305 Factor $x^2 + 2x - 15$. _______________ .

$(x + 5)(x - 3)$

306 Factor $x^2 - x - 20$. _______________ .

$(x - 5)(x + 4)$

307 $3x^2 + 5x - 2$ has a coefficient of 3 for its x^2 term. To factor $3x^2 + 5x - 2$, first multiply 3 and $^-2$. $(3 \cdot {^-2} = {^-6})$. Then find factors of $^-6$ that have a sum of 5. Complete the factorization of $3x^2 + 5x - 2$.

$$3x^2 + 5x - 2$$

$$3x^2 + 6x - x - 2$$

$$3x(x + 2) - 1(x + 2)$$

________________ .

$(x + 2)(3x - 1)$

308 The coefficient of x^2 in the expression $4x^2 + 3x - 1$ is not 1. Therefore, the first step is to multiply 4 and $^-1$; then find factors of $^-4$ with a sum of 3. Complete the factorization of $4x^2 + 3x - 1$:

$$4x^2 + 3x - 1$$

$$4x^2 + 4x - x - 1$$

________________ .

$(x + 1)(4x - 1)$

309 Factor $2x^2 + 7x + 6$ by using the method necessary when the coefficient of x^2 is not 1. ________________ .

$(2x + 3)(x + 2)$

310 Factor $3x^2 - x - 4$. ________________ .

$(3x - 4)(x + 1)$

311 Factor $2x^2 + 5x - 3$. ________________ .

$(2x - 1)(x + 3)$

312 Factor $5x^2 + 18x - 8$. ________________ .

$(5x - 2)(x + 4)$

313 Factor $3x^2 + 5x - 12$. ________________ .

$(3x - 4)(x + 3)$

Three methods of factoring have been shown in this chapter.

1. The common factor method.

$$5x - 10 = 5(x - 2)$$

2. The method when the coefficient of the x^2 term is 1.

$$x^2 + 7x - 30 = (x + 10)(x - 3)$$

3. The method when the coefficient of the x^2 term is not 1.

$$3x^2 + 7x - 6 \qquad 3 \cdot {}^-6 = {}^-18$$

Factors of $^-18$ that have a sum of 7 are 9 and $^-2$.

$$3x^2 + 9x - 2x - 6$$

$$3x(x + 3) - 2(x + 3)$$

$$(x + 3)(3x - 2)$$

In the following frames each polynomial can be factored by using one of the above methods.

---

314 Factor $12x - 8$. ______________ . $\quad$ $4(3x - 2)$

315 Factor $x^2 + 11x + 30$. ______________ . $\quad$ $(x + 5)(x + 6)$

316 Factor $5x^2 - 25x$. ______________ . $\quad$ $5x(x - 5)$

317 Factor $4x^2 + 5x - 6$. ______________ . $\quad$ $(x + 2)(4x - 3)$

318 Factor $x^2 - 16$. ______________ . $\quad$ $(x + 4)(x - 4)$

319 Factor $x^2 - 14x + 40$. ______________ . $\quad$ $(x - 4)(x - 10)$

320 Factor $8x^2 - 20x - 2$. ______________ . $\quad$ $2(4x^2 - 10x - 1)$

321 Factor $25x^2 - 1$. ______________ . $\quad$ $(5x + 1)(5x - 1)$

322 Factor $2x^2 - x - 10$. ______________ . $\quad$ $(2x - 5)(x + 2)$

323 Factor $4x^3 - 2x^2$. ______________ . $\quad$ $2x^2(2x - 1)$

324 Factor $x^2 - 3x + 2$. ______________ . $(x - 2)(x - 1)$

325 Factor $x^2 - 9x + 18$. ______________ . $(x - 3)(x - 6)$

326 Factor $2x^2 + 3x - 5$. ______________ . $(2x + 5)(x - 1)$

327 Factor $7x^2 - 7x$. ______________ . $7x(x - 1)$

328 Factor $x^2 - 3x - 10$. ______________ . $(x + 2)(x - 5)$

329 Factor $x^2 - 9x + 8$. ______________ . $(x - 8)(x - 1)$

The first attempt at factoring any polynomial is to find a common factor. If a common factor is found, the remaining polynomial must be checked to determine whether or not it can be factored. This procedure is shown below:

$$6x^3 + 22x^2 - 8x$$
$$2x(3x^2 + 11x - 4)$$
$$3x^2 + 12x - 1x - 4$$
$$3x(x + 4) - 1(x + 4)$$
$$2x(x + 4)(3x - 1)$$

Each of the following polynomials must be factored twice: first by the common factor method and second by factoring the remaining polynomial.

---

330 Factor $3x^2 - 15x + 12$. ______________ . $3(x - 4)(x - 1)$

331 Factor $6x^2 - 54$. ______________ . $6(x + 3)(x - 3)$

332 Factor $2x^3 - 9x^2 + 9x$. ______________ . $x(2x - 3)(x - 3)$

333 Factor $5x^2 - 35x - 90$. ______________ . $5(x - 9)(x + 2)$

334 Factor $24x^2 - 6$. ______________ . $6(2x + 1)(2x - 1)$

335 Factor x^3-2x^2-3x. ________________ . $x(x-3)(x+1)$

336 Factor $2x^2-72$. ________________ . $2(x+6)(x-6)$

337 Factor $3x^2+21x-24$. ________________ . $3(x+8)(x-1)$

338 Factor $2x^3-2x^2-40x$. ________________ . $2x(x-5)(x+4)$

339 Factor x^3-2x^2-15x. ________________ . $x(x-5)(x+3)$

340 Factor $36x^3-9x$. ________________ . $9x(2x+1)(2x-1)$

341 Factor $5x^2-10x+5$. ________________ . $5(x-1)(x-1)$

342 Factor $3x^3-15x^2-42x$. ________________ . $3x(x-7)(x+2)$

343 Factor $5x^2-10x-15$. ________________ . $5(x+1)(x-3)$

344 Factor x^3+3x^2-10x. ________________ . $x(x+5)(x-2)$

345 Factor $3x^2-12$. ________________ . $3(x+2)(x-2)$

346 Factor $4x^2+8x-12$. ________________ . $4(x+3)(x-1)$

347 Factor $7x^3-7x$. ________________ . $7x(x+1)(x-1)$

348 Factor $3x^2-21x+30$. ________________ . $3(x-5)(x-2)$

349 Factor $2x^3-8x^2+6x$. ________________ . $2x(x-3)(x-1)$

350 Factor $3x^3-75x$. ________________ . $3x(x+5)(x-5)$

351 Factor $3x^2-18x+15$. ________________ . $3(x-1)(x-5)$

352 Factor x^3+4x^2-5x. ______________ . $x(x+5)(x-1)$

353 Factor $3x^2-12$. ______________ . $3(x+2)(x-2)$

354 Factor $^-2x^2-10x-12$. ______________ . $^-2(x+2)(x+3)$

355 Factor $3x^3-6x^2-24x$. ______________ . $3x(x-4)(x+2)$

Self-Quiz # 5

The following questions test the objectives of the preceding section. 100% mastery is desired.

Factor:

1. $3x - 24$. ________ .
2. $x^2 + 3x - 10$. ________ .
3. $3x^2 - 9x$. ________ .
4. $x^2 - 25$. ________ .
5. $6x^2 + 7x - 3$. ________ .
6. $x^2 - 13x + 40$. ________ .
7. $6x^2 - 3x - 45$. ________ .
8. $3x^2 + 11x - 4$. ________ .
9. $2x^2 + 5x - 1$. ________ .
10. $3x^3 - 48x$. ________ .
11. $2x^2 + x - 15$. ________ .
12. $6x^2 - x - 12$. ________ .

Chapter Summary

The factorization of polynomials is an important skill in many algebra problems.

In this chapter three different methods of factoring were explained:

1. The common factor method:

$$14x - 21 = 7(2x - 3)$$

2. Factoring trinomials when the coefficient of x^2 is 1:

$$x^2 - 8x - 33 = (x - 11)(x + 3)$$

3. Factoring trinomials when the coefficient of x^2 is not 1:

$$6x^2 + 13x - 8$$

$$6x^2 + 16x - 3x - 8$$

$$2x(3x + 8) - 1(3x + 8)$$

$$(3x + 8)(2x - 1)$$

CHAPTER POST-TEST

The following questions test the objectives of this chapter. A score of 90% indicates sufficient mastery, and the student may proceed to the next chapter.

I Factor:

1. $7x - 35$
2. $6x^3 - 5x$
3. $^-9x + 12$
4. $4x + 4$

II Factor:

1. $x^2 + 4x + 3$
2. $x^2 - 81$
3. $x^2 - 3x - 28$
4. $x^2 - 11x + 24$

III Factor:

1. $6x^2 - 11x + 4$
2. $5x^2 - 8x - 4$
3. $3x^2 - 11x - 4$
4. $4x^2 + 4x + 1$

IV Factor:

1. $x^2 - 5x - 36$
2. $2x^3 - 9x^2 - x$
3. $25x^2 - 1$
4. $3x^2 - 22x + 7$
5. $x^2 + 14x + 24$
6. $3x^2 - 15x + 12$
7. $x^2 - 1$
8. $6x^2 - 16x - 30$

ALGEBRA

Polynomial Fractions

CHAPTER PRE-TEST

The following questions indicate the objectives of this chapter. A score of 90% indicates sufficient mastery, and the student may immediately take the Chapter Post-Test.

I Simplify:

1. $\dfrac{10x^5}{15x}$

2. $\dfrac{x+4}{x^2-5x-36}$

3. $\dfrac{x^2-49}{x-7}$

II Multiply:

1. $\dfrac{4x^3}{x-3} \cdot \dfrac{x-3}{14x^6}$

2. $\dfrac{x^2+10x+9}{3x-6} \cdot \dfrac{x^2-6x+8}{x^2-3x-4}$

3. $\dfrac{7}{5-x} \cdot \dfrac{x-5}{x+7}$

III Add:

1. $\dfrac{x+8}{5} - \dfrac{2x-5}{3}$

2. $\dfrac{7}{12x} + \dfrac{3}{8x}$

3. $\dfrac{x+8}{x^2-6x+5} + \dfrac{x-3}{x^2-25}$

The following section is devoted to simplifying polynomial fractions. $\frac{x^2 + 5x + 6}{x^2 + 7x + 12}$ is a polynomial fraction. It can be simplified to $\frac{x + 2}{x + 4}$, as will be shown in the following frames.

---

1 $\frac{6}{8}$ can be simplified to $\frac{3}{4}$ because 2 is a common factor for 6 and 8.

$$\frac{6}{8} = \frac{2 \cdot 3}{2 \cdot 4} = \frac{\not{2} \cdot 3}{\not{2} \cdot 4} = \frac{3}{4}$$

Simplify $\frac{8}{12}$ by dividing out the common factor 4.

_______ .

$\frac{2}{3}$.

2 $2x^2$ and $5x$ have a common factor x, and $\frac{2x^2}{5x}$ is simplified as follows:

$$\frac{2x^2}{5x} = \frac{2x \cdot x}{5 \cdot x} = \frac{2x \cdot \not{x}}{5 \cdot \not{x}} = \frac{2x}{5}$$

Simplify $\frac{4x^3}{3x^2}$ by dividing out the common factor x^2.

_______ .

$\frac{4x}{3}$

3 $\frac{7}{8}$ cannot be simplified because the only common factors of 7 and 8 are 1 and $^-1$. Can $\frac{3x^2}{2y}$ be simplified?

_______ .

No

4 $\frac{7}{8}$ cannot be simplified. $\frac{7}{8}$ can be written as $\frac{5 + 2}{5 + 3}$, but since 5 is *not* a common factor, the 5's cannot be divided out. If $\frac{5}{7}$ is written as $\frac{4 + 1}{4 + 3}$, can the 4's be divided out? _______________ .

No, 4 is not a factor

5 Only common factors can be divided out of any fraction. Can $\frac{x + 7}{x + 3}$ be simplified by dividing out the x's?

____________________ .

No, x is not a common factor

6 $2x + 7$ is prime. $x + 5$ is also prime. $\frac{2x + 7}{x + 5}$ cannot be simplified because the only common factors are 1 and $^{-}1$. Can $\frac{x + 7}{2x - 3}$ be simplified? ____________________ .

No, there is no common factor

7 $\frac{(x + 2)(x + 3)}{(x + 4)(x + 3)}$ can be simplified because $(x + 3)$ is a common factor.

$$\frac{(x + 2)\cancel{(x + 3)}}{(x + 4)\cancel{(x + 3)}} = \frac{x + 2}{x + 4}$$

Simplify $\frac{(x + 7)(x - 5)}{(x + 7)(x + 10)}$ by dividing out the common factor $(x + 7)$. ________ .

$\frac{x - 5}{x + 10}$

8 Simplify $\frac{(x - 2)(x + 5)}{(x + 8)(x - 2)}$ by dividing out the common factor $(x - 2)$. ________ .

$\frac{x + 5}{x + 8}$

9 Simplify $\frac{3(x - 2)}{6(x + 8)}$ by dividing out the common factor 3.

________ .

$\frac{(x - 2)}{2(x + 8)}$

10 Simplify $\frac{x(x + 5)}{(x + 7)(x + 5)}$ by dividing out the common factor $(x + 5)$. ______ .

$\frac{x}{(x + 7)}$

11 To simplify $\frac{x^2 + 5x + 6}{x^2 + 7x + 12}$, first factor both numerator and denominator. Finish the following simplification.

$$\frac{x^2 + 5x + 6}{x^2 + 7x + 12} = \frac{(x + 3)(x + 2)}{(x + 3)(x + 4)} = \underline{\qquad\qquad}.$$

$\frac{x + 2}{x + 4}$

12 Complete the following simplification:

$$\frac{5x - 10}{4x - 8} = \frac{5(x - 2)}{4(x - 2)} = \underline{\qquad\qquad}.$$

$\frac{5}{4}$

13 Complete the following simplification:

$$\frac{3x - 12}{x^2 - 7x + 12} = \frac{3(x - 4)}{(x - 3)(x - 4)} = \underline{\qquad\qquad}.$$

$\frac{3}{x - 3}$

14 Complete the following simplification:

$$\frac{2x - 6}{4x + 4} = \frac{2(x - 3)}{4(x + 1)} = \underline{\qquad\qquad}.$$

$\frac{x - 3}{2(x + 1)}$

15 Complete the following simplification:

$$\frac{x^2 - 49}{x^2 + 8x + 7} = \frac{(x + 7)(x - 7)}{(x + 1)(x + 7)} = \underline{\qquad\qquad}.$$

$\frac{x - 7}{x + 1}$

16 Simplify $\frac{x^2 + 6x + 8}{x^2 + 3x + 2}$. $\underline{\qquad\qquad}$.

$\frac{x + 4}{x + 1}$

17 Simplify $\frac{x^2 + 5x - 14}{3x - 6}$. $\underline{\qquad\qquad}$.

$\frac{x + 7}{3}$

18 Simplify $\frac{x^2 - 3x}{x^2 - 10x + 21}$. $\underline{\qquad\qquad}$.

$\frac{x}{x - 7}$

19 Simplify $\frac{3x - 24}{6x}$. $\underline{\qquad\qquad}$.

$\frac{x - 8}{2x}$

20 Simplify $\frac{x^2 - 4}{3x + 6}$. $\underline{\qquad\qquad}$.

$\frac{x - 2}{3}$

21 Simplify $\dfrac{x^2 + 6x + 9}{2x + 6}$. ______ .

$\dfrac{x + 3}{2}$

22 Simplify $\dfrac{x^2 + 5x - 24}{x^2 + 10x + 16}$. ______ .

$\dfrac{x - 3}{x + 2}$

23 Simplify $\dfrac{{}^{-}3x^2 + 12x}{x^2 - 4x}$. ______ .

$\dfrac{{}^{-}3}{1}$ or ${}^{-}3$

24 Simplify $\dfrac{x^2 + 9x + 8}{x^2 - 64}$. ______ .

$\dfrac{x + 1}{x - 8}$

25 Simplify $\dfrac{x^2 - 5x + 4}{x^2 - 1}$. ______ .

$\dfrac{x - 4}{x + 1}$

Self-Quiz # 1

The following questions test the objectives of the preceding section. 100% mastery is desired.

Simplify each of the following:

1. $\dfrac{10x^3}{2x}$. ______ .
2. $\dfrac{2x - 6}{x^2 + 2x - 15}$. ______ .
3. $\dfrac{x^2 - 2x - 8}{x^2 + 5x + 6}$. ______ .
4. $\dfrac{x^2 + 10x + 24}{x^2 - 16}$. ______ .
5. $\dfrac{x^2 - 5x - 14}{x^2 - 10x + 21}$. ______ .
6. $\dfrac{x^2 - 7x - 8}{x^2 + x}$. ______ .

The multiplication of rational numbers is accomplished by multiplying numerators, and also by multiplying denominators, as the following example shows:

$$\frac{3}{4} \cdot \frac{5}{7} = \frac{3 \cdot 5}{4 \cdot 7} = \frac{15}{28}$$

The multiplication of polynomial fractions is accomplished in the same manner and is studied in the following frames.

---

26 $\frac{3}{8} \cdot \frac{5}{7} = \frac{15}{56}$. $\frac{4}{5} \cdot \frac{2}{3} =$ ________ .

$\frac{8}{15}$

27 $\frac{2x^2}{y} \cdot \frac{3x^5}{5y} = \frac{6x^7}{5y^2}$. $\frac{4x}{3y^2} \cdot \frac{5x}{7y^3} =$ ________ .

$\frac{20x^2}{21y^5}$

28 $\frac{x+2}{x-3} \cdot \frac{x+1}{x-4} = \frac{(x+2)(x+1)}{(x-3)(x-4)} = \frac{x^2+3x+2}{x^2-7x+12}$.

$\frac{x-5}{x-1} \cdot \frac{x+1}{x+3} =$ ________________ .

$\frac{x^2-4x-5}{x^2+2x-3}$

29 The multiplication of $\frac{2}{3} \cdot \frac{5}{4}$ can be simplified because 2 and 4 have a common factor of 2, which can be divided out, as shown in the following example:

$$\frac{2}{3} \cdot \frac{5}{4} = \frac{1 \cdot \not{2}}{3} \cdot \frac{5}{\not{2} \cdot 2} = \frac{1}{3} \cdot \frac{5}{2} = \frac{5}{6}$$

Multiply $\frac{2}{7}$ and $\frac{3}{4}$ by first dividing out the common factor 2. ________ .

$\frac{3}{14}$

30 3 is a common factor for 6 and 9. The multiplication of $\frac{9}{7} \cdot \frac{5}{6}$ can be done by dividing out the common factor 3 in the numerator and denominator. Multiply $\frac{8}{9} \cdot \frac{7}{12}$. 4 is a common factor of 8 and 12 . ______ .

$\frac{14}{27}$

31 x is a common factor of $2x^2$ and $3x$. The multiplication of $\frac{2x^2}{7} \cdot \frac{5}{3x}$ is done as follows:

$$\frac{2x^2}{7} \cdot \frac{5}{3x} = \frac{2x \cdot \cancel{x}}{7} \cdot \frac{5}{3\cancel{x}} = \frac{2x}{7} \cdot \frac{5}{3} = \frac{10x}{21}$$

Multiply $\frac{4x^3}{5} \cdot \frac{2}{3x^2}$ by first dividing out the common factor x^2. ________ .

$\frac{8x}{15}$

32 Multiply $\frac{3x^2}{5} \cdot \frac{4}{9x}$ by dividing out the common factor $3x$. ________ .

$\frac{4x}{15}$

33 Multiply $\frac{5x}{2y^2} \cdot \frac{4y}{7}$ by dividing out the common factor $2y$. ________ .

$\frac{10x}{7y}$

34 Multiply $\frac{7x^2}{5} \cdot \frac{10}{3x^2}$ by dividing out 5 and x^2, which are both common factors. ________ .

$\frac{14}{3}$

35 Multiply $\frac{3y^3}{7x} \cdot \frac{14x^2}{9y}$. Both $3y$ and $7x$ are common factors. ________ .

$\frac{2xy^2}{3}$

36 Multiply $\frac{4x}{3y^2} \cdot \frac{6y}{7x^3}$ by dividing out any common factors of the numerator and denominator. ________ .

$\frac{8}{7x^2y}$

37 Multiply $\frac{2x^5}{5y^2} \cdot \frac{5y^2}{4x^3}$. ________ .

$\frac{x^2}{2}$

38 Multiply $\frac{^-3x^2}{5} \cdot \frac{7y}{3x}$. ________ .

$\frac{^-7xy}{5}$

39 Multiply $\frac{^-4x}{5y^3} \cdot \frac{^-15y^4}{16x^3}$. ________.

$\frac{3y}{4x^2}$

40 Common factors may be divided out when multiplying monomial or polynomial fractions.

$$\frac{x-3}{5} \cdot \frac{7}{x-3} = \frac{(\cancel{x-3})}{5} \cdot \frac{7}{(\cancel{x-3})} = \frac{1}{5} \cdot \frac{7}{1} = \frac{7}{5}$$

[*Note*: $(x - 3)$ was the common factor that could be divided out of the numerator and denominator.] Multiply $\frac{x+2}{4} \cdot \frac{3}{x+2}$ by dividing out the common factor $(x + 2)$. ________.

$\frac{3}{4}$

41 Multiply $\frac{2(x-5)}{x+7} \cdot \frac{3}{(x-5)}$ by dividing out the common factor $(x-5)$. ________.

$\frac{6}{x+7}$

42 Multiply $\frac{(x+4)(x-1)}{x+3} \cdot \frac{5}{x+4}$. ________.
[*Note*: The common factor is $(x + 4)$.]

$\frac{5(x-1)}{x+3}$ or $\frac{5x-5}{x+3}$

43 Multiply $\frac{x+2}{x-1} \cdot \frac{3(x-1)}{2(x+4)}$. ________.
[*Note*: The common factor is $(x - 1)$.]

$\frac{3(x+2)}{2(x+4)}$ or $\frac{3x+6}{2x+8}$

44 Multiply $\frac{5(x-1)}{x+2} \cdot \frac{x+2}{10x}$. ________.
[*Note*: $(x + 2)$ and 5 are *both* common factors.]

$\frac{x-1}{2x}$

45 Multiply $\frac{3x^2}{x-7} \cdot \frac{x-7}{6x}$. ________.

$\frac{x}{2}$

46 To multiply $\frac{3x-6}{4} \cdot \frac{5}{x^2-4}$, first factor wherever possible, as follows:

$$\frac{3x-6}{4} \cdot \frac{5}{x^2-4} = \frac{3(\cancel{x-2})}{4} \cdot \frac{5}{(x+2)(\cancel{x-2})}$$

$$= \frac{15}{4(x+2)}$$

Multiply $\frac{8x+16}{5} \cdot \frac{1}{x^2-4}$ by first factoring wherever possible. ________ .

$\frac{8}{5(x-2)}$

Dividing out common factors from numerators and denominators is possible because any nonzero number divided by itself is 1.

$$\frac{6}{6} = 1, \quad \frac{^-7}{^-7} = 1, \quad \frac{48}{48} = 1$$

For the same reason,

$$\frac{x-5}{x-5} = 1, \quad \frac{2x+7}{2x+7} = 1, \quad \frac{x^2+5x+18}{x^2+5x+18} = 1$$

Whenever the numerator and denominator of a polynomial fraction are the same, the fraction is equivalent to 1.

---

47 $\frac{5}{5}$ = ________ .

1

48 $\frac{^-19}{^-19}$ = ________ .

1

49 $\frac{x+18}{x+18}$ = ________ .

1

50 $\frac{x-23}{x-23}$ = ________ .

1

51 $\frac{2x-19}{2x-19}$ = ________ .

1

52 $\dfrac{2 - x}{2 - x} = $ ________ . 1

53 $\dfrac{x^2 + 3x - 17}{x^2 + 3x - 17} = $ ________ . 1

54 $\dfrac{3x^3 - 15x + 2}{3x^3 - 15x + 2} = $ ________ . 1

55 $\dfrac{4x - 7}{4x - 7} = $ ________ . 1

56 Note the concellation of the common factor $(x - 2)$ as used in the following multiplication problem:

$$\frac{x - 2}{6} \cdot \frac{7}{x - 2}$$

$$\frac{\cancel{x - 2}}{6} \cdot \frac{7}{\cancel{x - 2}}$$

$$\frac{1}{6} \cdot \frac{7}{1}$$

$$\frac{7}{6}$$

Multiply $\dfrac{2x - 7}{4} \cdot \dfrac{3}{2x - 7} \cdot$ ________ . $\dfrac{3}{4}$

57 Note the manner in which $\frac{x + 3}{x + 3} = 1$ and $\frac{x - 7}{x - 7} = 1$ are used in simplifying the following multiplication problem:

$$\frac{2x + 6}{x^2 - 9x + 14} \cdot \frac{x - 7}{x + 3}$$

$$\frac{2(x + 3)}{(x - 7)(x - 2)} \cdot \frac{x - 7}{x + 3}$$

$$\frac{2(\cancel{x + 3})}{(\cancel{x - 7})(x - 2)} \cdot \frac{\cancel{x - 7}}{\cancel{x + 3}}$$

$$\frac{2}{x - 2} \cdot \frac{1}{1}$$

$$\frac{2}{x - 2}$$

Multiply $\frac{3x + 12}{x^2 - 6x + 5} \cdot \frac{x - 5}{x + 4}$: ________ . $\frac{3}{x - 1}$

58 Multiply $\frac{x + 4}{x - 2} \cdot \frac{5x - 10}{x^2 + 6x + 8}$: ________ . $\frac{5}{x + 2}$

59 $\frac{x^2 + x - 12}{x - 7} \cdot \frac{x - 7}{3x - 9}$: ________ . $\frac{x + 4}{3}$

60 When every factor, *except 1,* is divided out of the numerator, the product is indicated as in the following:

$$\frac{x^2 + 7x + 6}{x^2 + 4x + 3} \cdot \frac{x + 3}{x^2 - 36}$$

$$\frac{(x + 6)(x + 1)}{(x + 1)(x + 3)} \cdot \frac{(x + 3)}{(x + 6)(x - 6)}$$

$$\frac{\cancel{(x + 6)}\cancel{(x + 1)}}{\cancel{(x + 1)}\cancel{(x + 3)}} \cdot \frac{\cancel{(x + 3)}}{\cancel{(x + 6)}(x - 6)}$$

$$\frac{1}{1} \cdot \frac{1}{(x - 6)}$$

$$\frac{1}{x - 6}$$

Multiply $\frac{x + 5}{x - 8} \cdot \frac{x - 8}{x^2 - 25}$: ________ .

$\frac{1}{x - 5}$

61 Multiply $\frac{x - 8}{x^2 - 64} \cdot \frac{x + 8}{x + 1}$: ________ .

$\frac{1}{x + 1}$

62 $\frac{2x + 8}{x - 6} \cdot \frac{x - 6}{2x^2 + 8x} =$ ________ .

$\frac{1}{x}$

63 When every factor, *except 1,* is divided out of the denominator, the product is indicated as in the following:

$$\frac{x^2 + 10x + 9}{x - 7} \cdot \frac{x - 7}{x + 9}$$

$$\frac{(x + 9)(x + 1)}{x - 7} \cdot \frac{x - 7}{x + 9}$$

$$\frac{\cancel{(x + 9)}(x + 1)}{\cancel{x - 7}} \cdot \frac{\cancel{x - 7}}{\cancel{x + 9}}$$

$$\frac{(x + 1)}{1} \cdot \frac{1}{1}$$

$$\frac{x + 1}{1} = x + 1.$$

Multiply $\frac{x - 8}{x + 3} \cdot \frac{x^2 + 10x + 21}{x - 8} =$ ____________ .

$\frac{x + 7}{1} = x + 7$

64 $\dfrac{2x - 6}{2x + 14} \cdot \dfrac{x^2 + 9x + 14}{x - 3} =$ ________________.

$\dfrac{x + 2}{1} = x + 2$

65 $\dfrac{x^2 + 9x + 18}{x + 3} \cdot \dfrac{x^2 - 3x - 4}{x^2 + 2x - 24} =$ ________________.

$\dfrac{x + 1}{1} = x + 1$

66 When every factor, *except 1,* is divided out of both the numerator and the denominator, the product is indicated as in the following:

$$\frac{x^2 + 7x + 12}{3x + 12} \cdot \frac{3x - 9}{x^2 - 9}$$

$$\frac{(x + 3)(x + 4)}{3(x + 4)} \cdot \frac{3(x - 3)}{(x - 3)(x + 3)}$$

$$\frac{\cancel{(x + 3)}\cancel{(x + 4)}}{\cancel{3}\cancel{(x + 4)}} \cdot \frac{\cancel{3}\cancel{(x - 3)}}{\cancel{(x - 3)}\cancel{(x + 3)}}$$

$$\frac{1}{1} \cdot \frac{1}{1}$$

$$\frac{1}{1} = 1$$

Multiply $\dfrac{x^2 + 8x + 15}{x^2 - 25} \cdot \dfrac{2x - 10}{2x + 6} \cdot$ ________.

1

67 $\dfrac{x^2 + 5x + 6}{x^2 - 9} \cdot \dfrac{x^2 - 5x + 6}{x^2 - 4} =$ ________.

1

68 $\dfrac{3x + 12}{x^2 + 8x + 16} \cdot \dfrac{x^2 - 5x - 36}{3x - 27} =$ ________.

1

69 $\dfrac{2x + 6}{x - 1} \cdot \dfrac{2x - 2}{4x + 12} =$ ________.

1

$$\frac{1}{^-1} = {^-1}, \quad \frac{^-5}{5} = {^-1}, \quad \frac{12}{^-12} = {^-1}$$

$$\frac{^-37}{37} = \frac{^-1 \cdot 37}{37} = {^-1} \cdot \frac{37}{37} = {^-1} \cdot 1 = {^-1}$$

---

70 $\frac{^-7}{7} = $ ________ . $^-1$

71 $\frac{18}{^-18} = $ ________ . $^-1$

72 $\frac{x}{^-x} = \frac{1x}{^-1x} = $ ________ . $^-1$

73 To simplify $\frac{x - 7}{7 - x}$, first note that $7 - x$ is equivalent to $7 + {^-x}$. Is $7 + {^-x}$ equivalent to ${^-x} + 7$? ________ . Yes

74 ${^-x} + 13$ can be factored as $^-(x - 13)$. Factor ${^-x} + 7$.
________ . $^-(x - 7)$

75 To simplify $\frac{x - 7}{7 - x}$, the following steps are used:

$$\frac{x - 7}{7 - x} = \frac{x - 7}{^-x + 7} = \frac{x - 7}{^-(x - 7)} = \frac{1}{^-1} = {^-1}$$

Simplify $\frac{x - 12}{12 - x}$. ________ . $^-1$

76 Simplify $\frac{x - 17}{17 - x}$. ________ . $^-1$
[*Note:* The denominator is the opposite of the numerator.]

77 Simplify $\frac{9 - x}{x - 9}$. ________ . $^-1$

78 Simplify $\frac{x^2 - 8x + 13}{^-x^2 + 8x - 13}$. ________ . $^-1$
[*Note:* The denominator is the opposite of the numerator.]

79 Multiply $\frac{2 - x}{x - 3} \cdot \frac{x^2 - 9}{x - 2}$. ________________ . $^-(x + 3)$ or ${^-x} - 3$
[*Note*: $x - 2$ is one common factor to be divided out of the numerator and denominator because $2 - x = {^-(x - 2)}$.]

80 Multiply $\dfrac{^{-}3x + 18}{x^2 - 8x + 12} \cdot \dfrac{x - 2}{3x + 12} \cdot$ ________ .

$\dfrac{^{-}1}{x + 4}$

81 Multiply $\dfrac{x^2 - 3x + 2}{x^2 + 4x - 5} \cdot \dfrac{x + 5}{2 - x} \cdot$ ________ .

$^{-}1$

82 Multiply $\dfrac{x^2 + 5x + 6}{3} \cdot \dfrac{2}{5x + 10}$ by first factoring wherever possible. ________ .

$\dfrac{2(x + 3)}{15}$

83 Multiply $\dfrac{x^2 + 4x - 5}{3x - 6} \cdot \dfrac{5x - 10}{x + 5} \cdot$ ________ .
[*Note*: Both $(x + 5)$ and $(x - 2)$ are common factors.]

$\dfrac{5(x - 1)}{3}$

84 Multiply $\dfrac{3x + 12}{2x} \cdot \dfrac{4}{x^2 - 16} \cdot$ ________ .
[*Note*: Both $(x + 4)$ and 2 are common factors.]

$\dfrac{6}{x(x - 4)}$

85 Multiply $\dfrac{2x^2 - 3x}{5x} \cdot \dfrac{10}{4x - 6} \cdot$ ________ .

1

86 Multiply $\dfrac{x^2 - 4}{x^2 + 8x + 12} \cdot \dfrac{3x + 18}{5x - 10} \cdot$ ________ .

$\dfrac{3}{5}$

87 $\dfrac{x^2 - 5x}{x^2 + 3x + 2} \cdot \dfrac{x^2 + 7x + 10}{x^2 - 25} =$ ________ .

$\dfrac{x}{x + 1}$

88 $\dfrac{x + 7}{x^2 + 3x - 28} \cdot \dfrac{x^2 - 16}{x^2 + 5x + 4} =$ ________ .

$\dfrac{1}{x + 1}$

89 $\dfrac{4x^2}{x^2 + 4x - 12} \cdot \dfrac{x^2 - 5x + 6}{x^2 - 3x} =$ ________ .

$\dfrac{4x}{x + 6}$

90 $\dfrac{x^2 - 36}{x^2 + 7x + 6} \cdot \dfrac{3x + 3}{x^2 - 8x + 12} =$ ________ .

$\dfrac{3}{x - 2}$

91 $\dfrac{2x^2 + 5x + 3}{3x - 6} \cdot \dfrac{4x - 8}{4x + 4} =$ ________ .

$\dfrac{2x + 3}{3}$

92 $\frac{5x^2 - 15x + 10}{x^2 - 4} \cdot \frac{3x + 6}{10x} =$ ________ .

$\frac{3(x - 1)}{2x}$

93 $\frac{4x^2 - 9}{x^2 + 3x} \cdot \frac{5x + 15}{2x + 3} =$ ________________ .

$\frac{5(2x - 3)}{x}$

94 $\frac{4x + 12}{x^2 - 9} \cdot \frac{3 - x}{8} =$ ________ .

[*Note*: $3 - x$ is the opposite of $x - 3$.]

$\frac{{}^{-}1}{2}$

95 $\frac{x^2 - 4}{5x} \cdot \frac{10}{2 - x} =$ ________________ .

$\frac{{}^{-}2(x + 2)}{x}$

Self-Quiz # 2

The following questions test the objectives of the preceding section. 100% mastery is desired.

Multiply:

1. $\frac{x - 3}{x + 7} \cdot \frac{x - 1}{x - 3} =$ ________ .

2. $\frac{3x^2}{x + 2} \cdot \frac{x + 2}{6x} =$ ________ .

3. $\frac{x^2 - x - 6}{x^2 - 9} \cdot \frac{x^2 + 8x + 15}{x^2 + 9x + 14} =$ ________ .

4. $\frac{2x^2 - 5x - 3}{x^2 + 2x - 15} \cdot \frac{x + 5}{2x^2 + 5x + 2} =$ ________ .

5. $\frac{x^2 - 4}{x + 2} \cdot \frac{x + 3}{2 - x} =$ ________ .

The addition of polynomial fractions with the same denominator is done as follows:

$$\frac{5}{x} + \frac{3}{x} = \frac{5 + 3}{x} = \frac{8}{x}$$

This simple addition is possible only because the denominators are the same. Whenever the denominators are the same, the addition can be accomplished by simply adding the numerators. Addition of polynomial fractions with the same denominators will be studied in the following frames.

96 $\frac{4}{x} + \frac{11}{x} = \frac{4 + 11}{x} = \frac{15}{x}$.

$\frac{3}{y} + \frac{5}{y} =$ ________.

$\frac{8}{y}$

97 $\frac{2}{x} - \frac{3}{x} = \frac{2}{x} + \frac{^{-}3}{x} = \frac{^{-}1}{x}$.

$\frac{8}{y} - \frac{6}{y} =$ ________.

$\frac{2}{y}$

98

$$\frac{5}{3x} + \frac{2}{3x} = \frac{5 + 2}{3x} = \frac{7}{3x}$$

Since the denominators are the same, the addition of $\frac{5}{3x} + \frac{2}{3x}$ is accomplished by adding the numerators and keeping the same denominator.

$\frac{8}{5y} + \frac{4}{5y} =$ ________.

$\frac{12}{5y}$

99

$$\frac{8}{x - 2} + \frac{3}{x - 2} = \frac{8 + 3}{x - 2} = \frac{11}{x - 2}$$

[*Note*: The denominators are the same; therefore, the sum is obtained by adding the numerators.]

$\frac{9}{x + 5} + \frac{8}{x + 5} =$ ________.

$\frac{17}{x + 5}$

100 $\dfrac{3x}{x-7} + \dfrac{4x}{x-7} = \dfrac{3x+4x}{x-7} = \dfrac{7x}{x-7}$.

$\dfrac{5y}{x+1} + \dfrac{3y}{x+1} = $ ______ .

$\dfrac{8y}{x+1}$

101 $\dfrac{x+5}{x-3} + \dfrac{8}{x-3} = $ ______ .

$\dfrac{x+13}{x-3}$

102 $\dfrac{9}{x^2+5x+2} + \dfrac{5}{x^2+5x+2} = $ ______ .

[*Note:* The denominators of both fractions are the same.]

$\dfrac{14}{x^2+5x+2}$

103 $\dfrac{x+7}{5y} + \dfrac{x-3}{5y} = \dfrac{x+7+x-3}{5y} = \dfrac{2x+4}{5y}$.

$\dfrac{x-3}{6y} + \dfrac{3x-2}{6y} = $ ______ .

$\dfrac{4x-5}{6y}$

104 $\dfrac{x+9}{x-3} + \dfrac{x-8}{x-3} = \dfrac{x+9+x-8}{x-3} = \dfrac{2x+1}{x-3}$.

$\dfrac{x+5}{x-6} + \dfrac{x+7}{x-6} = $ ______ .

$\dfrac{2x+12}{x-6}$

105 The following example shows the way a minus sign must be used:

$$\frac{x+5}{3x} - \frac{x+3}{3x} = \frac{(x+5)-(x+3)}{3x} =$$

$$\frac{x+5-x-3}{3x} = \frac{2}{3x}.$$

$\dfrac{2x+1}{5x} - \dfrac{x+3}{5x} = $ ______ .

$\dfrac{x-2}{5x}$

106 Complete the following:

$\dfrac{3x-7}{4x} - \dfrac{x-2}{4x} = \dfrac{(3x-7)-(x-2)}{4x} =$

$\dfrac{3x-7-x+2}{4x} = $ ______ .

$\dfrac{2x-5}{4x}$

107 $\dfrac{x^2 + 5x + 18}{2x} - \dfrac{x^2 + 3x - 1}{2x} = $ ________ .

$\dfrac{2x + 19}{2x}$

108 $\dfrac{x^2 + 3}{4x} - \dfrac{x + 1}{4x} = $ ________________ .

$\dfrac{x^2 - x + 2}{4x}$

109 Complete the following addition:

$$\frac{5(x + 2)}{2x} + \frac{3(x - 4)}{2x} = \frac{5(x + 2) + 3(x - 4)}{2x} =$$

$\dfrac{5x + 10 + 3x - 12}{2x} = $ ________ .

$\dfrac{8x - 2}{2x}$ or $\dfrac{4x - 1}{x}$

110 Complete the following addition:

$$\frac{x(x + 3)}{4x^2} + \frac{2(x - 7)}{4x^2} = \frac{x(x + 3) + 2(x - 7)}{4x^2} =$$

$\dfrac{x^2 + 3x + 2x - 14}{4x^2} = $ ________________ .

$\dfrac{x^2 + 5x - 14}{4x^2}$

111 Complete the following addition:

$$\frac{4(x - 3)}{x - 8} - \frac{2(x + 3)}{x - 8} = \frac{4(x - 3) - 2(x + 3)}{x - 8} =$$

$\dfrac{4x - 12 - 2x - 6}{x - 8} = $ ________ .

$\dfrac{2x - 18}{x - 8}$

112 Complete the following:

$$\frac{3x + 7}{x^2 + 7x + 12} - \frac{x + 3}{x^2 + 7x + 12} =$$

$\dfrac{(3x + 7) - (x + 3)}{x^2 + 7x + 12} = $ ________________ .

$\dfrac{2x + 4}{x^2 + 7x + 12}$

113 Complete:

$$\frac{x^2 + 6x + 1}{x^2 + 8x - 1} - \frac{x^2 + 2x - 2}{x^2 + 8x - 1} =$$

$\dfrac{(x^2 + 6x + 1) - (x^2 + 2x - 2)}{x^2 + 8x - 1} = $ ________________ .

$\dfrac{4x + 3}{x^2 + 8x - 1}$

114 $\frac{x^2 + 3x + 2}{x + 7} + \frac{x - 3}{x + 7} =$ ________________ .

$\frac{x^2 + 4x - 1}{x + 7}$

115 $\frac{x^2 - 3}{x + 9} - \frac{2x + 7}{x + 9} =$ ________________ .

$\frac{x^2 - 2x - 10}{x + 9}$

116 $\frac{x^2 + 3x - 5}{2x} + \frac{3x - 6}{2x} =$ ________________ .

$\frac{x^2 + 6x - 11}{2x}$

117 $\frac{8x - 5}{x + 9} - \frac{3x + 1}{x + 9} =$ ________________ .

$\frac{5x - 6}{x + 9}$

118 $\frac{4x + 7}{x^2 + 9x - 1} + \frac{3x - 8}{x^2 + 9x - 1} =$ ________________ .

$\frac{7x - 1}{x^2 + 9x - 1}$

119 $\frac{x^2 - 3x + 2}{x - 6} - \frac{x^2 + 3x + 2}{x - 6} =$ ________________ .

$\frac{{}^{-}6x}{x - 6}$

120 $\frac{3x + 7}{x^2 - 80} + \frac{5x - 11}{x^2 - 80} =$ ________________ .

$\frac{8x - 4}{x^2 - 80}$

121 $\frac{x^2 + 5x + 3}{x^2 - 11x + 7} + \frac{2x^2 - 16}{x^2 - 11x + 7} =$ ________________ .

$\frac{3x^2 + 5x - 13}{x^2 - 11x + 7}$

122 $\frac{5x - 1}{2x + 3} - \frac{2x + 5}{2x + 3} =$ ________________ .

$\frac{3x - 6}{2x + 3}$

123 $\frac{x^2 - 5x + 7}{x - 11} - \frac{x^2 - 9x - 3}{x - 11} =$ ________________ .

$\frac{4x + 10}{x - 11}$

124 $\frac{7x - 5}{3x} - \frac{2x - 4}{3x} =$ ________________ .

$\frac{5x - 1}{3x}$

Self-Quiz # 3

The following questions test the objectives of the preceding section. 100% mastery is desired.

1. $\frac{6}{x} + \frac{9}{x} =$ ________ .

2. $\frac{3}{2y} - \frac{6}{2y} =$ ________ .

3. $\frac{5}{x-3} + \frac{2}{x-3} =$ ________ .

4. $\frac{x-3}{x+14} + \frac{x+5}{x+14} =$ ________ .

5. $\frac{2x-1}{5x} - \frac{x+3}{5x} =$ ________ .

6. $\frac{4x}{x^2+3x-7} + \frac{2x-3}{x^2+3x-7} =$ ________ .

7. $\frac{5x-3}{x^2-19} - \frac{2x+7}{x^2-19} =$ ________ .

8. $\frac{x^2-19x+3}{x+2} - \frac{x^2+14x-1}{x+2} =$ ________ .

The following section will explain the addition of polynomial fractions in which the denominators are not the same.

Finding a common denominator is dependent upon finding the least common multiple of the polynomial denominators.

---

125 To determine the least common multiple of 6 and 4,
(a) Find their highest common factor, which is 2.
(b) Multiply 6 and 4, and divide the product by the highest common factor, 2.

Hence, the least common multiple of 6 and 4 is $\frac{6 \cdot 4}{2}$ or 12. Find the least common multiple of 6 and 8.
________ .

24

126 Find the least common multiple of 10 and 15.
________ .

30

127 Find the least common multiple of 9 and 12.
________ .

36

128 To find the least common multiple of $5x$ and $6x$,
(a) Find the highest common factor x.
(b) Multiply $5x$ and $6x$, and divide the product by the highest common factor, x.

Hence, the least common multiple of $5x$ and $6x$ is $\frac{5x \cdot 6x}{x}$ or $30x$. What is the least common multiple of $7x$ and $2x$? ________ .

$14x$

129 What is the least common multiple of $2x^2$ and $4x$?
$\frac{2x^2 \cdot 4x}{2x} =$ ________ .

$4x^2$

130 What is the least common multiple of $3x^2$ and $9x$?
$\frac{3x^2 \cdot 9x}{3x} =$ ________ .

$9x^2$

131 What is the least common multiple of $2x$ and $3y$?
[*Note:* The highest common factor is 1.]
$\frac{2x \cdot 3y}{1} =$ ________ .

$6xy$

132 What is the least common multiple of $4x$ and $5y$?
________ .

$20xy$

133 What is the least common multiple of 3 and 6?

$\frac{3 \cdot 6}{3} =$ ________ . 6

134 Find the least common multiple of 4 and 12.

________ . 12

135 Find the least common multiple of 8 and 10.

$\frac{8 \cdot 10}{2} =$ ________ . 40

136 Find the least common multiple of 6 and 9. ________ . 18

[*Note:* 3 is the highest common factor of 6 and 9.]

137 Find the least common multiple of $8x^3$ and $12x^2$.

($4x^2$ is the highest common factor.)

$\frac{8x^3 \cdot 12x^2}{4x^2} =$ ________ . $24x^3$

138 Find the least common multiple of 8 and 7. ________ . 56

[*Note:* 1 is the highest common factor of 8 and 7.]

139 Find the least common multiple of $7x$ and $14x^2$.

________ . $14x^2$

140 Find the least common multiple of $6x^2$ and $10x$.

________ . $30x^2$

141 Find the least common multiple of $6x^3$ and $8x^2$.

________ . $24x^3$

To find the least common multiple of $x^2 - 3x + 2$ and $4x - 4$:

1. Factor each polynomial.

$$x^2 - 3x + 2 = (x - 1)(x - 2)$$
$$4x - 4 = 4 \cdot (x - 1)$$

2. $(x - 1)$ is the common factor for $(x - 1)(x - 2)$ and $4(x - 1)$.
3. Multiply $(x - 1)(x - 2) \cdot 4(x - 1)$ and divide this product by the largest common factor, $(x - 1)$.

$$\frac{\cancel{(x - 1)}(x - 2) \cdot 4(x - 1)}{\cancel{(x - 1)}} = 4(x - 2)(x - 1)$$

The expression $4(x - 2)(x - 1)$ is the least common multiple of $x^2 - 3x + 2$ and $4x - 4$.

142

$$x^2 + 9x + 20 = (x + 5)(x + 4)$$
$$x^2 - 25 = (x + 5)(x - 5)$$

What is the highest common factor of $x^2 + 9x + 20$ and $x^2 - 25$? ________ .

$(x + 5)$

143 Factor each of the following polynomials and find their highest common factor. ________ .

$$x^2 + 6x + 8, \qquad 3x + 12$$

$x + 4$

144 The first step in finding the least common multiple for $x^2 + 5x + 6$ and $3x + 9$ is to factor both polynomials:
$x^2 + 5x + 6 = (x + 2)(x + 3)$.
$3x + 9 =$ ________ .

$3(x + 3)$

145 The second step in finding the least common multiple of $x^2 + 5x + 6$ and $3x + 9$ is to find the highest common factor for the two polynomials.

$$x^2 + 5x + 6 = (x + 2)(x + 3)$$
$$3x + 9 = 3(x + 3)$$

What is the highest common factor for the two polynomials? ________ .

$x + 3$

146 To find the least common multiple for $x^2 + 5x + 6$ and $3x + 9$, multiply the polynomials and divide their product by their highest common factor, $x + 3$.

$$\frac{(x + 2)\cancel{(x + 3)} \cdot 3(x + 3)}{\cancel{x + 3}} = 3(x + 2)(x + 3)$$

What is the least common multiple for $x^2 + 5x + 6$ and $3x + 9$? ______________ .

$3(x + 2)(x + 3)$

147 To find the least common multiple for $x^2 - 16$ and $x^2 - 6x + 8$,

(a) Factor both polynomials

$$x^2 - 16 = (x + 4)(x - 4)$$

$$x^2 - 6x + 8 = (x - 4)(x - 2)$$

(b) The highest common factor is $(x - 4)$.

(c) Multiply the two polynomials, and divide by the highest common factor, $(x - 4)$, as follows:

$$\frac{(x + 4)\cancel{(x - 4)} \cdot (x - 4)(x - 2)}{\cancel{x - 4}} =$$

$$(x + 4)(x - 4)(x - 2)$$

The least common multiple of $x^2 - 16$ and $x^2 - 6x + 8$ is ______________ .

$(x + 4)(x - 4)(x - 2)$

148 To find the least common multiple of $x^2 + 9x + 18$ and $7x + 21$,

(a) Factor both polynomials.

$$x^2 + 9x + 18 = (x + 3)(x + 6)$$

$$7x + 21 = 7(x + 3)$$

(b) $(x + 3)$ is the highest common factor.

(c) Multiply the two polynomials, and divide their product by the highest common factor, $x + 3$.

$$\frac{\cancel{(x + 3)}(x + 6) \cdot 7(x + 3)}{\cancel{x + 3}}$$

What is the least common multiple for $x^2 + 9x + 18$ and $7x + 21$? ______________ .

$7(x + 3)(x + 6)$

149 Find the least common multiple for $x^2 + 10x - 11$ and $5x - 5$ by completing the following:

$$\frac{(x^2 + 10x - 11) \cdot (5x - 5)}{\text{HCF}} = \frac{(x + 11)(x - 1) \cdot 5(x - 1)}{x - 1} =$$

________________ .

$5(x + 11)(x - 1)$

[*Note:* The highest common factor of the two polynomials is $x - 1$.] (Leave your answer in factor form.)

150 Find the least common multiple for $x^2 + 7x + 6$ and $3x + 3$. ________________ .

$3(x + 6)(x + 1)$

$$x^2 + 7x + 6 = (x + 6)(x + 1)$$
$$3x + 3 = 3(x + 1)$$

151 Find the least common multiple for $x^2 - 49$ and $5x - 35$. ________________ .

$5(x + 7)(x - 7)$

152 Find the least common multiple for $x^2 + 5x + 4$ and $6x + 24$. ________________ .

$6(x + 1)(x + 4)$

153 Find the least common multiple for $8x - 16$ and $x^2 + 5x - 14$. ________________ .

$8(x - 2)(x + 7)$

154 Find the least common multiple for $x^2 - 7x - 18$ and $x^2 - 10x + 9$. ________________ .

$(x - 9)(x + 2)(x - 1)$

155 Find the least common multiple for $x^2 - 64$ and $x^2 - x - 56$. ________________ .

$(x + 8)(x - 8)(x + 7)$

Self-Quiz # 4

The following questions test the objectives of the preceding section. 100% mastery is desired.

Find the least common multiple for all of the following problems:

1. 18 and 8. ______ .

2. 30 and 4. ______ .

3. $8x$ and $5x$. ______ .

4. $3x^2$ and $12x$. ______ .

5. $x^2 - 9$ and $5x + 15$. ______ .

6. $x^2 - 5x - 24$ and $x^2 - 64$. ______ .

7. $5x + 10$ and $x^2 - 4$. ______ .

8. $x^2 + 10x + 24$ and $x^2 + 3x - 18$. ______ .

156 12 is the least common multiple of 4 and 6. What number can be multiplied by 4 to give 12?
$4 \cdot$ ______ $= 12$. — 3

157 $6 \cdot$ ______ $= 12$. — 2

158 $\frac{5}{5}, \frac{19}{19}, \frac{47}{47},$ and $\frac{916}{916}$ are all equal to 1. How may 1 be written in the following problem to make a true statement?
$\frac{1}{4} \cdot$ ______ $= \frac{3}{12}$. — $\frac{3}{3}$

159 How may 1 be written in the following problem to make a true statement?
$\frac{5}{6} \cdot$ ______ $= \frac{10}{12}$. — $\frac{2}{2}$

160 How may 1 be written in the following problem to make a true statement?

$$\frac{2}{3} \cdot \underline{\hspace{2cm}} = \frac{10}{15}.$$

$\frac{5}{5}$

161 How may 1 be written in the following problem to make a true statement?

$$\frac{1}{6} \cdot \underline{\hspace{2cm}} = \frac{3}{18}.$$

$\frac{3}{3}$

162 How may 1 be written in the following problem to make a true statement?

$$\frac{2}{5} \cdot \underline{\hspace{2cm}} = \frac{16}{40}.$$

$\frac{8}{8}$

163 To add $\frac{5}{3x^2} + \frac{7}{6x}$, the first step is to find the least common multiple of $3x^2$ and $6x$. What is the least common multiple of $3x^2$ and $6x$?

$$\frac{3x^2 \cdot 6x}{3x} = \underline{\hspace{2cm}}.$$

$6x^2$

164 The least common multiple for $3x^2$ and $6x$ is $6x^2$. The second step in adding $\frac{5}{3x^2} + \frac{7}{6x}$ is to write both fractions so that the denominator is $6x^2$.

$$\frac{5}{3x^2} \cdot \frac{2}{2} = \frac{10}{6x^2}.$$

$$\frac{7}{6x} \cdot \frac{x}{x} = \frac{?}{6x^2}.$$

$7x$

165 Complete the following addition:

$$\frac{5}{3x^2} + \frac{7}{6x} = \frac{5}{3x^2} \cdot \frac{2}{2} + \frac{7}{6x} \cdot \frac{x}{x} =$$

$$\frac{10}{6x^2} + \frac{7x}{6x^2} = \underline{\hspace{2cm}}.$$

$\frac{10 + 7x}{6x^2}$

166 In adding $\frac{x+5}{4x} + \frac{x-1}{2x^2}$, the first step is to find the least common multiple of $4x$ and $2x^2$. What is the least common multiple?

$\frac{4x \cdot 2x^2}{2x} =$ ______ .

$4x^2$

167 The least common multiple of $2x^2$ and $4x$ is $4x^2$. In adding $\frac{x+5}{4x} + \frac{x-1}{2x^2}$, the second step is to write each fraction with a denominator of $4x^2$.

$$\frac{x+5}{4x} \cdot \frac{x}{x} + \frac{x-1}{2x^2} \cdot \frac{2}{2}$$

Complete the addition:

$\frac{(x+5) \cdot x}{4x^2} + \frac{(x-1) \cdot 2}{4x^2} =$ ______________ .

$\frac{x^2 + 7x - 2}{4x^2}$

168 What is the least common multiple for $5x^2$ and $2x^2$?

$10x^2$

169 $5x^2 \cdot$ ______ $= 10x^2$.

2

170 $2x^2 \cdot$ ______ $= 10x^2$.

5

171 Complete the following addition:

$\frac{4}{5x^2} + \frac{3}{2x^2} = \frac{4}{5x^2} \cdot \frac{2}{2} + \frac{3}{2x^2} \cdot \frac{5}{5} =$

$\frac{8}{10x^2} + \frac{15}{10x^2} =$ ______ .

$\frac{23}{10x^2}$

172 What is the least common multiple for $6x$ and $8x^2$?

______ .

$24x^2$

173 Complete the following addition:

$\frac{x-3}{6x} + \frac{3}{8x^2} = \frac{(x-3)}{6x} \cdot \frac{4x}{4x} + \frac{3}{8x^2} \cdot \frac{3}{3} =$

$\frac{4x^2 - 12x}{24x^2} + \frac{9}{24x^2} =$ ______________ .

$\frac{4x^2 - 12x + 9}{24x^2}$

174 What is the least common multiple for 15 and 10?

________ .

30

175 Complete the following addition:

$$\frac{x-7}{15}+\frac{x+2}{10}=\frac{(x-7)}{15}\cdot\frac{2}{2}+\frac{(x+2)}{10}\cdot\frac{3}{3}=$$

$$\frac{2x-14}{30}+\frac{3x+6}{30}=\_\_\_\_\_\_\_\_ .$$

$$\frac{5x-8}{30}$$

176 What is the least common multiple for $3x$ and $4x^2$?

________ .

$12x^2$

177 Complete the following addition:

$$\frac{x-6}{3x}+\frac{x+1}{4x^2}=\frac{(x-6)}{3x}\cdot\frac{4x}{4x}+\frac{(x+1)}{4x^2}\cdot\frac{3}{3}=$$

$$\frac{4x^2-24x}{12x^2}+\frac{3x+3}{12x^2}=\_\_\_\_\_\_\_\_\_\_\_\_\_\_ .$$

$$\frac{4x^2-21x+3}{12x^2}$$

178 Find the least common multiple of $6x$ and $2x^2$ and add:

$$\frac{x+4}{6x}+\frac{x-2}{2x^2}=\_\_\_\_\_\_\_\_\_\_\_\_ .$$

$$\frac{x^2+7x-6}{6x^2}$$

179 Find the least common multiple for $2x$ and $7x$ and add:

$$\frac{x+1}{2x}+\frac{x-6}{7x}=\_\_\_\_\_\_\_\_ .$$

$$\frac{9x-5}{14x}$$

180 Find the least common multiple for $2x$ and $4x$ and add:

$$\frac{x+6}{2x}+\frac{x-3}{4x}=\_\_\_\_\_\_\_\_ .$$

$$\frac{3x+9}{4x}$$

181 Add $\frac{8}{3x}+\frac{5}{2x}$ by first finding the least common multiple of $3x$ and $2x$. ________ .

$$\frac{31}{6x}$$

182 Add $\frac{2}{3x} + \frac{1}{6x^2}$ by first finding the least common multiple for $3x$ and $6x^2$. ________ .

$\frac{4x + 1}{6x^2}$

183 $\frac{x + 4}{3x^2} + \frac{5}{9x} =$ ________ .

$\frac{8x + 12}{9x^2}$

184 $\frac{x + 3}{5x} + \frac{x - 1}{2x} =$ ________ .

$\frac{7x + 1}{10x}$

185 $\frac{x + 7}{12x} - \frac{3}{4x^2} =$ ________ .

$\frac{x^2 + 7x - 9}{12x^2}$

186 $\frac{5}{3x} - \frac{2}{x^3} =$ ________ .

$\frac{5x^2 - 6}{3x^3}$

187 $\frac{x + 1}{4x} - \frac{x - 2}{3x} =$ ________ .

$\frac{{}^{-}x + 11}{12x}$

188 $\frac{x + 5}{2x^2} - \frac{x + 1}{3x} =$ ________ .

$\frac{{}^{-}2x^2 + x + 15}{6x^2}$

189 $\frac{x + 9}{6} - \frac{x - 2}{5} =$ ________ .

$\frac{{}^{-}x + 57}{30}$

190 $\frac{5x - 4}{8x} - \frac{x + 1}{2x} =$ ________ .

$\frac{x - 8}{8x}$

191 To find the least common multiple for $3x - 6$ and $2x - 4$,

(a) Factor each binomial.

$$3x - 6 = 3(x - 2)$$
$$2x - 4 = 2(x - 2)$$

(b) $(x - 2)$ is the highest common factor of $3(x - 2)$ and $2(x - 2)$.

(c) Multiply $3(x - 2) \cdot 2(x - 2)$ and divide the product by $(x - 2)$ to find their least common multiple.

$$\frac{3\cancel{(x - 2)} \cdot 2(x - 2)}{\cancel{(x - 2)}} = 6(x - 2)$$

The least common multiple for $3x - 6$ and $2x - 4$ is ________ .

$6(x - 2)$

192 To find the least common multiple for $x^2 - 16$ and $x^2 + 4x$,

(a) Factor each binomial.

$$x^2 - 16 = (x + 4)(x - 4)$$
$$x^2 + 4x = x(x + 4)$$

(b) $(x + 4)$ is the highest common factor.

(c) Multiply $(x + 4)(x - 4) \cdot x(x + 4)$ and divide the product by $(x + 4)$ to find the least common multiple.

$$\frac{\cancel{(x + 4)}(x - 4) \cdot x(x + 4)}{\cancel{(x + 4)}} = x(x - 4)(x + 4)$$

The least common multiple for $x^2 - 16$ and $x^2 + 4x$ is ________________ .

$x(x - 4)(x + 4)$

193 To find the least common multiple for $x^2 - 16$ and $5x - 20$, first find their highest common factor. Find the highest common factor of $x^2 - 16$ and $5x - 20$ by first factoring each expression. ________ .

$x - 4$

194 The highest common factor for $x^2 - 16$ and $5x - 20$ is $(x - 4)$. To find the least common multiple for $x^2 - 16$ and $5x - 20$, divide the product of the two expressions by $(x - 4)$, their highest common factor:

$$\frac{(x^2 - 16) \cdot (5x - 20)}{(x - 4)} = \frac{(x - 4)(x + 4) \cdot 5(x - 4)}{(x - 4)} =$$

________________ .

$5(x + 4)(x - 4)$

(Leave the answer in factored form.)

195 Find the least common multiple for $x^2 + 5x + 6$ and $x^2 - 9$ by dividing their product by their highest common factor. ____________________ .

$(x + 2)(x + 3)(x - 3)$

196 Find the least common multiple for $x^2 + 7x + 12$ and $x^2 - 16$ by dividing their product by their highest common factor. ____________________ .

$(x + 3)(x + 4)(x - 4)$

197 Find the least common multiple for $x^2 - 5x - 14$ and $x^2 + 6x + 8$. ____________________ .

$(x - 7)(x + 2)(x + 4)$

198 Find the least common multiple for $x^2 + 10x + 9$ and $x^2 - 1$. ____________________ .

$(x + 9)(x + 1)(x - 1)$

199 The least common multiple for $x^2 + 10x + 9$ and $x^2 - 1$ is $(x + 9)(x + 1)(x - 1)$. To add

$$\frac{5}{x^2 + 10x + 9} + \frac{3}{x^2 - 1}$$

first write both of the fractions with the denominator $(x + 9)(x + 1)(x - 1)$.

$$\frac{5}{x^2 + 10x + 9} = \frac{5}{(x + 9)(x + 1)} \cdot \frac{(x - 1)}{(x - 1)} =$$

____________________ .

$$\frac{5x - 5}{(x + 9)(x + 1)(x - 1)}$$

200 Write $\frac{3}{x^2 - 1}$ with the denominator

$$(x + 9)(x + 1)(x - 1)$$

$$\frac{3}{x^2 - 1} = \frac{3}{(x + 1)(x - 1)} \cdot \frac{x + 9}{x + 9} =$$

________________ .

$$\frac{3(x + 9)}{(x + 9)(x + 1)(x - 1)}$$

201 Complete the following addition:

$$\frac{5}{x^2 + 10x + 9} + \frac{3}{x^2 - 1} =$$

$$\frac{5}{(x + 9)(x + 1)} + \frac{3}{(x + 1)(x - 1)} =$$

$$\frac{5}{(x + 9)(x + 1)} \cdot \frac{(x - 1)}{(x - 1)} + \frac{3}{(x + 1)(x - 1)} \cdot \frac{(x + 9)}{(x + 9)} =$$

$$\frac{5x - 5}{(x + 9)(x + 1)(x - 1)} + \frac{3x + 27}{(x + 9)(x + 1)(x - 1)} =$$

________________ .

$$\frac{8x + 22}{(x + 9)(x + 1)(x - 1)}$$

202 Complete the following addition:

$$\frac{4}{x^2 + 3x + 2} + \frac{7}{x^2 + 5x + 6} =$$

$$\frac{4}{(x + 2)(x + 1)} + \frac{7}{(x + 2)(x + 3)} =$$

$$\frac{4}{(x + 2)(x + 1)} \cdot \frac{x + 3}{x + 3} + \frac{7}{(x + 2)(x + 3)} \cdot \frac{x + 1}{x + 1} =$$

$$\frac{4x + 12}{(x + 2)(x + 1)(x + 3)} + \frac{7x + 7}{(x + 2)(x + 1)(x + 3)} =$$

________________ .

$$\frac{11x + 19}{(x + 2)(x + 1)(x + 3)}$$

The following steps are used to combine the fractions.

$$\frac{x + 3}{x^2 - 7x + 12} + \frac{x - 2}{x^2 - 3x - 4}$$

1. Factor both denominators:

$$\frac{x + 3}{(x - 4)(x - 3)} + \frac{x - 2}{(x - 4)(x + 1)}$$

2. Find the highest common factor of the denominators:

$$(x - 4)$$

3. Multiply the two denominators and divide their product by the highest common factor, $(x - 4)$:

$$\frac{(x - 4)(x - 3)(x - 4)(x + 1)}{(x - 4)} = (x - 3)(x - 4)(x + 1)$$

4. Write both fractions with the least common multiple as the common denominator:

$$\frac{(x + 3)}{(x - 4)(x - 3)} \cdot \frac{(x + 1)}{(x + 1)} + \frac{(x - 2)}{(x - 4)(x + 1)} \cdot \frac{(x - 3)}{(x - 3)} =$$

$$\frac{(x + 3)(x + 1)}{(x - 4)(x - 3)(x + 1)} + \frac{(x - 2)(x - 3)}{(x - 4)(x - 3)(x + 1)}$$

5. Do the multiplication as indicated in the numerator of each fraction:

$$\frac{x^2 + 4x + 3}{(x - 4)(x - 3)(x + 1)} + \frac{x^2 - 5x + 6}{(x - 4)(x - 3)(x + 1)}$$

6. Combine the numerators over the least common denominator:

$$\frac{(x^2 + 4x + 3) + (x^2 - 5x + 6)}{(x - 4)(x - 3)(x + 1)}$$

7. Simplify the numerator:

$$\frac{2x^2 - x + 9}{(x - 4)(x - 3)(x + 1)}$$

---

203 Using the steps shown in the previous example, combine the following fractions:

$$\frac{x - 2}{x^2 - 2x - 15} + \frac{x + 7}{x^2 + 7x + 12}.$$

__________________ .

$$\frac{2x^2 + 4x - 43}{(x - 5)(x + 3)(x + 4)}$$

204 Combine the following fractions:

$$\frac{x + 1}{x^2 - 4x + 3} + \frac{x - 2}{x^2 + x - 2}.$$

____________________ .

$$\frac{2x^2 - 2x + 8}{(x - 3)(x - 1)(x + 2)}$$

205 Combine the following fractions:

$$\frac{x + 7}{x^2 + 3x + 2} + \frac{x - 2}{x^2 + 5x + 6}.$$

____________________ .

$$\frac{2x^2 + 9x + 19}{(x + 1)(x + 2)(x + 3)}$$

206 Combine the following fractions:

$$\frac{x - 1}{x^2 - 3x - 10} + \frac{x + 4}{x^2 - 8x + 15}.$$

____________________ .

$$\frac{2x^2 + 2x + 11}{(x + 2)(x - 5)(x - 3)}$$

207 Combine the following fractions:

$$\frac{x + 1}{x^2 - 4} - \frac{x + 2}{x^2 + 3x - 10}.$$

____________________ .

$$\frac{2x + 1}{(x + 2)(x - 2)(x + 5)}$$

208 Combine the following fractions:

$$\frac{x - 2}{x^2 - 6x + 9} - \frac{x + 5}{x^2 - x - 6}.$$

____________________ .

$$\frac{{}^{-}2x + 11}{(x - 3)(x - 3)(x + 2)}$$

209 Combine the following fractions:

$$\frac{x - 6}{2x^2 - 5x - 3} + \frac{x + 2}{2x^2 - x - 1}.$$

____________________ .

$$\frac{2x^2 - 8x}{(2x + 1)(x - 3)(x - 1)}$$

210 Combine the following fractions:

$$\frac{2}{x^2 + 2x - 15} - \frac{x}{2x^2 + 9x - 5}.$$

____________________ .

$$\frac{-x^2 + 7x - 2}{(x - 3)(x + 5)(2x - 1)}$$

211 Combine the following fractions:

$$\frac{x + 3}{2x^2 - 7x - 15} - \frac{x - 2}{2x^2 + 5x + 3}.$$

____________________.

$$\frac{11x - 7}{(x - 5)(2x + 3)(x + 1)}$$

212 Combine the following fractions:

$$\frac{5}{x^2 - 9} - \frac{3}{x^2 - 2x - 3}.$$

____________________.

$$\frac{2x - 4}{(x + 3)(x - 3)(x + 1)}$$

213 Combine the following fractions: $\frac{2x - 3}{5} - \frac{x + 2}{4}$. ________.

$$\frac{3x - 22}{20}$$

214 Combine the following fractions:

$$\frac{x - 1}{x^2 + 5x + 6} - \frac{x - 2}{x^2 + 2x - 3}.$$

____________________.

$$\frac{^{-}2x + 5}{(x + 3)(x + 2)(x - 1)}$$

215 Combine the following fractions: $\frac{2x - 1}{3x} + \frac{x + 7}{4x}$. ________.

$$\frac{11x + 17}{12x}$$

216 Combine the following fractions:

$$\frac{2x - 1}{x^2 - x - 2} + \frac{x + 3}{x^2 - 5x + 6}.$$

____________________.

$$\frac{3x^2 - 3x + 6}{(x + 1)(x - 2)(x - 3)}$$

217 Combine the following fractions: $\frac{x - 5}{2x^2} - \frac{x + 2}{6x}$. ________________.

$$\frac{^{-}x^2 + x - 15}{6x^2}$$

218 Combine the following fractions:

$$\frac{x + 4}{x^2 + 6x + 5} - \frac{x - 2}{x^2 - 4x - 5}.$$

____________________.

$$\frac{^{-}4x - 10}{(x + 5)(x + 1)(x - 5)}$$

219 Combine the following fractions:

$$\frac{x+4}{x^2-x-2}+\frac{x+2}{x^2-2x-3}.$$

______________ .

$$\frac{2x^2+x-16}{(x-2)(x+1)(x-3)}$$

220 Combine the following fractions:

$$\frac{5}{x^2-4}-\frac{7}{x^2+5x-14}.$$

______________ .

$$\frac{{}^{-}2x+21}{(x+2)(x-2)(x+7)}$$

221 Combine the following fractions:

$$\frac{x-3}{x^2-x-12}+\frac{2x+1}{x^2+5x+6}.$$

______________ .

$$\frac{3x^2-8x-10}{(x-4)(x+3)(x+2)}$$

222 Combine the following fractions:

$$\frac{4x-7}{3}-\frac{9x+16}{6}.$$

______________ .

$$\frac{{}^{-}x-30}{6}$$

223 Combine the following fractions:

$$\frac{2x-9}{x^2-3x-10}-\frac{5x}{x^2-25}.$$

______________ .

$$\frac{{}^{-}3x^2-9x-45}{(x+2)(x-5)(x+5)}$$

224 Combine the following fractions:

$$\frac{3x+1}{7x}+\frac{x-5}{3x}.$$

______________ .

$$\frac{16x-32}{21x}$$

225 Combine the following fractions:

$$\frac{4x-3}{x+7}-\frac{3x-7}{x-1}.$$

______________ .

$$\frac{x^2-21x+52}{(x+7)(x-1)}$$

Self-Quiz # 5

The following questions test the objectives of the preceding section. 100% mastery is desired.

Combine the following fractions:

1. $\dfrac{x-3}{6x}+\dfrac{x+2}{2x}$. ________ .

2. $\dfrac{x+5}{4}-\dfrac{2x+1}{3}$. ________ .

3. $\dfrac{x-3}{x^2+x-12}+\dfrac{x+1}{x^2-x-6}$. ________ .

4. $\dfrac{2x-1}{x^2+x-20}-\dfrac{3x+2}{x^2-7x+12}$. ________ .

5. $\dfrac{x-3}{x^2-1}+\dfrac{x+2}{x^2+x-2}$. ________ .

Chapter Summary

Three types of problems with polynomial fractions were explained in this chapter.

The first type involved the simplification of a single polynomial fraction. To simplify such a fraction, both numerator and denominator are factored and any factor common to both numerator and denominator is divided out.

The multiplication of polynomial fractions is accomplished by factoring all polynomials, dividing out any factors that appear in both a numerator and a denominator, and, finally, multiplying numerators and multiplying denominators.

The addition of polynomial fractions is dependent upon finding the least common multiple (LCM) of the denominators. After each fraction is written with the LCM as its denominator the addition is completed by combining the numerators.

CHAPTER POST-TEST

The following questions test the objectives of this chapter. A score of 90% indicates sufficient mastery, and the student may proceed to the next chapter.

I Simplify:

1. $\dfrac{18x^4}{14x}$

2. $\dfrac{x + 10}{x^2 + 7x - 30}$

3. $\dfrac{x^2 - 9x + 20}{x - 4}$

II Multiply:

1. $\dfrac{6x^6}{2x + 7} \cdot \dfrac{2x + 7}{16x^4}$

2. $\dfrac{x^2 + 6x - 55}{x^2 - 25} \cdot \dfrac{4x + 20}{x^2 + 13x + 22}$

III Add:

1. $\dfrac{3x + 4}{6} - \dfrac{x - 9}{5}$

2. $\dfrac{5}{9x} + \dfrac{7}{12x}$

3. $\dfrac{x - 6}{x^2 + 3x - 28} + \dfrac{x + 5}{x^2 - 16}$

I'M GONNA
WIN

6
Solving Linear Equations

CHAPTER PRE-TEST

The following questions indicate the objectives of this chapter. A score of 90% indicates sufficient mastery, and the student may immediately take the Chapter Post-Test.

Using the set of rational numbers, solve and check each of the following equations:

1. $x + 12 = 5$
2. $^{-}7x = 16$
3. $4x - 9 = 6$
4. $8x + 5 = 7$
5. $3x + 7 = 2x - 8$
6. $5x - 9 = 14 - 3x$
7. $4(x - 3) - 2x = 7$
8. $5 - 3(3x - 2) = 7$
9. $x - \frac{5}{4} = \frac{2}{3}$
10. $\frac{17}{3}x = \frac{^{-}9}{4}$
11. $\frac{3}{4}x + \frac{4}{7} = \frac{1}{5}$
12. $\frac{6}{5}x - \frac{9}{2} = \frac{2}{3}$
13. $\frac{6}{5x} - \frac{3}{4} = \frac{1}{2x}$
14. $\frac{4}{5x} + \frac{2}{15} = \frac{3}{10}$

In this chapter the solution of equations is studied. To solve an equation such as $3x+2=7$ or $x^2+5x+6=0$ means to find all the number replacements for x that will result in a true statement.

Equations of the form $ax + b = c$ are called *linear equations.* $3x - 7 = 10$, $4x + 5 = 8$, and $^-2x + 4 = {}^-6$ are examples of linear equations. The first section of this chapter is a review of the solution of linear equations.

1 An equation is solved with respect to a set of numbers that may serve as possible replacements for the letter. Using the set of counting numbers, $\{1, 2, 3, \ldots\}$, the truth set of $x + 3 = 5$ is $\{2\}$. What is the truth set of $x + 7 = 11$? ________ .

$\{4\}$

2 Using the set of counting numbers, what is the truth set of $x + 13 = 18$? ________ .

$\{5\}$

3 The truth set of $4x = 24$ is $\{6\}$ because $4 \cdot 6 = 24$ is a true statement. Using $\{1, 2, 3, \ldots\}$, what is the truth set of $5x = 15$? ________ .

$\{3\}$

4 Using $\{1, 2, 3, \ldots\}$, what is the truth set of $9x = 18$? ________ .

$\{2\}$

5 Using $\{1, 2, 3, \ldots\}$, what is the truth set of $3x = 12$? ________ .

$\{4\}$

6 Using $\{1, 2, 3, \ldots\}$, what is the truth set of $x + 17 = 23$? ________ .

$\{6\}$

7 Using $\{1, 2, 3, \ldots\}$, what is the truth set of $9x = 45$? ________ .

$\{5\}$

8 Some linear equations have no solution in the set of counting numbers. Is there a counting-number solution for $x + 5 = 2$? ________ .

No

9 Using $\{1, 2, 3, \ldots\}$, the truth set of $x + 5 = 2$ is $\{\ \}$. The truth set does not contain any counting number. Using $\{1, 2, 3, \ldots\}$, what is the truth set of $x + 13 = 5$? ________ .

$\{\ \}$

10 Using $\{1, 2, 3, \ldots\}$, what is the truth set of $2x = 3$? ________ .

$\{\ \}$

11 Which of the following equations do not have a solution in the set of counting numbers? ______________ .

(b), (c), (e)

(There may be more than one.)
(a) $x + 9 = 26$.
(b) $3x = 10$.
(c) $x + 4 = 1$.
(d) $x + 8 = 10$.
(e) $x + 16 = 6$.

12 Does every linear equation have a solution in the set of counting numbers ________ .

No

13 $\{\ldots, {}^{-}2, {}^{-}1, 0, 1, 2, \ldots\}$ is the set of integers. 7 is the opposite of ${}^{-}7$ because $7 + {}^{-}7 = 0$. What is the opposite of ${}^{-}10$? ________ .

10

14 What is the opposite of 8? ________ .

${}^{-}8$

15 $\{\ldots, {}^{-}2, {}^{-}1, 0, 1, 2, \ldots\}$ is the set of integers. The set of integers includes all the counting numbers, zero, and an opposite of every counting number. Does every integer have an opposite? ________ .

Yes

16 ${}^{-}9$ is the opposite of 9. 73 is the opposite of ${}^{-}73$. 0 is its own opposite. What is the sum of any integer and its opposite? ________ .

0

17 Using the set of integers $\{\ldots, {}^{-}2, {}^{-}1, 0, 1, 2, \ldots\}$, the truth set of $x + 3 = 1$ is $\{{}^{-}2\}$. What is the truth set of $x + 8 = 3$? ________ .

$\{{}^{-}5\}$

18 Using the set of integers, what is the truth set of $x + 13 = 6$? ______ . {⁻7}

19 Using {. . . , ⁻2, ⁻1, 0, 1, 2, . . .}, what is the truth set of $x + 9 = 15$? ______ . {6}

20 Using the set of integers, what is the truth set of $5x = {}^{-}20$? ______ . {⁻4}

21 Using the set of integers, what is the truth set of ${}^{-}6x = 18$? ______ . {⁻3}

22 Using the set of integers, what is the truth set of ${}^{-}3x = 12$? ______ . {⁻4}

23 Using the set of integers, what is the truth set of $4x = 24$? ______ . {6}

24 Some linear equations do not have a solution in the set of integers. Using the set of integers, the truth set of $2x = 5$ is { }. Using the set of integers, what is the truth set of $3x = 7$? ______ . { }

25 Using the set of integers, what is the truth set of ${}^{-}3x = {}^{-}8$? ______ . { }

26 Which of the following equations do not have a solution in the set of integers? ______ . (e)

(a) $x + 9 = {}^{-}2$.
(b) $3x = {}^{-}18$.
(c) $x - 7 = {}^{-}16$.
(d) ${}^{-}5x = {}^{-}50$.
(e) $2x = 3$.

The truth set of an equation is dependent upon the set used as replacements for the letter as shown in the following examples:

For the equation $x + 1 = 13$:
  Using the set of counting numbers, the truth set is {12};
  Using the set of integers, the truth set is {12}.

For the equation $x + 5 = 3$:
  Using the set of counting numbers, the truth set is { };
  Using the set of integers, the truth set is {$^-2$}.

For the equation $2x = 7$:
  Using the set of counting numbers, the truth set is { };
  Using the set of integers, the truth set is { }.

---

27 Using the set of counting numbers {1, 2, 3, . . .}, what is the truth set of $3x = 12$? ________ .

{4}

28 Using the set of integers {. . . , $^-2$, $^-1$, 0, 1, 2, . . .}, what is the truth set of $x + 3 = 1$? ________ .

{$^-2$}

29 Using the set of counting numbers, what is the truth set of $x + 11 = 2$? ________ .

{ }

30 Using the set of integers, what is the truth set of $3x = 10$? ________ .

{ }

31 Using the set of counting numbers, what is the truth set of $4x = 20$? ________ .

{5}

32 Using the set of integers, what is the truth set of $x + 5 = 12$? ________ .

{7}

33 Using the set of counting numbers, what is the truth set of $5x = 7$? ________ .

{ }

34 Using the set of integers, what is the truth set of $x + 9 = 5$? ________ .

{$^-4$}

35 Every integer has an opposite. $^{-}8$ is the opposite of 8 because $8 + {}^{-}8 = 0$. What number may be added to both sides of $3x + 8 = 29$ to obtain the simpler equation $3x = 21$? ______________ .

$^{-}8$

$$\begin{array}{rcl} 3x + 8 & = & 29 \\ {}^{-}8 & & {}^{-}8 \\ \hline 3x & = & 21 \end{array}$$

36 What integer can be added to both sides of $2x - 7 = 11$ to obtain the simpler equation $2x = 18$?

______________ .

[*Note:* 7 is the opposite of $^{-}7$ because $^{-}7 + 7 = 0$.]

7

$$\begin{array}{rcl} 2x - 7 & = & 11 \\ {}^{+}7 & & {}^{+}7 \\ \hline 2x & = & 18 \end{array}$$

37 What integer can be added to both sides of $7x + 5 = 47$ to obtain the simpler equation $7x = 42$? ______________ .

$^{-}5$

$$\begin{array}{rcl} 7x + 5 & = & 47 \\ {}^{-}5 & & {}^{-}5 \\ \hline 7x & = & 42 \end{array}$$

38 What integer can be added to both sides of $8x - 3 = 19$ to obtain the simpler equation $8x = 22$? ______________ .

3

$$\begin{array}{rcl} 8x - 3 & = & 19 \\ {}^{+}3 & & {}^{+}3 \\ \hline 8x & = & 22 \end{array}$$

39 To find the truth set of $2x + 5 = 13$, first add the opposite of 5 to both sides of the equation to obtain $2x = 8$. Using the set of integers, the truth set of $2x = 8$ is ________ .

$\{4\}$

40 To find the truth set of $3x - 2 = 19$, first add the opposite of $^{-}2$ to both sides of the equation to obtain $3x = 21$. Using the set of integers, the truth set of $3x = 21$ is ________ .

$\{7\}$

41 Using the set of integers, find the truth set of $7x + 6 = 27$? ________ .

$\{3\}$

42 Using the set of integers, find the truth set of $5x - 2 = 13$. ________ .

$\{3\}$

43 Using the set of integers, find the truth set of $5x + 2 = {}^{-}13$. ________ .

$\{{}^{-}3\}$

44 Using the set of integers, the truth set of $4 - 3x = 10$ is $\{^-2\}$ because $4 - 3 \cdot {^-2} = 10$ is a true statement. Using the set of integers, find the truth set of $2 - 7x = 23$. ________ .

$\{^-3\}$

45 Using the set of integers, find the truth set of $10x + 7 = 47$. ________ .

$\{4\}$

46 Using the set of integers, find the truth set of $8x - 17 = 71$. ________ .

$\{11\}$

47 Using the set of integers, find the truth set of $4x + 13 = 13$. ________ .

$\{0\}$

48 Using the set of integers, find the truth set of $^-3x - 4 = 11$. ________ .

$\{^-5\}$

49 Every integer has an opposite. The equation $3x + 5 = x + 1$ can be simplified by adding ^-x to both sides to obtain $2x + 5 = 1$. Using the set of integers, find the truth set of $2x + 5 = 1$. ________ .

$\{^-2\}$

50 Using the set of integers, find the truth set of $5x - 4 = 2x + 8$ by first adding ^-2x to both sides of the equation. ________________ .

$\{4\}$

$$\begin{array}{rcr} 5x - 4 & = & 2x + 8 \\ {^-2x} \quad & & {^-2x} \quad \\ \hline 3x - 4 & = & 8 \\ {^+4} & & {^+4} \\ \hline 3x \quad & = & 12 \end{array}$$

51 Using the set of integers, find the truth set of $10x - 3 = 4x + 15$ by first adding ^-4x to both sides of the equation. ________________ .

$\{3\}$

$$\begin{array}{rcr} 10x - 3 & = & 4x + 15 \\ {^-4x} \quad & & {^-4x} \quad \\ \hline 6x - 3 & = & 15 \\ {^+3} & & {^+3} \\ \hline 6x \quad & = & 18 \end{array}$$

52 Using the set of integers, find the truth set of $^{-}3x - 2 = 12 + 4x$ by first adding $^{-}4x$ to both sides of the equation. ________ .

$\{^{-}2\}$

$$\begin{array}{rcl} ^{-}3x - 2 &=& 12 + 4x \\ ^{-}4x & & ^{-}4x \\ \hline ^{-}7x - 2 &=& 12 \\ ^{+}2 & & ^{+}2 \\ \hline ^{-}7x &=& 14 \end{array}$$

53 Using the set of integers, find the truth set of $4x + 7 = 25 - 2x$ by first adding $2x$ to both sides of the equation. ________ .

$\{3\}$

54 By adding $2x$ to both sides of the equation $10x + 2 = {}^{-}2x + 26$, the simpler equation $12x + 2 = 26$ can be obtained. $12x + 2 = 26$ can be further simplified by adding $^{-}2$ to both sides of the equation to obtain $12x = 24$. The truth set of $10x + 2 = {}^{-}2x + 26$ is $\{2\}$. Using the set of integers, find the truth set of $7x - 2 = 2x + 13$.

________ .

$\{3\}$

$$\begin{array}{rcl} 7x - 2 &=& 2x + 13 \\ ^{-}2x & & ^{-}2x \\ \hline 5x - 2 &=& 13 \\ ^{+}2 & & ^{+}2 \\ \hline 5x &=& 15 \end{array}$$

55 By adding ^{-}x to both sides of $4x - 5 = x + 4$, the simpler equation $3x - 5 = 4$ can be obtained. $3x - 5 = 4$ can be further simplified by adding 5 to both sides of the equation to obtain $3x = 9$. The truth set of $4x - 5 = x + 4$ is $\{3\}$. Using the set of integers, find the truth set of $5x - 6 = 2x + 15$.

________ .

$\{7\}$

$$\begin{array}{rcl} 5x - 6 &=& 2x + 15 \\ ^{-}2x & & ^{-}2x \\ \hline 3x - 6 &=& 15 \\ ^{+}6 & & ^{+}6 \\ \hline 3x &=& 21 \end{array}$$

56 Using the set of integers, find the truth set of $2x + 5 = x - 4$. ______ . $\{^{-}9\}$

57 Using the set of integers, find the truth set of $x + 5 = 2x - 3$. ______ . $\{8\}$

58 Using the set of integers, find the truth set of $2x - 6 = 5x + 6$. ______ . $\{^{-}4\}$

59 The equation $3x - 6 = x + 7$ can be simplified to $2x = 13$. Using the set of integers, the truth set is { }. Using the set of integers, what is the truth set of $5x + 2 = x - 3$? ______ . { }

60 Using the set of integers, find the truth set of $4x + 2 = x - 5$. ______ . { }

61 Using the set of integers, find the truth set of $3x - 7 = x + 11$. ______ . $\{9\}$

62 Using the set of integers, find the truth set of $4x + 6 = x - 5$. ______ . { }

Self-Quiz # 1

The following questions test the objectives of the preceding section. 100% mastery is desired.

1. Using the set of counting numbers, {1, 2, 3, . . .}, find the truth set of each equation.
(a) $x + 8 = 20$. ________ .
(b) $x + 13 = 5$. ________ .
(c) $3x = 33$. ________ .
(d) $4x = 11$. ________ .

2. Using the set of integers, {. . . , $^-2$, $^-1$, 0, 1, 2, . . .}, find the truth set of each equation.
(a) $x + 5 = 1$. ________ .
(b) $3x + 10 = 1$. ________ .
(c) $2x = 13$. ________ .
(d) $5x - 7 = 13$. ________ .
(e) $3x + 4 = x - 8$. ________ .
(f) $2x - 3 = 5 - 2x$. ________ .

63 $\frac{5}{7}$, $\frac{^-9}{4}$, $\frac{0}{6}$, $\frac{^-4}{4}$, $\frac{^-5}{3}$, and $\frac{6}{6}$ are rational numbers. Any integer may be used as the numerator of a rational number. Any integer, except 0, may be used as a denominator. Is $\frac{10}{21}$ a rational number? ________ .

Yes

64 Every rational number has an opposite. $\frac{2}{3}$ is the opposite of $\frac{^-2}{3}$ because $\frac{2}{3} + \frac{^-2}{3} = 0$. $\frac{^-8}{5}$ is the opposite of $\frac{8}{5}$ because $\frac{^-8}{5} + \frac{8}{5} = 0$. What is the opposite of $\frac{9}{5}$? ________ .

$\frac{^-9}{5}$

65 What is the opposite of $\frac{5}{17}$? ________ . $\frac{^-5}{17}$

66 What is the opposite of $\frac{^-8}{5}$? ________ . $\frac{8}{5}$

67 The truth set of $x - \frac{1}{2} = \frac{3}{4}$ can be found by adding $\frac{1}{2}$ to both sides of the equation. Using the set of rational numbers, the truth set of $x - \frac{1}{2} = \frac{3}{4}$ is ________ . $\left\{\frac{5}{4}\right\}$

$\left[\textit{Note: } \frac{3}{4} + \frac{1}{2} = \frac{3}{4} + \frac{2}{4} = \frac{5}{4}.\right]$

68 The truth set of $x + \frac{3}{5} = \frac{7}{10}$ can be found by adding $\frac{^-3}{5}$ to both sides of the equation. Using the set of rational numbers, find the truth set of $x + \frac{3}{5} = \frac{7}{10}$. ________ . $\left\{\frac{1}{10}\right\}$

69 Using the set of rational numbers, find the truth set of $x + \frac{2}{3} = \frac{1}{6}$ by first adding $\frac{^-2}{3}$ to both sides of the equation. ________ . $\left\{\frac{^-1}{2}\right\}$

70 Using the set of rational numbers, find the truth set of $x - \frac{3}{5} = \frac{1}{4}$ by first adding $\frac{3}{5}$ to both sides of the equation. ________ . $\left\{\frac{17}{20}\right\}$

71 The truth set of $x + \frac{1}{3} = \frac{^-1}{6}$ is $\left\{\frac{^-1}{2}\right\}$ because $\frac{^-1}{2} + \frac{1}{3} = \frac{^-1}{6}$ is a true statement. Using the set of rational numbers, find the truth set of $x + \frac{5}{8} = \frac{1}{4}$.

________ . $\left\{\frac{^-3}{8}\right\}$

72 Using the set of rational numbers, find the truth set of $x - \frac{8}{3} = \frac{^-2}{5}$. ________ .

$\left\{\frac{34}{15}\right\}$

73 $\frac{2}{9}$ is the reciprocal of $\frac{9}{2}$ because $\frac{2}{9} \cdot \frac{9}{2} = 1$. $\frac{^-5}{3}$ is the reciprocal of $\frac{^-3}{5}$ because $\frac{^-3}{5} \cdot \frac{^-5}{3} = 1$. What is the reciprocal of $\frac{3}{8}$? ________ .

$\frac{8}{3}$

74 $\frac{^-4}{7}$ is the reciprocal of $\frac{^-7}{4}$ because $\frac{^-4}{7} \cdot \frac{^-7}{4} = 1$. $\frac{^-5}{9}$ is the reciprocal of ________ .

$\frac{^-9}{5}$

75 What is the reciprocal of $\frac{^-17}{3}$? ________ .

$\frac{^-3}{17}$

76 8 can be written as $\frac{8}{1}$. What is the reciprocal of 8? ________ .

$\frac{1}{8}$

77 $^-5$ can be written as $\frac{^-5}{1}$. What is the reciprocal of $^-5$? ________ .

$\frac{^-1}{5}$

78 What is the reciprocal of $^-12$? ________ .

$\frac{^-1}{12}$

79 What is the reciprocal of $\frac{17}{20}$? ________ .

$\frac{20}{17}$

80 What is the reciprocal of 15? ________ .

$\frac{1}{15}$

81 What is the reciprocal of $\frac{7}{8}$? ________ .

$\frac{8}{7}$

82 The product of any rational number and its reciprocal is 1. Zero does not have a reciprocal because 0 multiplied by any rational number gives 0, not 1. Does every rational number except 0 have a reciprocal? ______ .

Yes

83 Which of the following rational numbers does not have a reciprocal? ______ .

(d)

(a) $\frac{^-1}{3}$.
(b) 5.
(c) $\frac{8}{19}$.
(d) 0.
(e) $^-4$.

For the remainder of this chapter use the set of rational numbers as possible replacements for all equation solutions.

---

84 The reciprocal of $\frac{2}{5}$ is $\frac{5}{2}$. To find the truth set of $\frac{2}{5}x = \frac{7}{3}$, first multiply both sides of the equation by $\frac{5}{2}$.

$$\frac{2}{5}x = \frac{7}{3}$$

$$\frac{5}{2}\left(\frac{2}{5}x\right) = \frac{5}{2} \cdot \frac{7}{3}$$

$$1x = \frac{35}{6}$$

What is the truth set of $\frac{2}{5}x = \frac{7}{3}$? ______ .

$\left\{\frac{35}{6}\right\}$

85 The reciprocal of $\frac{5}{7}$ is $\frac{7}{5}$. Find the truth set of $\frac{5}{7}x = \frac{2}{3}$ by multiplying both sides of the equation by $\frac{7}{5}$.

________ .

$\left\{\frac{14}{15}\right\}$

86 The reciprocal of $\frac{^-10}{9}$ is $\frac{^-9}{10}$. To find the truth set of $\frac{^-10}{9}x = \frac{7}{4}$, multiply both sides of the equation by $\frac{^-9}{10}$.

$$\frac{^-10}{9}x = \frac{7}{4}$$

$$\frac{^-9}{10}\left(\frac{^-10}{9}x\right) = \frac{^-9}{10} \cdot \frac{7}{4}$$

$$1x = \frac{^-63}{40}$$

What is the truth set of $\frac{^-10}{9}x = \frac{7}{4}$? ________ .

$\left\{\frac{^-63}{40}\right\}$

87 The reciprocal of $\frac{^-2}{3}$ is $\frac{^-3}{2}$. Find the truth set of $\frac{^-2}{3}x = \frac{5}{9}$. ________ .

$\left\{\frac{^-5}{6}\right\}$

88 Find the truth set of $\frac{4}{3}x = \frac{^-2}{7}$. ________ .

$\left\{\frac{^-3}{14}\right\}$

89 Find the truth set of $\frac{8}{5}x = \frac{1}{2}$. ________ .

$\left\{\frac{5}{16}\right\}$

90 Find the truth set of $5x = \frac{^-2}{3}$. ________ .

$\left[\textit{Note: } \frac{1}{5} \text{ is the reciprocal of 5.}\right]$

$\left\{\frac{^-2}{15}\right\}$

91 Find the truth set of $^-3x = \frac{14}{5}$. ________ .

$\left\{\frac{^-14}{15}\right\}$

92 The truth set of $5x = {}^-3$ is $\left\{\frac{^-3}{5}\right\}$, because $5 \cdot \frac{^-3}{5} = {}^-3$ is a true statement. Find the truth set of $^-2x = 9$.

________ .

$\left\{\frac{^-9}{2}\right\}$

In solving $^-2x = 9$ the number $\frac{^-9}{2}$ is found for x. To check that $\left\{\frac{^-9}{2}\right\}$ is the truth set of $^-2x = 9$, the x in the equation is replaced by $\frac{^-9}{2}$ as follows:

$$^-2x = 9 \qquad x = \frac{^-9}{2}$$

$$^-2 \cdot \frac{^-9}{2} = 9$$

The numerical sentence $^-2 \cdot \frac{^-9}{2} = 9$ is *true*. Therefore, $\left\{\frac{^-9}{2}\right\}$ is the truth set of $^-2x = 9$.

---

93 Solve and check $8x = 11$. ________________ .
(To solve an equation means to find its truth set.)

$\left\{\frac{11}{8}\right\}$

$8 \cdot \frac{11}{8} = 11$ is true

94 Solve and check $\frac{1}{2}x = \frac{3}{4}$. ________________ .

$\left\{\frac{3}{2}\right\}$

$\frac{1}{2} \cdot \frac{3}{2} = \frac{3}{4}$ is true

95 Is $\frac{4}{7}$ in the truth set of $\frac{2}{3}x = \frac{8}{21}$?

________________ .

Yes,
$\frac{2}{3} \cdot \frac{4}{7} = \frac{8}{21}$ is true

96 Is $\frac{^-5}{3}$ in the truth set of $\frac{4}{7}x = \frac{^-20}{33}$?

__________________ .

No, $\frac{4}{7} \cdot \frac{^-5}{3} = \frac{^-20}{33}$ is false

97 Solve and check $^-11x = {}^-6$. ________ .

$\left\{\frac{6}{11}\right\}$

98 Solve and check $45x = 3$. ________ .

$\left\{\frac{1}{15}\right\}$

99 To solve $2x - 3 = 11$, the following steps are used:

$$\begin{array}{rcl} 2x - 3 &=& 11 \\ {}^+3 && {}^+3 \\ \hline 2x &=& 14 \\ \frac{1}{2}(2x) &=& \frac{1}{2}(14) \\ x &=& 7 \end{array}$$

To check, the x in the original equation is replaced by 7. Is $2 \cdot 7 - 3 = 11$ a true statement? ________ .

Yes

100 To solve $5x - 3 = {}^-6$, the following steps are used:

$$\begin{array}{rcl} 5x - 3 &=& {}^-6 \\ {}^+3 && {}^+3 \\ \hline 5x &=& {}^-3 \\ \frac{1}{5}(5x) &=& \frac{1}{5}({}^-3) \\ x &=& \frac{^-3}{5} \end{array}$$

Check by replacing x in the original equation by $\frac{^-3}{5}$. Is $\frac{^-3}{5}$ in the truth set of $5x - 3 = {}^-6$? ________ .

Yes, $5 \cdot \frac{^-3}{5} - 3 = {}^-6$ is true

101 Solve and check $8x + 5 = 11$. ________ .

$\left\{\frac{3}{4}\right\}$

102 Solve and check $2x - 7 = 13$. ________ .

$\{10\}$

103 Solve and check $^{-}3x + 5 = {}^{-}9$. ________ .

$\left\{\frac{14}{3}\right\}$

104 Solve and check $4x - 9 = 6$. ________ .

$\left\{\frac{15}{4}\right\}$

105 Solve and check $5x + 7 = {}^{-}1$. ________ .

$\left\{\frac{^{-}8}{5}\right\}$

106 Solve $4x - 3 = x + 5$ by first adding ^{-}x to both sides of the equation. ________ .

$\left\{\frac{8}{3}\right\}$

107 Solve $5x + 7 = {}^{-}4x + 34$. ________ .

$\{3\}$

108 Solve $2x + 4 = 5x - 1$. ________ .

$\left\{\frac{5}{3}\right\}$

109 Solve $5x + 3 = x + 15$. ________ .

$\{3\}$

110 Solve $6x - 4 = 3x + 7$. ________ .

$\left\{\frac{11}{3}\right\}$

111 Solve $7x - 1 = 5x - 17$. ________ .

$\{^{-}8\}$

112 The truth set of $2x + 9 = 7x - 3$ is $\left\{\frac{12}{5}\right\}$ because $2 \cdot \frac{12}{5} + 9 = 7 \cdot \frac{12}{5} - 3$ is a true statement. Solve and check $x + 7 = 3x + 10$. ________ .

$\left\{\frac{^{-}3}{2}\right\}$

$\frac{^{-}3}{2} + 7 = 3 \cdot \frac{^{-}3}{2} + 10$ is true

113 To solve $2(x - 7) + 4 = 8$, first simplify $2(x - 7) + 4$ by removing the parentheses as follows:

$$2(x - 7) + 4 = 8$$
$$2x - 14 + 4 = 8$$
$$2x - 10 = 8$$

Solve and check $2(x - 7) + 4 = 8$.

______________ .

$\{9\}$
$2(9 - 7) + 4 = 8$
is true

114 To solve $6 - 3(x - 1) = 12$, first remove the parentheses as follows:

$$6 - 3(x - 1) = 12$$
$$6 - 3x + 3 = 12$$

Simplify the left side of the equation to $9 - 3x$ and solve $6 - 3(x - 1) = 12$. ________ .

$\{^{-}1\}$

115 Solve $4(x + 2) - 1 = 6$ by first simplifying the left side of the equation. ________ .

$\left\{\frac{^{-}1}{4}\right\}$

116 Solve and check $8 - 3(x + 2) = 7$. ________ .

$\left\{\frac{^{-}5}{3}\right\}$

117 Solve and check $9 - 4(2x - 3) = 8$. ________ .

$\left\{\frac{13}{8}\right\}$

118 Solve and check $8 + (7x - 6) = 3$. ________ .

[*Note:* $+(7x - 6)$ is equivalent to $1(7x - 6)$.]

$\left\{\frac{1}{7}\right\}$

119 The truth set of $5x - (2x + 7) = 6$ is $\left\{\frac{13}{3}\right\}$ because $5 \cdot \frac{13}{3} - \left(2 \cdot \frac{13}{3} + 7\right) = 6$ is a true statement. Solve and check $2x + (3x - 2) = 38$.

______________ .

$\{8\}$
$2 \cdot 8 + (3 \cdot 8 - 2) = 38$
is true

120 Solve and check $5 - (x + 3) = 4$. ________ . $\{^{-}2\}$
[*Note:* $-(x + 3)$ is equivalent to $^{-}1(x + 3)$.]

121 Solve and check $6 - (2x + 7) = 5$. ________ . $\{^{-}3\}$

122 Solve and check $4(x + 2) - 2(x - 3) = 14$.
________ . $\{0\}$

123 Solve and check $3(x + 4) - x = 4(x - 7)$.
________ . $\{20\}$

124 Solve and check $5 - 3(x + 4) = 6$. ________ . $\left\{\frac{^{-}13}{3}\right\}$

Self-Quiz # 2

The following questions test the objectives of the preceding section. 100% mastery is desired.

Solve and check each equation.

1. $x + \frac{1}{2} = \frac{2}{5}$. ________ .
2. $x - \frac{2}{3} = \frac{1}{7}$. ________ .
3. $\frac{2}{3}x = \frac{5}{8}$. ________ .
4. $^{-}3x = 18$. ________ .
5. $\frac{5}{7}x = \frac{^{-}2}{3}$. ________ .
6. $4x + 5 = 2$. ________ .
7. $6x - 3 = 15$. ________ .
8. $5x - 2 = x + 13$. ________ .
9. $2x + 3(x - 5) = 7$. ________ .
10. $11x - (^{-}2x + 5) = 21$. ________ .

Finding truth sets of fractional equations is studied in the following frames.

An example of a fractional equation is

$$\frac{2}{3}x + \frac{5}{6} = \frac{1}{2}$$

---

125 One method of solving $\frac{2}{3}x + \frac{5}{6} = \frac{1}{2}$ would be to first add $\frac{^-5}{6}$ to both sides of the equation. Add $\frac{^-5}{6}$ to both sides of $\frac{2}{3}x + \frac{5}{6} = \frac{1}{2}$. What equation is obtained? ________.

$\frac{2}{3}x = \frac{^-1}{3}$

126 $\frac{2}{3}x + \frac{5}{6} = \frac{1}{2}$ can be simplified to $\frac{2}{3}x = \frac{^-1}{3}$. Find the truth set of $\frac{2}{3}x = \frac{^-1}{3}$ by multiplying both sides of the equation by $\frac{3}{2}$. ______.

$\left\{\frac{^-1}{2}\right\}$

127 Find the truth set of $\frac{5}{6}x + \frac{1}{3} = \frac{2}{3}$ by first adding $\frac{^-1}{3}$ to both sides of the equation. ______.

$\left\{\frac{2}{5}\right\}$

128 Find the truth set of $\frac{2}{5}x + \frac{1}{4} = \frac{7}{8}$ by first adding $\frac{^-1}{4}$ to both sides of the equation. ______.

$\left\{\frac{25}{16}\right\}$

129 Solve $\frac{3}{4}x - \frac{2}{3} = \frac{1}{5}$. ______.

$\left\{\frac{52}{45}\right\}$

130 Solve $\frac{4}{3}x + \frac{1}{2} = \frac{3}{4}$. ______.

$\left\{\frac{3}{16}\right\}$

131 Solve $2x - \frac{2}{5} = \frac{3}{4}$. ________ .

$\left\{\frac{23}{40}\right\}$

132 Solve $5x - \frac{1}{2} = \frac{5}{8}$. ________ .

$\left\{\frac{9}{40}\right\}$

133 Solve $\frac{2}{5}x - 3 = \frac{1}{5}$. ________ .

$\{8\}$

134 Solve $\frac{7}{8}x + \frac{1}{2} = 4$. ________ .

$\{4\}$

Equations such as $\frac{2}{3}x + \frac{5}{6} = \frac{1}{2}$ have been solved in the preceding frames. In the following frames another method for solving $\frac{2}{3}x + \frac{5}{6} = \frac{1}{2}$ will be explained.

---

135 The denominators, in order, of the equation $\frac{2}{5}x + \frac{1}{4} = \frac{1}{6}$ are 5, 4, and 6. The first step in the solution is to multiply both sides of the equation by the first denominator, 5, as follows:

$$5\left(\frac{2}{5}x + \frac{1}{4}\right) = 5 \cdot \frac{1}{6}$$

$$5 \cdot \frac{2}{5}x + 5 \cdot \frac{1}{4} = 5 \cdot \frac{1}{6}$$

Complete the first step of this solution by simplifying each term of the equation. ________________ .

$2x + \frac{5}{4} = \frac{5}{6}$

136 The first step in solving $\frac{2}{5}x + \frac{1}{4} = \frac{1}{6}$ is to multiply both sides of the equation by the denominator 5 to obtain

$$2x + \frac{5}{4} = \frac{5}{6}$$

The second step is to multiply both sides of the equation by the denominator 4 to obtain

$$4\left(2x + \frac{5}{4}\right) = 4 \cdot \frac{5}{6}$$

After simplifying, what equation is obtained?

________________ .

$8x + 5 = \frac{10}{3}$

137

$$\frac{2}{5}x + \frac{1}{4} = \frac{1}{6}$$

$$5\left(\frac{2}{5}x + \frac{1}{4}\right) = 5 \cdot \frac{1}{6}$$

$$2x + \frac{5}{4} = \frac{5}{6}$$

$$4\left(2x + \frac{5}{4}\right) = 4 \cdot \frac{5}{6}$$

$$8x + 5 = \frac{10}{3}$$

The third step in solving $\frac{2}{5}x + \frac{1}{4} = \frac{1}{6}$ is to multiply both sides of $8x + 5 = \frac{10}{3}$ by the denominator 3.

$$3(8x + 5) = 3 \cdot \frac{10}{3}$$

After simplifying, what equation is obtained?

________________ .

$24x + 15 = 10$

138 After multiplying both sides of $8x + 5 = \frac{10}{3}$ by 3, the equation $24x + 15 = 10$ is obtained. Complete the solution of $\frac{2}{5}x + \frac{1}{4} = \frac{1}{6}$ by solving $24x + 15 = 10$.

_______ .

$\left\{\frac{^-5}{24}\right\}$

The solution of $\frac{3}{4}x - \frac{2}{3} = \frac{5}{12}$ may be shortened by multiplying both sides of the equation by 12 because 12 is the least common multiple (LCM) of the denominators.

To multiply both sides of $\frac{3}{4}x - \frac{2}{3} = \frac{5}{12}$ by 12, each term on both sides of the equation is multiplied by 12.

---

139 The least common multiple of 4, 3, and 12 is 12 because 12 is the smallest counting number that is a multiple of 4, 3, and 12. Multiply each term of $\frac{3}{4}x - \frac{2}{3} = \frac{5}{12}$ by 12. What equation is obtained?

___________ .

$9x - 8 = 5$

140 To solve $\frac{3}{4}x - \frac{2}{3} = \frac{5}{12}$, multiply each side of the equation by 12, which is the least common multiple of the denominators.

$$\frac{3}{4}x - \frac{2}{3} = \frac{5}{12}$$

$$12\left(\frac{3}{4}x - \frac{2}{3}\right) = 12\left(\frac{5}{12}\right)$$

$$9x - 8 = 5$$

Complete the solution of $\frac{3}{4}x - \frac{2}{3} = \frac{5}{12}$ by solving $9x - 8 = 5$. _______ .

$\left\{\frac{13}{9}\right\}$

141 To solve $\frac{2}{5}x - \frac{3}{4} = \frac{7}{10}$, each term of the equation should be multiplied by the least common multiple (LCM) of the denominators. What is the LCM of 5, 4, and 10? ________ .

20

142 Multiply each term of $\frac{2}{5}x - \frac{3}{4} = \frac{7}{10}$ by 20. What equation is obtained? ________________ .

$8x - 15 = 14$

143 Complete the solution of $\frac{2}{5}x - \frac{3}{4} = \frac{7}{10}$:

$$\frac{2}{5}x - \frac{3}{4} = \frac{7}{10}$$

$$20 \cdot \frac{2}{5}x - 20 \cdot \frac{3}{4} = 20 \cdot \frac{7}{10}$$

$$8x - 15 = 14$$

________ .

$\left\{\frac{29}{8}\right\}$

144 To solve $\frac{7}{8}x + \frac{5}{6} = \frac{1}{12}$, each term of the equation should be multiplied by the least common multiple of the denominators. What is the LCM of the denominators 8, 6, and 12? ________ .

24

145 Multiply each term of $\frac{7}{8}x + \frac{5}{6} = \frac{1}{12}$ by the LCM of the denominators. What equation is obtained?

________________ .

$21x + 20 = 2$

146 Complete the solution of $\frac{7}{8}x + \frac{5}{6} = \frac{1}{12}$:

$$\frac{7}{8}x + \frac{5}{6} = \frac{1}{12}$$

$$24 \cdot \frac{7}{8}x + 24 \cdot \frac{5}{6} = 24 \cdot \frac{1}{12}$$

$$21x + 20 = 2$$

________ .

$\left\{\frac{^-6}{7}\right\}$

147 Complete the solution for the following equation by multiplying each term by the LCM of the denominators:

$$\frac{3}{4}x + \frac{9}{10} = \frac{2}{5}$$

$$20 \cdot \frac{3}{4}x + 20 \cdot \frac{9}{10} = 20 \cdot \frac{2}{5}$$

________ .

$\left\{\frac{^-2}{3}\right\}$

148 Solve the following equation by first multiplying each term by the LCM of the denominators:
$\frac{5}{3}x + \frac{1}{4} = \frac{7}{9}$. ________ .

$\left\{\frac{19}{60}\right\}$

149 Solve the following equation by first multiplying each term by the LCM of the denominators:
$\frac{5}{6}x + \frac{7}{12} = \frac{2}{3}$. ________ .

$\left\{\frac{1}{10}\right\}$

150 Solve the following equation by first multiplying each term by the LCM of the denominators:
$\frac{2}{3}x + \frac{1}{4} = \frac{5}{12}$. ________ .

$\left\{\frac{1}{4}\right\}$

151 Solve the following equation by first multiplying each term by the LCM of the denominators:
$\frac{3}{8}x - \frac{5}{12} = \frac{1}{6}$. ________ .

$\left\{\frac{14}{9}\right\}$

152 Solve $\frac{5}{3}x - \frac{2}{7} = \frac{1}{3}$. ________ . $\left\{\frac{13}{35}\right\}$

153 Solve $\frac{1}{4}x - \frac{3}{8} = \frac{1}{2}$. ________ . $\left\{\frac{7}{2}\right\}$

154 Solve $\frac{5}{12}x + \frac{2}{3} = \frac{1}{4}$. ________ . $\{{}^{-}1\}$

155 Solve $\frac{2}{3}x - \frac{2}{5} = \frac{3}{5}x - \frac{1}{3}$. ________ . $\{1\}$

156 Solve $\frac{5}{6}x - \frac{1}{3} = \frac{1}{4}x + \frac{1}{2}$. ________ . $\left\{\frac{10}{7}\right\}$

157 Solve $\frac{3}{4}x - \frac{1}{2} = \frac{1}{4}x + \frac{11}{2}$. ________ . $\{12\}$

158 Solve $\frac{1}{3}x + \frac{3}{4} = \frac{1}{2} - \frac{2}{3}x$. ________ . $\left\{\frac{{}^{-}1}{4}\right\}$

159 The truth set of $\frac{1}{3}x + \frac{3}{4} = \frac{1}{2} - \frac{2}{3}x$ is $\left\{\frac{{}^{-}1}{4}\right\}$ because $\frac{1}{3} \cdot \frac{{}^{-}1}{4} + \frac{3}{4} = \frac{1}{2} - \frac{2}{3} \cdot \frac{{}^{-}1}{4}$ is a true statement. Solve and check $\frac{2}{3}x + 1 = \frac{5}{9}x - \frac{1}{6}$. ________ . $\left\{\frac{{}^{-}21}{2}\right\}$

160 To solve $\frac{5}{x} + \frac{2}{3} = \frac{4}{x}$, the following steps are used to obtain a simpler equation:

$$\frac{5}{x} + \frac{2}{3} = \frac{4}{x}$$

$$3x \cdot \frac{5}{x} + 3x \cdot \frac{2}{3} = 3x \cdot \frac{4}{x}$$

$$15 + 2x = 12$$

Complete the solution of $\frac{5}{x} + \frac{2}{3} = \frac{4}{x}$ by solving

$15 + 2x = 12$. ________ . $\left\{\frac{{}^{-}3}{2}\right\}$

161 Complete the solution for the following equation:

$$\frac{4}{3x} - \frac{1}{2} = \frac{5}{2x}$$

$$6x \cdot \frac{4}{3x} - 6x \cdot \frac{1}{2} = 6x \cdot \frac{5}{2x}$$

$$8 - 3x = 15$$

________ .

$\left\{\frac{^-7}{3}\right\}$

162 The LCM of 5, x, and 10 is $10x$. Complete the solution for the following equation:

$$\frac{3}{5} - \frac{2}{x} = \frac{1}{10}$$

$$10x \cdot \frac{3}{5} - 10x \cdot \frac{2}{x} = 10x \cdot \frac{1}{10}$$

________________ .

$6x - 20 = x$

$\{4\}$

163 Solve $\frac{1}{6} - \frac{2}{x} = \frac{1}{2}$ by multiplying each term of the equation by the LCM of 6, x, and 2. ________ .

$\{^-6\}$

164 Solve $\frac{5}{8} - \frac{3}{2x} = \frac{1}{4}$ by multiplying each term by the LCM of the denominators. ________ .

$\{4\}$

To check that 4 is in the truth set of $\frac{5}{8} - \frac{3}{2x} = \frac{1}{4}$, the following steps are used:

$$\frac{5}{8} - \frac{3}{2x} = \frac{1}{4} \quad \text{becomes}$$

$$\frac{5}{8} - \frac{3}{2 \cdot 4} = \frac{1}{4} \quad \text{when } x = 4$$

$$\frac{5}{8} - \frac{3}{8} = \frac{1}{4} \quad \text{is true.}$$

Therefore, 4 is in the truth set of $\frac{5}{8} - \frac{3}{2x} = \frac{1}{4}$.

To check $\frac{^-7}{3}$ as a solution of $\frac{4}{3x} - \frac{1}{2} = \frac{5}{2x}$, the following steps are used:

$$\frac{4}{3x} - \frac{1}{2} = \frac{5}{2x} \quad \text{becomes}$$

$$\frac{4}{3 \cdot \frac{^-7}{3}} - \frac{1}{2} = \frac{5}{2 \cdot \frac{^-7}{3}} \quad \text{when } x = \frac{^-7}{3}$$

$$\frac{4}{^-7} - \frac{1}{2} = \frac{5}{\frac{^-14}{3}}$$

$$\frac{^-4}{7} - \frac{1}{2} = \frac{^-15}{14} \quad \text{because} \quad \frac{5}{\frac{^-14}{3}} = 5 \cdot \frac{^-3}{14} = \frac{^-15}{14}$$

$$\frac{^-8}{14} - \frac{7}{14} = \frac{^-15}{14} \quad \text{is true.}$$

For each of the following frames, solve and check the equation.

---

165 $\frac{2}{5} - \frac{4}{x} = \frac{3}{10}$. ________. {40}

166 $\frac{2}{7} - \frac{3}{x} = \frac{^-5}{7}$. ________. {3}

167 $\frac{5}{x} - \frac{9}{4} = \frac{1}{2x}$. ________ . $\{2\}$

168 $\frac{2}{5} - \frac{1}{x} = \frac{3}{5x}$. ________ . $\{4\}$

169 $\frac{6}{x} - \frac{1}{2} = \frac{1}{4}$. ________ . $\{8\}$

170 $\frac{4}{x} - \frac{2}{3} = \frac{5}{6x}$. ________ . $\left\{\frac{19}{4}\right\}$

171 $\frac{3}{x} - \frac{1}{2} = \frac{5}{6}$. ________ . $\left\{\frac{9}{4}\right\}$

172 $\frac{5}{x} - \frac{2}{3} = \frac{3}{2x}$. ________ . $\left\{\frac{21}{4}\right\}$

173 $\frac{1}{x} - \frac{3}{5} = \frac{3}{x} - \frac{2}{5}$. ________ . $\{{}^{-}10\}$

174 $\frac{3}{4x} - \frac{1}{2} = \frac{2}{3} + \frac{3}{x}$. ________ . $\left\{\frac{{}^{-}27}{14}\right\}$

Self-Quiz # 3

The following questions test the objectives of the preceding section. 100% mastery is desired.

Solve and check.

1. $\frac{2}{3}x + \frac{5}{6} = \frac{1}{12}$. ________ .

2. $\frac{3}{4}x - \frac{1}{5} = \frac{1}{10}$. ________ .

3. $\frac{2}{3}x - \frac{3}{8} = \frac{1}{2}x + \frac{17}{8}$. ________ .

4. $\frac{5}{x} + \frac{1}{2} = \frac{3}{x} + \frac{7}{10}$. ________ .

5. $\frac{4}{3x} + \frac{1}{2} = \frac{2}{x}$. ________ .

Chapter Summary

This chapter reviewed and extended the skills for solving linear equations such as $3x + 8 = 13$ and $\frac{5}{6}x - \frac{1}{2} = \frac{3}{4}$.

The two most useful methods for solving equations are:

1. Add the same number or variable to both sides of the equation.
2. Multiply the same nonzero number or variable by both sides of the equation.

Every rational number has an opposite and every rational number, except zero, has a reciprocal. Opposites and reciprocals are used to generate simpler equations until the truth sets are obvious.

CHAPTER POST-TEST

The following questions test the objectives of this chapter. A score of 90% indicates sufficient mastery, and the student may proceed to the next chapter.

Using the set of rational numbers, solve and check each of the following equations:

1. $x + 18 = 5$
2. $^{-}9x = 13$
3. $3x + 13 = 2$
4. $7x - 3 = 19$
5. $4x - 3 = x - 7$
6. $3x + 2 = 5 - 7x$
7. $3(x + 5) + 4x = 9$
8. $6 - 2(5x - 2) = 4$
9. $x + \frac{6}{5} = \frac{1}{2}$
10. $\frac{13}{5}x = \frac{2}{3}$
11. $\frac{2}{3}x - \frac{7}{6} = \frac{1}{4}$
12. $\frac{7}{10}x + \frac{3}{5} = \frac{7}{15}$
13. $\frac{8}{x} - \frac{2}{7} = \frac{3}{2}$
14. $\frac{4}{3x} - \frac{5}{6} = \frac{7}{2x}$

ALGEBRA
ALGEBRA

Fractional and Quadratic Equations

1
2
3
4
5
6
7
8
9
10

CHAPTER PRE-TEST

The following questions indicate the objectives of this chapter. A score of 90% indicates sufficient mastery, and the student may immediately take the Chapter Post-Test.

Using the set of rational numbers, find the truth set for each of the following equations:

1. $\frac{3}{x-4} + 2 = \frac{7}{x-4}$

2. $\frac{4}{x+3} = \frac{3}{x-5}$

3. $\frac{3}{x-2} - \frac{5}{x^2-7x+10} = \frac{4}{x-5}$

4. $\frac{2}{x^2-4} + \frac{1}{x+2} = \frac{3}{x-2}$

5. $\frac{2}{x-5} + \frac{4}{x+3} = \frac{9}{x^2-2x-15}$

6. $x^2 + 6x - 16 = 0$

7. $x^2 - 13x = 0$

8. $x^2 - 64 = 0$

9. $x^2 + 11x + 28 = 0$

10. $x^2 - 12x + 32 = 0$

11. $5x^2 - 20 = 0$

12. $^{-}x^2 + 5x + 14 = 0$

13. $x^2 - 13x + 22 = 0$

14. $x^2 + 42 = 13x$

15. $3x^2 + 11x - 4 = 0$

In the preceding chapter fractional equations such as $\frac{3}{4}x - \frac{1}{7} = \frac{5}{2}$ were solved by first multiplying both sides of the equation by the least common multiple (LCM) of the denominators.

This same procedure is used to find the solution for equations such as $\frac{^-5}{x+2} + 3 = \frac{4}{x+2}$.

1 The first step in solving

$$\frac{^-5}{x+2} + 3 = \frac{4}{x+2}$$

is to multiply both sides of the equation by the denominator $(x + 2)$.

$$\frac{^-5}{x+2} + 3 = \frac{4}{x+2}$$

$$\frac{^-5}{(x+2)} \cdot \underline{(x+2)} + 3\underline{(x+2)} = \frac{4}{(x+2)} \cdot \underline{(x+2)}$$

$$\frac{^-5(x+2)}{(x+2)} + 3(x+2) = \frac{4(x+2)}{(x+2)}$$

The first term of the equation $\frac{^-5(x+2)}{(x+2)}$ can be simplified as follows:

$$\frac{^-5(x+2)}{(x+2)} = \frac{^-5\cancel{(x+2)}}{\cancel{(x+2)}} = {^-5}$$

Simplify $\frac{4(x+2)}{(x+2)}$. ________.

4

2 Complete the solution of the following equation:

$$\frac{^{-}5}{x + 2} + 3 = \frac{4}{x + 2}$$

$$\frac{^{-}5}{(x + 2)} \cdot (x + 2) + 3(x + 2) = \frac{4}{(x + 2)} \cdot (x + 2)$$

$$\frac{^{-}5(x + 2)}{(x + 2)} + 3(x + 2) = \frac{4(x + 2)}{(x + 2)}$$

$$^{-}5 + 3x + 6 = 4$$

$$3x + 1 = 4$$

________ .

$\{1\}$

3 To solve $\frac{4}{x - 5} + 2 = \frac{3}{x - 5}$, first multiply both sides of the equation by the denominator, $(x - 5)$.

$$\frac{4}{x - 5} + 2 = \frac{3}{x - 5}$$

$$\frac{4}{(x - 5)} \cdot (\underline{x - 5}) + 2(\underline{x - 5}) = \frac{3}{(x - 5)} \cdot (\underline{x - 5})$$

$$\frac{4(x - 5)}{(x - 5)} + 2(x - 5) = \frac{3(x - 5)}{(x - 5)}$$

$$4 + 2x - 10 = 3$$

$$2x - 6 = 3$$

Complete the solution by solving $2x - 6 = 3$.

________ .

$\left\{\frac{9}{2}\right\}$

4 Complete the solution of the following equation.

$$\frac{8}{x + 5} + 7 = \frac{^{-}3}{x + 5}$$

$$\frac{8(x + 5)}{(x + 5)} + 7(x + 5) = \frac{^{-}3(x + 5)}{(x + 5)}$$

$$8 + 7x + 35 = {}^{-}3$$

$$7x + 43 = {}^{-}3$$

________ .

$\left\{\frac{^{-}46}{7}\right\}$

5 Complete the solution of the following equation.

$$\frac{2}{x-6} - 3 = \frac{4}{x-6}$$

$$\frac{2(x-6)}{(x-6)} - 3(x-6) = \frac{4(x-6)}{(x-6)}$$

$$2 - 3x + 18 = 4$$

$$^{-}3x + 20 = 4$$

________ . $\left\{\frac{16}{3}\right\}$

6 Solve $\frac{4}{x+1} + 2 = \frac{6}{x+1}$ by first multiplying both sides of the equation by $(x+1)$. ________ . $\{0\}$

7 Solve $\frac{3}{x-2} + 2 = \frac{1}{x-2}$ by first multiplying both sides of the equation by $(x-2)$. ________ . $\{1\}$

8 Solve $\frac{8}{x-1} + 3 = \frac{20}{x-1}$. ________ . $\{5\}$

9 Solve $\frac{5}{x-6} - 3 = \frac{^{-}1}{x-6}$. ________ . $\{8\}$

10 Solve $\frac{3}{x+1} + 5 = \frac{^{-}12}{x+1}$. ________ . $\{^{-}4\}$

11 Solve $\frac{4}{x+2} + 3 = \frac{28}{x+2}$. ________ . $\{6\}$

12 Solve $\frac{3}{x-4} - 1 = \frac{1}{x-4}$. ________ . $\{6\}$

Numbers such as $\frac{5}{0}$, $\frac{^-2}{0}$, and $\frac{17}{0}$ do *not* name rational numbers because division by zero is *not* possible.

In checking an equation such as

$$\frac{^-5}{x + 2} + 3 = \frac{4}{x + 2}$$

if x is replaced by $^-2$, the following steps are used:

$$\frac{^-5}{x + 2} + 3 = \frac{4}{x + 2} \quad \text{becomes}$$

$$\frac{^-5}{^-2 + 2} + 3 = \frac{4}{^-2 + 2} \quad \text{when } x = {^-2}$$

$$\frac{^-5}{0} + 3 = \frac{4}{0} \quad \text{is } not \text{ true}$$

because $\frac{^-5}{0}$ and $\frac{4}{0}$ are not rational numbers.

In solving equations such as

$$\frac{^-5}{x + 2} + 3 = \frac{4}{x + 2}$$

when there is a variable in the denominator, the check is of great importance because it is possible to arrive at a value for the variable, which makes the denominator *zero*, and is not acceptable as an element of the truth set.

---

13 For the equation $\frac{5}{x - 3} + 2 = \frac{6}{x - 3}$, what number cannot be used as a replacement for x because it would result in a zero denominator? ________ .

3

14 What number cannot be used as a replacement for x in the following equation?

$$\frac{4}{x - 7} + 1 = \frac{6}{x - 7}$$

(What replacement for x would result in a zero denominator?) ________ .

7

15 What number cannot be used as a replacement for x in the following equation?

$$\frac{2}{x+4} + 5 = \frac{^-7}{x+4}. \underline{\qquad\qquad}.$$

$^-4$

16 There are two numbers that cannot be used as replacements for x in the following equation:

$$\frac{6}{x-5} = \frac{4}{x+2}$$

The two replacements for x that would result in zero denominators are 5 and ______ .

$^-2$

17 What two numbers cannot be used as replacements for x in the following equation?

$$\frac{4}{x-8} = \frac{5}{x+2}. \underline{\qquad\qquad}.$$

8 and $^-2$

18 What numbers cannot be used as replacements for x in the following equation?

$$\frac{5}{x+6} + \frac{3}{x-4} = \frac{^-4}{x-1}. \underline{\qquad\qquad\qquad\qquad}.$$

$^-6$, 4, and 1

19 What numbers cannot be used as replacements for x in the following equation?

$$\frac{4}{x+8} - \frac{2}{x+3} = \frac{6}{x-5}. \underline{\qquad\qquad\qquad\qquad}.$$

$^-8$, $^-3$, and 5

20 What number cannot be used as a replacement for x in the following equation?

$$\frac{4}{2x-7} + 1 = \frac{3}{2x-7}. \underline{\qquad\qquad}.$$

[*Note*: Solve $2x - 7 = 0$ to find the number that would result in a zero denominator.]

$\frac{7}{2}$

21 What numbers cannot be used as replacements for x in the following equation?

$$\frac{4}{3x+2} + 7 = \frac{5}{x-2}. \underline{\qquad\qquad}.$$

[*Note*: Solve $3x + 2 = 0$ and $x - 2 = 0$.]

$\frac{^-2}{3}$ and 2

22 What numbers cannot be used as replacements for x in the following equation?

$$\frac{4}{x-5} + 6 = \frac{5}{x}. \underline{\qquad\qquad}.$$

5 and 0

$\left[\textit{Note}: \frac{5}{x} \text{ is not a number if } x = 0.\right]$

23 What numbers cannot be used as replacements for x in the following equation?

$$\frac{2x}{4x-3} + \frac{3}{x} = 6. \underline{\qquad\qquad}.$$

$\frac{3}{4}$ and 0

In solving fractional equations it is possible to obtain values that are meaningless because they would result in zero denominators.

Do not include numbers in a truth set that would result in a zero denominator for any fraction.

---

24 In solving $\frac{1}{x-5} + 2 = \frac{1}{x-5}$ the following steps are used:

$$\frac{1}{x-5} + 2 = \frac{1}{x-5}$$

$$\frac{1(x-5)}{(x-5)} + 2(x-5) = \frac{1(x-5)}{(x-5)}$$

$$1 + 2x - 10 = 1$$

$$2x = 10$$

$$x = 5$$

Can 5 be used as a replacement for x in the following equation?

$$\frac{1}{x-5} + 2 = \frac{1}{x-5}. \underline{\qquad\qquad}.$$

No

25 { } is the truth set of

$$\frac{1}{x-5} + 2 = \frac{1}{x-5}$$

because 5 is the only replacement for x that makes $1 + 2(x - 5) = 1$ a true statement, but 5 cannot be used as a replacement for x in

$$\frac{1}{x-5} + 2 = \frac{1}{x-5}.$$

Solve and check.

$\frac{7}{x+2} + 4 = \frac{7}{x+2}$. ________ . { }

26 Solve and check.

$\frac{5}{x-3} + 4 = \frac{5}{x-3}$. ________ . { }

27 Solve $\frac{3}{x-1} + 6 = \frac{2}{x-1}$ by multiplying each fraction of the equation by the denominator $(x - 1)$.
________ . [Check the answer.] $\left\{\frac{5}{6}\right\}$

28 Solve $\frac{5}{x-2} - 3 = \frac{5}{x-2}$ by multiplying each side of the equation by the denominator $(x - 2)$. ________ . { }
[Check the answer.]

29 Solve and check.

$\frac{3}{x+2} - 7 = \frac{10}{x+2}$. ________ . $\{^{-}3\}$

30 Solve and check.

$\frac{1}{x-5} + 2 = \frac{1}{x-5}$. ________ . { }

31 To solve $\frac{5}{x - 3} = \frac{10}{x - 2}$, it is necessary to multiply both sides of the equation by each denominator or by the least common multiple (LCM) of the denominators.

$$\frac{5}{x - 3} = \frac{10}{x - 2}$$

$$(x - 3)(x - 2)\left(\frac{5}{x - 3}\right) = (x - 3)(x - 2)\left(\frac{10}{x - 2}\right)$$

$$\frac{5(x - 3)(x - 2)}{x - 3} = \frac{10(x - 3)(x - 2)}{x - 2}$$

$$\frac{5(\cancel{x - 3})(x - 2)}{\cancel{x - 3}} = \frac{10(x - 3)(\cancel{x - 2})}{\cancel{x - 2}}$$

$$5(x - 2) = 10(x - 3)$$

$$5x - 10 = 10x - 30$$

Complete the solution of $\frac{5}{x - 3} = \frac{10}{x - 2}$ by solving $5x - 10 = 10x - 30$. ________ . {4}

32 Complete the solution of the following equation:

$$\frac{3}{x - 7} = \frac{2}{x - 5}$$

$$(x - 7)(x - 5)\left(\frac{3}{x - 7}\right) = (x - 7)(x - 5)\left(\frac{2}{x - 5}\right)$$

$$\frac{3(x - 7)(x - 5)}{x - 7} = \frac{2(x - 7)(x - 5)}{x - 5}$$

$$\frac{3(\cancel{x - 7})(x - 5)}{\cancel{x - 7}} = \frac{2(x - 7)(\cancel{x - 5})}{\cancel{x - 5}}$$

$$3(x - 5) = 2(x - 7)$$

$$3x - 15 = 2x - 14$$

________ . {1}

33 Solve $\frac{4}{x + 1} = \frac{3}{x - 1}$ by multiplying both sides of the equation by the LCM of the denominators $(x + 1)(x - 1)$. ________ . {7}

34 Solve $\dfrac{10}{x + 3} = \dfrac{6}{x + 1}$ by multiplying both sides of the equation by the LCM of the denominators $(x + 3)(x + 1)$. ________.

{2}

35 Solve and check.

$\dfrac{4}{x + 3} = \dfrac{2}{x - 3}$. ________.

{9}

36 Solve and check.

$\dfrac{3}{x + 6} = \dfrac{5}{x + 10}$. ________.

{0}

37 Solve and check.

$\dfrac{5}{x - 3} = \dfrac{3}{x - 5}$. ________.

{8}

38 The truth set of $\dfrac{5}{x - 3} = \dfrac{3}{x - 5}$ is {8} because

$\dfrac{5}{8 - 3} = \dfrac{3}{8 - 5}$ is a true statement. Solve and check

$\dfrac{5}{x + 1} = \dfrac{8}{x - 2}$. ________.

$\{^{-}6\}$

$$\frac{5}{^{-}6 + 1} = \frac{8}{^{-}6 - 2}$$

$\dfrac{5}{^{-}5} = \dfrac{8}{^{-}8}$ is true

Self-Quiz # 1

The following questions test the objectives of the preceding section. 100% mastery is desired.

Solve and check the following equations:

1. $\dfrac{5}{x+1} + 3 = \dfrac{^{-}1}{x+1}$. ________.

2. $\dfrac{3}{x-5} - 4 = \dfrac{7}{x-5}$. ________.

3. $\dfrac{8}{x-2} = \dfrac{4}{x-3}$. ________.

4. $\dfrac{1}{x-3} + 5 = \dfrac{1}{x-3}$. ________.

5. $\dfrac{4}{x-8} = \dfrac{7}{x-14}$. ________.

39 Complete the solution of the following equation.

$$\frac{3}{x-2} + \frac{5}{(x-2)(x+3)} = \frac{7}{x+3}$$

$$\frac{3(x-2)(x+3)}{x-2} + \frac{5(x-2)(x+3)}{(x-2)(x+3)} = \frac{7(x-2)(x+3)}{x+3}$$

$$\frac{3(\cancel{x-2})(x+3)}{\cancel{x-2}} + \frac{5(\cancel{x-2})(\cancel{x+3})}{(\cancel{x-2})(\cancel{x+3})} = \frac{7(x-2)(\cancel{x+3})}{\cancel{x+3}}$$

$$3(x+3) + 5 = 7(x-2)$$

________. {7}

40 Complete the solution of the following equation by first multiplying both sides of the equation by the LCM of the denominators $(x + 1)(x - 7)$ and cancelling out like factors in each term.

$$\frac{4}{x + 1} + \frac{3}{(x - 7)(x + 1)} = \frac{5}{x - 7}$$

$$4(x - 7) + 3 = 5(x + 1)$$

______ . $\{^{-}30\}$

41 Complete the solution of the following equation in which each term on both sides of the equation has been multiplied by the LCM of the denominators.

$$\frac{3}{x - 1} - \frac{5}{(x + 4)(x - 1)} = \frac{8}{x + 4}$$

$$\frac{3(x - 1)(x + 4)}{x - 1} - \frac{5(x - 1)(x + 4)}{(x + 4)(x - 1)} = \frac{8(x - 1)(x + 4)}{x + 4}$$

$$3(x + 4) - 5 = 8(x - 1)$$

______ . $\{3\}$

42 Solve $\frac{4}{x + 4} - \frac{12}{(x + 2)(x + 4)} = \frac{6}{x + 2}$ by multiplying each fraction on both sides of the equation by the LCM of the denominators. ______ . $\{^{-}14\}$

43 Solve and check.

$\frac{8}{x + 2} + \frac{15}{(x - 1)(x + 2)} = \frac{7}{x - 1}$. ______ . $\{7\}$

44 Solve and check.

$\frac{5}{x + 4} + \frac{12}{(x - 2)(x + 4)} = \frac{2}{x - 2}$. ______ . $\{\ \}$ (because x cannot be replaced by 2)

45 To solve $\frac{5}{x-2} - \frac{8}{x^2+2x-8} = \frac{3}{x+4}$, first factor $x^2 + 2x - 8$ as follows:

$$\frac{5}{x-2} - \frac{8}{(x+4)(x-2)} = \frac{3}{x+4}$$

Complete the solution. ________ . $\{^-9\}$

46 Solve $\frac{8}{x+1} + \frac{1}{x^2-2x-3} = \frac{5}{x-3}$ by first factoring $x^2 - 2x - 3$ as follows:

$$\frac{8}{x+1} + \frac{1}{(x-3)(x+1)} = \frac{5}{x-3}$$

Complete the solution. ________ . $\left\{\frac{28}{3}\right\}$

47 Solve $\frac{3}{x+1} + \frac{1}{x^2+5x+4} = \frac{8}{x+4}$ by first factoring $x^2 + 5x + 4$. ________ . $\{1\}$

48 Solve and check.
$\frac{3}{x+1} - \frac{2}{x^2+7x+6} = \frac{4}{x+6}$. ________ . $\{12\}$

49 Solve and check.
$\frac{3}{x-5} + \frac{2}{x^2-25} = \frac{5}{x+5}$. ________ . $\{21\}$

50 The truth set of $\frac{3}{x-5} + \frac{2}{x^2-25} = \frac{5}{x+5}$ is $\{21\}$ because $\frac{3}{21-5} + \frac{2}{21^2-25} = \frac{5}{21+5}$ is a true statement. Solve and check.
$\frac{4}{x-5} - \frac{12}{x^2-7x+10} = \frac{5}{x-2}$. ________ . $\{\ \}$

51 Solve and check.
$\frac{9}{x+3} - \frac{2}{x^2-x-12} = \frac{4}{x-4}$. ________ . $\{10\}$

Self-Quiz # 2

The following questions test the objectives of the preceding section. 100% mastery is desired.

Solve and check the following equations:

1. $\dfrac{6}{x+1} + 2 = \dfrac{12}{x+1}$. ________ .

2. $\dfrac{3}{x-2} - 1 = \dfrac{^-2}{x-2}$. ________ .

3. $\dfrac{2}{x+1} + \dfrac{6}{(x+1)(x+2)} = \dfrac{4}{x+2}$. ________ .

4. $\dfrac{7}{x-2} - \dfrac{18}{(x-2)(x+4)} = \dfrac{3}{x+4}$. ________ .

5. $\dfrac{5}{x+1} + \dfrac{1}{x^2-1} = \dfrac{2}{x-1}$. ________ .

6. $\dfrac{4}{x-3} - \dfrac{15}{x^2+2x-15} = \dfrac{5}{x+5}$. ________ .

$5x - 3 = 8$ and $2x + 7 = 9$ are linear equations.

In the following section equations such as $x^2 - 5x + 6 = 0$ and $x^2 - 49 = 0$ will be solved.

$x^2 - 5x + 6 = 0$ and $x^2 - 49 = 0$ are quadratic equations.

---

52 What is the truth set of $3x = 0$? ________ . {0}

53 What is the truth set of $^-5x = 0$? ________ . {0}

54 If $5 \cdot$ ________ $= 0$, the blank must be filled by 0 to have a true statement. $^-3 \cdot$ ________ $= 0$. ________ . 0

55 ________ $\cdot\ 10 = 0$. 0

56 If $3x = 0$, then x must be replaced by ______ to obtain a true statement. 0

57 If $y \cdot 6 = 0$, then y must be replaced by ______ to obtain a true statement. 0

58 If $a \cdot b = 0$, and if a is replaced by any number except zero, b must be replaced by ______ to obtain a true statement. 0

59 If $x \cdot y = 0$, and if y is not replaced by zero, then what number must replace x to give a true statement? ______ . 0

60 If $w \cdot x = 0$, and if w is not replaced by zero, then what number must replace x to give a true statement? ______ . 0

61 The number zero has a very useful property in multiplication. Whenever two factors have a product of zero, then *one of the factors must be zero.* If $xy = 0$, what number must replace either x or y to obtain a true statement? ______ . 0

62 If $x \cdot z = 0$, then $x = 0$ or $z = 0$. If $w \cdot y = 0$, then $w = 0$ or ______ . $y = 0$

63 If $a \cdot b = 0$, then $a = 0$ or $b =$ ______ . 0

64 For the equation $x(x + 7) = 0$, the factors are x and $(x + 7)$. To have a product of 0, one of the factors must be ______ . 0

65 If $x(x - 4) = 0$, then either $x = 0$ or $(x - 4) = 0$. If $y(y + 5) = 0$, then either $y = 0$ or ______ . $y + 5 = 0$

66 If $x(x - 6) = 0$, then $x = 0$ or $x - 6 = 0$. If $y(y + 2) = 0$, then $y + 2 = 0$ or ______ . $y = 0$

67 For the equation $(x + 9)(x - 6) = 0$, the factors are $(x + 9)$ and $(x - 6)$. To have the product zero, one of the factors must be ________ .

0

68 If $(x - 3)(x - 5) = 0$, then either $(x - 3) = 0$ or $(x - 5) = 0$. If $(x + 8)(x + 4) = 0$, then either $(x + 8) = 0$ or ______________ .

$(x + 4) = 0$

69 If $w(w + 19) = 0$, then either $(w + 19) = 0$ or $w = 0$. If $x(x - 17) = 0$, then either $(x - 17) = 0$ or ________ .

$x = 0$

70 In the equation $(x - 2)(x + 5) = 0$, $(x - 2)$ and $(x + 5)$ are factors that have a product of zero. What number must be represented by either $(x - 2)$ or $(x + 5)$ to obtain a true statement for the equation $(x - 2)(x + 5) = 0$? ________ .

0

71 For the equation $(x + 5)(x - 2) = 0$, either $(x + 5) = 0$ or ______________ .

$(x - 2) = 0$

72 For the equation $(x + 17)(x - 3) = 0$, the factors are $(x + 17)$ and $(x - 3)$. Either $x + 17 = 0$ or ______________ .

$x - 3 = 0$

73 For the equation $x(x + 9) = 0$, the factors are x and $(x + 9)$. Either $x = 0$ or ______________ .

$x + 9 = 0$

74 To make $(x - 2)(x + 5) = 0$ a true statement, either $(x - 2)$ or $(x + 5)$ must be zero. What is the truth set of $x - 2 = 0$? ________ .

$\{2\}$

75 To make $(x - 2)(x + 5) = 0$ a true statement, either $(x - 2)$ or $(x + 5)$ must be zero. What is the truth set of $x + 5 = 0$? ________ .

$\{{}^{-}5\}$

76 {2} is the truth set of $x - 2 = 0$. Replacing x by 2 in $(x - 2)(x + 5) = 0$ gives $(2 - 2)(2 + 5) = 0$ or $0 \cdot 7 = 0$, which is a true statement. $\{^{-}5\}$ is the truth set of $x + 5 = 0$. Replace x by $^{-}5$ in $(x - 2)(x + 5) = 0$. Is $(^{-}5 - 2)(^{-}5 + 5) = 0$ a true statement? ________ .

Yes

77 {2} is the truth set of $x - 2 = 0$. $\{^{-}5\}$ is the truth set of $x + 5 = 0$. The truth set of $(x - 2)(x + 5) = 0$ is $\{2, ^{-}5\}$ because one of the factors, $(x - 2)$ or $(x + 5)$, must be ________ .

0

78 For the equation $(x - 4)(x - 3) = 0$, either $(x - 4)$ or $(x - 3)$ must be 0. What is the truth set of $x - 4 = 0$? ________ .

{4}

79 For the equation $(x - 4)(x - 3) = 0$, either $(x - 4)$ or $(x - 3)$ must be 0. What is the truth set of $x - 3 = 0$? ________ .

{3}

80 To solve $(x - 4)(x - 3) = 0$, at least one of the factors must be zero. Solve:

$$x - 4 = 0 \qquad x - 3 = 0$$

Indicate these solutions as the truth set for the equation $(x - 4)(x - 3) = 0$. ______________ .

{3, 4} or {4, 3}

81 {4} is the truth set of $x - 4 = 0$. {3} is the truth set of $x - 3 = 0$. {3, 4} is the truth set of $(x - 4)(x - 3) = 0$ because one of the factors, $(x - 4)$ or $(x - 3)$, must be ________ .

0

82 To find the truth set of $(x + 7)(x - 5) = 0$, either $(x + 7)$ or $(x - 5)$ must be zero. To find the truth set of $(x + 7)(x - 5) = 0$, solve:

$$x + 7 = 0 \qquad x - 5 = 0$$

________________ .

$\{^{-}7, 5\}$ or $\{5, ^{-}7\}$

83 $\{^-7\}$ is the truth set of $x + 7 = 0$. $\{5\}$ is the truth set of $x - 5 = 0$. $\{5, ^-7\}$ is the truth set of $(x + 7)(x - 5) = 0$ because one of the factors, $(x + 7)$ or $(x - 5)$, must be ________ . — 0

84 To solve $(x - 8)(x - 1) = 0$, either $x - 8 = 0$ or $x - 1 = 0$. Solve:

$$x - 8 = 0 \qquad x - 1 = 0$$

________________ . — $\{8, 1\}$ or $\{1, 8\}$

85 Find the truth set of $(x - 3)(x - 5) = 0$ by writing $x - 3 = 0$ or $x - 5 = 0$, and solving both.

$$x - 3 = 0 \qquad x - 5 = 0$$

Complete the solution. ________ . — $\{3, 5\}$
[*Note*: The order in which the elements are listed in the truth set makes no difference.]

86 Find the truth set of $x(x + 7) = 0$ by writing $x = 0$ or $x + 7 = 0$ and solving both. ________ . — $\{0, ^-7\}$
[*Note*: The truth set of $x = 0$ is $\{0\}$.]

87 Find the truth set of $(x + 6)(x - 5) = 0$ by writing $x + 6 = 0$ and $x - 5 = 0$ and solving both. ________ . — $\{^-6, 5\}$

88 Find the truth set of $(2x + 3)(x - 4) = 0$ by solving both $2x + 3 = 0$ and $x - 4 = 0$. ________ . — $\left\{\frac{^-3}{2}, 4\right\}$

89 Find the truth set of $3x(4x - 5) = 0$ by writing $3x = 0$ and $4x - 5 = 0$ and solving both. ________ . — $\left\{0, \frac{5}{4}\right\}$

90 For the equation $(x - 6)(x - 6) = 0$, both factors are $(x - 6)$. The truth set of $(x - 6)(x - 6) = 0$ is $\{6\}$. It is not necessary to show 6 twice in the truth set. What is the truth set of $(x - 8)(x - 8) = 0$? ________ . — $\{8\}$

91 Find the truth set of $(x + 9)(x + 9) = 0$ by solving $x + 9 = 0$. ________ . — $\{^-9\}$

92 Find the truth set of $(x - 5)(x + 3) = 0$ by writing $x - 5 = 0$ and $x + 3 = 0$ and solving both. ______. $\{5, {}^{-}3\}$

The number zero has an important multiplication property that makes it possible to solve equations such as $x^2 - 7x - 18 = 0$.

If the product of two factors is zero, then at least one of the factors is zero.

$$x^2 - 7x - 18 = 0$$
$$(x - 9)(x + 2) = 0$$

If the factor $(x - 9)$ is zero, then $x = 9$.
If the factor $(x + 2)$ is zero, then $x = {}^{-}2$.

---

93 Solve ${}^{-}7x(x + 4) = 0$. ______. $\{0, {}^{-}4\}$

94 Solve $(x + 4)(x - 2) = 0$. ______. $\{{}^{-}4, 2\}$

95 Solve $(x + 4)(x + 4) = 0$. ______. $\{{}^{-}4\}$

96 Solve $(x + 11)(x - 1) = 0$. ______. $\{{}^{-}11, 1\}$

97 Solve $x(x - 8) = 0$. ______. $\{0, 8\}$

98 Solve $(2x - 5)(2x - 5) = 0$. ______. $\left\{\frac{5}{2}\right\}$

99 Solve $(2x + 5)(x + 4) = 0$. ______. $\left\{\frac{{}^{-}5}{2}, {}^{-}4\right\}$

100 Solve $x(x + 9) = 0$. ______. $\{0, {}^{-}9\}$

101 Solve $(x - 7)(x - 7) = 0$. ______. $\{7\}$

Self-Quiz # 3

The following questions test the objectives of the preceding section. 100% mastery is desired.

Solve:

1. $(x - 5)(x + 2) = 0$. ________.

2. $(x - 3)(x - 4) = 0$. ________.

3. $x(x + 7) = 0$. ________.

4. $(x + 6)(x + 1) = 0$. ________.

5. $(x - 4)(x - 4) = 0$. ________.

6. $5x(x - 8) = 0$. ________.

7. $(2x - 3)(x + 9) = 0$. ________.

8. $(x + 5)(x - 6) = 0$. ________.

9. $(2x + 5)(2x + 5) = 0$. ________.

10. $^-5x(3x + 2) = 0$. ________.

Three methods for factoring polynomials have been explained in an earlier chapter in the text. They are as follows:

1. The common factor method.

$$14x^2 - 21x = 7x(2x - 3)$$

2. The method for factoring a trinomial when the coefficient of x^2 is 1.

$$x^2 + 11x + 24 = (x + 8)(x + 3)$$

3. The method for factoring a trinomial when the coefficient of x^2 is *not* 1.

$$6x^2 + 5x - 4$$

(8 and $^-3$ have a product of $^-24$ and a sum of 5)

$$6x^2 + 8x - 3x - 4$$

$$2x(3x + 4) - 1(3x + 4)$$

$$(3x + 4)(2x - 1)$$

---

102 Each term of $15x^2 - 6x$ has $3x$ as a factor. To factor $15x^2 - 6x$, the common factor method is used. $15x^2 - 6x = 3x(5x - 2)$. Factor $20x^2 - 5x$.

______________ . $5x(4x - 1)$

103 To factor $x^2 + 6x - 7$, first note that the coefficient of x^2 is 1. Consequently, it is necessary to find factors of $^{-}7$ that have a sum of 6.

$$x^2 + 6x - 7 = (x + 7)(x - 1)$$

Factor $x^2 - 8x + 7$. ______________ . $(x - 7)(x - 1)$

104 To factor $5x^2 - 11x + 2$, first note that the coefficient of x^2 is *not* 1. Consequently, it is necessary to find factors of $5 \cdot 2$ or 10 that have a sum of $^{-}11$.

$$5x^2 - 11x + 2$$
$$5x^2 - 10x - 1x + 2$$
$$5x(x - 2) - 1(x - 2)$$
$$(x - 2)(5x - 1)$$

Factor $4x^2 - 8x + 3$. ______________ . $(2x - 3)(2x - 1)$

105 To solve $x^2 + 5x - 6 = 0$, first factor $x^2 + 5x - 6$.

$$x^2 + 5x - 6 = 0$$
$$(x + 6)(x - 1) = 0$$

Complete the solution of $x^2 + 5x - 6 = 0$. ________ . $\{1, {}^{-}6\}$

106 Solve $x^2 + 8x + 12 = 0$ by first factoring $x^2 + 8x + 12$. ________ . $\{{}^{-}6, {}^{-}2\}$

107 Solve $x^2 - 7x + 12 = 0$ by factoring. ________ . $\{3, 4\}$

108 Solve $x^2 + 11x - 12 = 0$. ________ . $\{1, {}^{-}12\}$

109 Solve $x^2 - 8x + 7 = 0$. ________ . $\{1, 7\}$

110 Solve $x^2 + 7x + 10 = 0$. ________ . $\{{}^{-}2, {}^{-}5\}$

111 Solve $x^2 - 4x - 21 = 0$. ________ . $\{^-3, 7\}$

112 To solve $x^2 - 49 = 0$, first factor $x^2 - 49$.

$$x^2 - 49 = 0$$
$$(x - 7)(x + 7) = 0$$

Complete the solution of $x^2 - 49 = 0$. ________ . $\{7, ^-7\}$

113 Solve $x^2 - 9 = 0$ by factoring $x^2 - 9$. ________ . $\{3, ^-3\}$

114 Solve $x^2 - 64 = 0$. ________ . $\{8, ^-8\}$

115 Solve $4x^2 - 9 = 0$. ________ . $\left\{\frac{3}{2}, \frac{^-3}{2}\right\}$

[*Note:* $4x^2 - 9 = 4x^2 + 0x - 9$.]

116 To solve $x^2 - 12x + 36 = 0$, first factor $x^2 - 12x + 36$.

$$x^2 - 12x + 36 = 0$$
$$(x - 6)(x - 6) = 0$$

Complete the solution of $x^2 - 12x + 36 = 0$.
________ . $\{6\}$

117 Solve $x^2 - 6x + 9 = 0$ by first factoring $x^2 - 6x + 9$.
________ . $\{3\}$

118 Solve $x^2 + 4x + 4 = 0$. ________ . $\{^-2\}$

119 To solve $x^2 - 4x = 0$, first factor $x^2 - 4x$.

$$x^2 - 4x = 0$$
$$x(x - 4) = 0$$

Complete the solution of $x^2 - 4x = 0$. ________ . $\{0, 4\}$

120 Solve $6x^2 - 12x = 0$ by first factoring $6x^2 - 12x$.
________ . $\{0, 2\}$

121 Solve $5x^2 - 20x = 0$. ________ . $\{0, 4\}$

122 Solve $x^2 + 8x + 15 = 0$. ________. $\{^-3, ^-5\}$

123 Solve $x^2 - 2x - 3 = 0$. ________. $\{3, ^-1\}$

124 Solve $x^2 + 20x + 100 = 0$. ________. $\{^-10\}$

125 Solve $x^2 - 81 = 0$. ________. $\{9, ^-9\}$

126 Solve $x^2 + 5x = 0$. ________. $\{0, ^-5\}$

127 The factors of $2x^2 - 11x + 15$ are $(2x - 5)(x - 3)$. Solve $2x^2 - 11x + 15 = 0$. ________. $\left\{\frac{5}{2}, 3\right\}$

128 Solve $3x^2 - 11x + 6 = 0$. ________. $\left\{\frac{2}{3}, 3\right\}$

[*Note*: $3x^2 - 11x + 6 = 3x^2 - 9x - 2x + 6$.]

129 Solve $2x^2 - 7x + 6 = 0$. ________. $\left\{\frac{3}{2}, 2\right\}$

To check $\left\{\frac{3}{2}, {}^{-}5\right\}$ as the truth set of $2x^2 + 7x - 15 = 0$, the values $\frac{3}{2}$ and ${}^{-}5$ are separately used to replace x in the equation.

$$2x^2 + 7x - 15 = 0 \quad \text{becomes}$$

$$2\left(\frac{3}{2}\right)^2 + 7\left(\frac{3}{2}\right) - 15 = 0 \quad \text{when } x = \frac{3}{2}$$

$$2\left(\frac{9}{4}\right) + 7\left(\frac{3}{2}\right) - 15 = 0$$

$$\frac{9}{2} + \frac{21}{2} - 15 = 0$$

$$\frac{30}{2} - 15 = 0$$

$$15 - 15 = 0 \quad \text{is true.}$$

$$2x^2 + 7x - 15 = 0 \quad \text{becomes}$$

$$2({}^{-}5)^2 + 7({}^{-}5) - 15 = 0 \quad \text{when } x = {}^{-}5$$

$$2 \cdot 25 + 7 \cdot {}^{-}5 - 15 = 0$$

$$50 + {}^{-}35 - 15 = 0$$

$$15 - 15 = 0 \quad \text{is true.}$$

---

130 Solve and check $x^2 - 7x + 10 = 0$. ________ . $\{2, 5\}$

131 Solve and check $x^2 - 4 = 0$. ________ . $\{{}^{-}2, 2\}$

132 Solve and check $x^2 - 12x = 0$. ________ . $\{0, 12\}$

133 Solve and check $x^2 - 8x + 16 = 0$. ________ . $\{4\}$

134 Solve and check $2x^2 - 3x - 2 = 0$. ________ . $\left\{2, \frac{{}^{-}1}{2}\right\}$

Self-Quiz # 4

The following questions test the objectives of the preceding section. 100% mastery is desired.

Solve and check the following equations:

1. $x^2 + 13x + 30 = 0$. ________ .

2. $x^2 - 36 = 0$. ________ .

3. $x^2 - 5x - 24 = 0$. ________ .

4. $x^2 + 7x = 0$. ________ .

5. $x^2 - 7x + 12 = 0$. ________ .

6. $3x^2 + 18x = 0$. ________ .

7. $x^2 - 25 = 0$. ________ .

8. $2x^2 - 11x + 5 = 0$. ________ .

The solution of quadratic equations such as $x^2 - 9x + 18 = 0$ and $x^2 - 25 = 0$ is dependent upon having a product of zero for the factors of the polynomials.

The number zero has the useful property that whenever two factors have a product of zero, then at least one of the factors must be zero.

---

135 To solve $x^2 + 4x = 12$, first add $^{-}12$ to both sides of the equation so that the right side of the equation will be zero:

$$x^2 + 4x = 12$$

$$x^2 + 4x - 12 = 12 - 12$$

$$x^2 + 4x - 12 = 0$$

Complete the solution of $x^2 + 4x = 12$. ________ . $\{^{-}6, 2\}$

136 To solve $x^2 + 10 = 7x$, first add $^{-}7x$ to both sides of the equation so that the right side of the equation will be zero:

$$x^2 + 10 = 7x$$
$$x^2 + 10 - 7x = 7x - 7x$$
$$x^2 - 7x + 10 = 0$$

Complete the solution of $x^2 + 10 = 7x$. ________ .

$\{2, 5\}$

137 Solve $x^2 + 6x = 16$ by first adding $^{-}16$ to both sides of the equation. ________ .

$\{2, {}^{-}8\}$

138 Solve $x^2 - 15 = 2x$ by first adding $^{-}2x$ to both sides of the equation. ________ .

$\{{}^{-}3, 5\}$

139 Solve $x^2 + 5x = {}^{-}6$ by first adding 6 to both sides of the equation. ________ .

$\{{}^{-}2, {}^{-}3\}$

140 Solve $x^2 + 2 = {}^{-}3x$. ________ .

$\{{}^{-}1, {}^{-}2\}$

141 Solve $x^2 + 10x = {}^{-}25$. ________ .

$\{{}^{-}5\}$

142 Solve $x^2 + 10x = 11$. ________ .

$\{1, {}^{-}11\}$

143 Solve $x^2 + 49 = 14x$. ________ .

$\{7\}$

144 Solve $x^2 - 3 = 2x$. ________ .

$\{{}^{-}1, 3\}$

145 Solve and check $x^2 - 7x = 8$. ________ .

$\{{}^{-}1, 8\}$

$$x^2 - 7x = 8$$
$$({}^{-}1)^2 - 7({}^{-}1) = 8$$
$$1 + 7 = 8$$

is true.

$$x^2 - 7x = 8$$
$$(8)^2 - 7(8) = 8$$
$$64 - 56 = 8$$

is true.

146 Solve $x^2 - 9x = 22$. ________ .

$\{{}^{-}2, 11\}$

147 Solve $x^2 - 6x = 27$. ________ .

$\{^-3, 9\}$

148 Solve and check $x^2 + 81 = 18x$. ________________ .

$\{9\}$

$(9)^2 + 81 = 18(9)$

$81 + 81 = 162$

is true.

149 Solve $x^2 = 2x + 24$. ________ .

$\{6, ^-4\}$

150 Solve $5x - 6 = {}^-x^2$. ________ .

$\{^-6, 1\}$

151 Solve and check $3x = 4 - x^2$. ________ .

$\{^-4, 1\}$

$3(^-4) = 4 - (^-4)^2$

$^-12 = 4 - 16$

is true.

$3(1) = 4 - (1)^2$

$3 = 4 - 1$

is true.

152 To solve $2x^2 - 10x - 48 = 0$, factor as follows:

$$2x^2 - 10x - 48 = 0$$
$$2(x^2 - 5x - 24) = 0$$
$$2(x - 8)(x + 3) = 0$$

One of these factors

$$2, (x - 8), (x + 3)$$

must be zero because the product is zero. Which factor cannot be zero? ________ .

2

153 To solve $2x^2 - 10x - 48 = 0$, factor as follows:

$$2x^2 - 10x - 48 = 0$$
$$2(x^2 - 5x - 24) = 0$$
$$2(x - 8)(x + 3) = 0$$

The factor 2 cannot be zero; therefore, both $x - 8 = 0$ and $x + 3 = 0$ must be solved to obtain the solution of $2x^2 - 10x - 48 = 0$. What is the truth set of $2x^2 - 10x - 48 = 0$? ________ .

$\{8, ^-3\}$

154 To solve $4x^2 - 16x + 12 = 0$, factor as follows:

$$4x^2 - 16x + 12 = 0$$
$$4(x^2 - 4x + 3) = 0$$
$$4(x - 3)(x - 1) = 0$$

One of the three factors must be zero, because the product is zero. Which factor

$$4, (x - 3), \text{ or } (x - 1)$$

cannot be zero? ________ .

4

155 To solve $4x^2 - 16x + 12 = 0$, factor as follows:

$$4x^2 - 16x + 12 = 0$$
$$4(x^2 - 4x + 3) = 0$$
$$4(x - 3)(x - 1) = 0$$

The factor 4 cannot be zero; therefore only $x - 3 = 0$ and $x - 1 = 0$ must be solved to obtain the solution for $4x^2 - 16x + 12 = 0$. What is the truth set of $4x^2 - 16x + 12 = 0$? ________ .

$\{1, 3\}$

156 $^{-}x^2 + 7x - 12$ can be factored as follows:

$$^{-}x^2 + 7x - 12$$
$$^{-}1(x^2 - 7x + 12)$$

This factoring makes the solution of $^{-}x^2 + 7x - 12 = 0$ much easier. Solve $^{-}x^2 + 7x - 12 = 0$. ________ .

$\{3, 4\}$

157 To solve $^{-}x^2 + 3x + 28 = 0$, first factor as follows:

$$^{-}x^2 + 3x + 28 = 0$$
$$^{-}1(x^2 - 3x - 28) = 0$$
$$^{-}1(x - 7)(x + 4) = 0$$

One of the three factors must be zero because the product is zero. Which factor

$$^{-}1, (x - 7), \text{ or } (x + 4)$$

cannot be zero? ________ .

$^{-}1$

158 To solve $^{-}x^2 + 3x + 28 = 0$, first factor as

$$^{-}1(x^2 - 3x - 28) = 0$$

$$^{-}1(x - 7)(x + 4) = 0$$

The factor $^{-}1$ cannot be zero; therefore, only $x - 7 = 0$ and $x + 4 = 0$ must be solved to obtain the solution for $^{-}x^2 + 3x + 28 = 0$. Find the truth set of $^{-}x^2 + 3x + 28 = 0$. ________ .

$\{7, {^{-}4}\}$

159 Solve $2x^2 + 12x + 16 = 0$. ________ .
[*Note*: First find the common factor.]

$\{{^{-}2}, {^{-}4}\}$

160 Solve $3x^2 - 15x - 18 = 0$. ________ .
[*Note*: First find the common factor.]

$\{{^{-}1}, 6\}$

161 Solve $^{-}x^2 + 7x - 12 = 0$. ________ .
[*Note*: First remove factor of $^{-}1$ to make the factoring of the trinomial expression easier.]

$\{3, 4\}$

162 Solve $5x^2 + 20x + 20 = 0$. ________ .

$\{{^{-}2}\}$

163 Solve and check $^{-}x^2 + 7x - 10 = 0$. ________ .

$\{2, 5\}$

$$^{-}x^2 + 7x - 10 = 0$$
$$^{-}(2)^2 + 7(2) - 10 = 0$$
$$^{-}4 + 14 - 10 = 0$$
$$10 - 10 = 0$$

is true.

$$^{-}x^2 + 7x - 10 = 0$$
$$^{-}(5)^2 + 7(5) - 10 = 0$$
$$^{-}25 + 35 - 10 = 0$$
$$10 - 10 = 0$$

is true.

164 Solve $4x^2 - 12x - 40 = 0$. ________ .

$\{{^{-}2}, 5\}$

165 Solve $^{-}x^2 + 36 = 0$. ________ .

$\{{^{-}6}, 6\}$

166 Solve and check $2x + 3 = x^2$. ________________.
[*Note*: Add $^{-}x^2$ to both sides of the equation.]

$\{3, {}^{-}1\}$
$2(3) + 3 = (3)^2$
$6 + 3 = 9$
is true.
$2({}^{-}1) + 3 = ({}^{-}1)^2$
${}^{-}2 + 3 = 1$
is true.

167 Solve $3x + 28 = x^2$. ________.

$\{{}^{-}4, 7\}$

168 Solve $2x^2 + 14x = 16$. ________.

$\{1, {}^{-}8\}$

169 Solve $x^2 - 12x + 36 = 0$. ________.

$\{6\}$

170 Solve $x^2 - 5x = 14$. ________.

$\{7, {}^{-}2\}$

171 Solve and check $2x^2 - 3x = 0$.
________________.

$\left\{0, \frac{3}{2}\right\}$
$2(0)^2 - 3(0) = 0$
$0 - 0 = 0$
is true.
$2\left(\frac{3}{2}\right)^2 - 3\left(\frac{3}{2}\right) = 0$
$2\left(\frac{9}{4}\right) - \frac{9}{2} = 0$
$\frac{9}{2} - \frac{9}{2} = 0$
is true.

172 Solve $2x^2 + 13x - 7 = 0$. ________.

$\left\{\frac{1}{2}, {}^{-}7\right\}$

173 Solve $x^2 - 9x - 22 = 0$. ________.

$\{{}^{-}2, 11\}$

174 Solve $x^2 + 6 = 7x$. ________.

$\{1, 6\}$

175 Solve $5x^2 - 20 = 0$. ________.

$\{2, {}^{-}2\}$

176 Solve $2x^2 + 7x - 4 = 0$. ________.

$\left\{{}^{-}4, \frac{1}{2}\right\}$

177 Solve and check $3x^2 - 6x + 3 = 0$. ________ .

$\{1\}$

178 Solve $x^2 - 100 = 0$. ________ .

$\{10, {}^-10\}$

179 Solve and check $2x^2 - 7x + 3 = 0$. ________ .

$\left\{\frac{1}{2}, 3\right\}$

$$2\left(\frac{1}{2}\right)^2 - 7\left(\frac{1}{2}\right) + 3 = 0$$
$$2\left(\frac{1}{4}\right) - \frac{7}{2} + 3 = 0$$
$$\frac{1}{2} - \frac{7}{2} + 3 = 0$$

is true.

$$2(3)^2 - 7(3) + 3 = 0$$
$$2(9) - 21 + 3 = 0$$
$$18 - 21 + 3 = 0$$

is true.

180 Solve $x^2 - 8x + 12 = 0$. ________ .

$\{2, 6\}$

181 Solve $2x^2 + 8x - 42 = 0$. ________ .

$\{{}^-7, 3\}$

182 Solve ${}^-x^2 - 4x + 45 = 0$. ________ .

$\{{}^-9, 5\}$

183 Solve $x^2 - 2x - 3 = 0$. ________ .

$\{{}^-1, 3\}$

184 Solve $x^2 - 49 = 0$. ________ .

$\{{}^-7, 7\}$

185 Solve $x^2 + 14x + 49 = 0$. ________ .

$\{{}^-7\}$

186 Solve $6x^2 - 7x - 3 = 0$. ________ .

$\left\{\frac{{}^-1}{3}, \frac{3}{2}\right\}$

187 Solve $4x^2 - 100 = 0$. ________ .

$\{{}^-5, 5\}$

188 Solve and check $x^2 + 3x - 28 = 0$. ________ .

$\{4, ^-7\}$

$(4)^2 + 3(4) - 28 = 0$

$16 + 12 - 28 = 0$

is true.

$(^-7)^2 + 3(^-7) - 28 = 0$

$49 - 21 - 28 = 0$

is true.

Self-Quiz # 5

The following questions test the objectives of the preceding section. 100% mastery is desired.

Solve and check the following equations:

1. $x^2 + 2x - 15 = 0$. ________ .
2. $x^2 - 7x + 6 = 0$. ________ .
3. $x^2 - 81 = 0$. ________ .
4. $3x^2 - 27 = 0$. ________ .
5. $x^2 - 4x + 4 = 0$. ________ .
6. $3x^2 - 7x + 2 = 0$. ________ .
7. $5x^2 - 3x = 0$. ________ .
8. $3x^2 - 3x = 36$. ________ .
9. $^-x^2 - 4x + 5 = 0$. ________ .
10. $x = 20 - x^2$. ________ .

Chapter Summary

In this chapter methods of solving equations have been studied.

Equations of the form $ax + b = c$ are linear equations. Methods have been shown for converting equations such as

$$\frac{{}^{-}3}{x + 5} + 4 = \frac{7}{x + 5} \quad \text{or} \quad \frac{3}{x - 2} + \frac{2}{x^2 - 5x + 6} = \frac{5}{x - 3}$$

to linear equations of the form $ax + b = c$.

The last section of this chapter was concerned with quadratic equations such as $x^2 - 8x + 12 = 0$, $x^2 - 7x = 0$, and $2x^2 - 3x = 2$. These equations were solved by factoring the polynomials.

$\{0, 7\}$ is checked as the truth set of $x^2 - 7x = 0$ by separately replacing x by 0 and 7 in the equation and showing that a true statement is obtained in both cases.

Equations such as $x^2 + 5x + 2 = 0$ were not solved in this chapter because $x^2 + 5x + 2$ is not factorable. In Chapter 9 methods for solving equations such as $x^2 + 5x + 2 = 0$ will be studied.

CHAPTER POST-TEST

The following questions test the objectives of this chapter. A score of 90% indicates sufficient mastery, and the student may proceed to the next chapter.

Using the set of rational numbers, find the truth set for each of the following equations:

1. $\dfrac{4}{x-2} + 5 = \dfrac{1}{x-2}$

2. $\dfrac{9}{x-6} = \dfrac{1}{x+1}$

3. $\dfrac{6}{x-6} - \dfrac{5}{x^2-7x+6} = \dfrac{2}{x-1}$

4. $\dfrac{5}{x^2-9} + \dfrac{3}{x-3} = \dfrac{2}{x+3}$

5. $\dfrac{2}{x-4} + \dfrac{5}{x-3} = \dfrac{4}{x^2-7x+12}$

6. $x^2 + 5x - 14 = 0$

7. $x^2 - 17x = 0$

8. $x^2 - 100 = 0$

9. $x^2 + 15x + 56 = 0$

10. $x^2 - 10x + 21 = 0$

11. $2x^2 - 8 = 0$

12. $^-x^2 - 3x + 10 = 0$

13. $x^2 - 9x + 14 = 0$

14. $x^2 + 35 = 12x$

15. $3x^2 - 10x + 8 = 0$

Rational and Irrational Numbers

1
2
3
4
5
6
7
8
9
10

CHAPTER PRE-TEST

The following questions indicate the objectives of this chapter. A score of 90% indicates sufficient mastery, and the student may immediately take the Chapter Post-Test.

I Which of the following are perfect square integers?
(a) 36 (b) $^{-}25$ (c) 27 (d) 0 (e) 100

II Which of the following are perfect squares?
(a) $\frac{25}{64}$ (b) $\frac{^{-}9}{25}$ (c) $\frac{12}{75}$ (d) $\frac{63}{7}$ (e) $\frac{23}{49}$

III Which of the following are rational numbers?

(a) $\sqrt{\frac{16}{49}}$ (b) $\sqrt{\frac{3}{25}}$ (c) $^{-}\sqrt{81}$ (d) $\sqrt{55}$ (e) $^{-}\sqrt{\frac{4}{9}}$

IV Simplify:

1. $3\sqrt{50}$

2. $^{-}7\sqrt{27}$

3. $^{-}\sqrt{64}$

4. $\frac{\sqrt{50}}{\sqrt{10}}$

5. $\frac{7}{\sqrt{11}}$

6. $\frac{\sqrt{35}}{\sqrt{7}}$

7. $\frac{3\sqrt{7}}{5\sqrt{21}}$

8. $\sqrt{\frac{2}{5}}$

9. $12 - 3\sqrt{13} + 4 + \sqrt{13}$

10. $3\sqrt{5} - 7\sqrt{20}$

11. $\sqrt{63} - 5\sqrt{7}$

12. $4\sqrt{27} + 3\sqrt{12}$

13. $3\sqrt{32} - 5\sqrt{8}$

14. $5\sqrt{7} + 3\sqrt{28}$

15. $\sqrt{72} - 2\sqrt{50}$

The quadratic equation $x^2+7x-2=0$ has no rational-number solutions. In this chapter another set of numbers is studied, so that solutions for equations such as $x^2+7x-2=0$ can be found.

1 When an integer is multiplied by itself, the product is called a *perfect square integer.* $2 \cdot 2 = 4$. 4 is a *perfect square integer.* $7 \cdot 7 = 49$. 49 is a ________ ________ integer.

perfect square

2 25 is a perfect square integer because $^{-}5 \cdot {}^{-}5 = 25$. $^{-}7 \cdot {}^{-}7 = 49$. Is 49 a perfect square integer? ________ .

Yes

3 64 is a perfect square integer because $^{-}8 \cdot {}^{-}8 = 64$. Is 81 a perfect square integer? ________ .

Yes

4 9 is a perfect square integer because $3 \cdot 3 = 9$. Is 16 a perfect square integer? ________ .

Yes

5 A positive product is obtained by multiplying any *positive* integer by itself. $8 \cdot 8 = 64$. A positive product is obtained by multiplying any *negative* integer by itself. $^{-}4 \cdot {}^{-}4 =$ ________ .

16

6 A positive integer multiplied by itself gives a positive product. $10 \cdot 10 = 100$. 100 is a perfect square integer. $12 \cdot 12 = 144$. Is 144 a perfect square integer? ________ .

Yes

7 A negative number multiplied by itself gives a positive product. $^{-}6 \cdot {}^{-}6 = 36$. 36 is a perfect square integer. $^{-}5 \cdot {}^{-}5 = 25$. Is 25 a perfect square integer? ________ .

Yes

8 $1 \cdot 1 = 1$. Is 1 a perfect square integer? ________ .

Yes

9 $^{-}9 \cdot {}^{-}9 = 81$. Is 81 a perfect square integer? ________ . — Yes

10 There is no integer that can be multiplied by itself to give 2. Is there any integer that can be multiplied by itself to give a product of 5? ________ . — No

11 2 is *not* a perfect square integer because no integer can be multiplied by itself to give a product of 2. Is 3 a perfect square integer? ________ . — No

12 Is 15 a perfect square integer? ________ . — No

13 25 is a perfect square integer because $5 \cdot 5 = 25$. Is 7 a perfect square integer? ________ . — No

14 Is 64 a perfect square integer? ________ . — Yes
[*Note:* $8 \cdot 8 = 64$.]

15 Is 10 a perfect square integer? ________ . — No

16 Is 9 a perfect square integer? ________ . — Yes

17 $3 \cdot 3 = 9$. (9 is positive.) $^{-}5 \cdot {}^{-}5 = 25$. (25 is positive.) A perfect square integer *cannot* be a negative integer because any number multiplied by itself *cannot* be negative. Is $^{-}9$ a perfect square integer? ________ . — No

18 Is $^{-}25$ a perfect square integer? ________ . — No
[*Note:* $5 \cdot 5 = 25$ and $^{-}5 \cdot {}^{-}5 = 25$.]

19 Is 1 a perfect square integer? ________ . — Yes

20 Is 36 a perfect square integer? ________ . — Yes

21 0 is a perfect square integer because $0 \cdot 0 = 0$. Is 49 a perfect square integer? ________ . — Yes

22 Is 17 a perfect square integer? ________ . No

23 Is 0 a perfect square integer? ________________ . Yes $(0 \cdot 0 = 0)$

24 Is $^{-}49$ a perfect square integer? ________ . No

25 Is 100 a perfect square integer? ________ . Yes

26 Which of the following are *not* perfect square integers? ________ . (c), (e)

(a) 64.
(b) 0.
(c) 17.
(d) 36.
(e) $^{-}9$.

27 Which of the following are *not* perfect square integers? ________ . (b), (e)

(a) 81.
(b) $^{-}36$.
(c) 1.
(d) 49.
(e) 21.

28 64 is the only perfect square integer between the integers 60 and 70. What is the only perfect square integer between 40 and 50? ________ . 49

29 What is the perfect square integer between the integers 5 and 12? ________ . 9

30 What is the perfect square integer between the integers 75 and 91? ________ . 81

31 What is the perfect square integer between the integers 2 and 7? ________ . 4

32 What is the perfect square integer between the integers 19 and 33? ________ . 25

33 What is the perfect square integer between the integers 90 and 106? ________ . 100

34 What is the perfect square integer between the integers 10 and 20? ________ . 16

35 0, 1, 4, 9, 16, 25, 36, 49, 64, 81, and 100 are perfect square integers. Is 73 a perfect square integer? ________ . No

36 Is 56 a perfect square integer? ________ . No

37 Is 36 a perfect square integer? ________ . Yes

38 {25, 64} is the set of perfect square integers obtained by multiplying each element of $\{^-5, 8\}$ by itself.

$$^-5 \cdot {^-5} = 25$$

$$8 \cdot 8 = 64$$

What is the set of perfect square integers obtained by multiplying each element of $\{^-1, 7\}$ by itself? ________ . {1, 49}

39 {16, 36, 49} is the set of perfect square integers obtained by multiplying each element of $\{^-4, 6, {^-7}\}$ by itself. What is the set of perfect square integers obtained by multiplying each element of $\{1, {^-8}, 9\}$ by itself? ________________ . {1, 64, 81}

40 What is the set of perfect square integers obtained by multiplying each element of $\{2, {^-6}, 9, 10\}$ by itself? ________________ . {4, 36, 81, 100}

41 Write the set of perfect square integers obtained by multiplying each element of $\{^-2, 7, {^-9}\}$ by itself. ________________ . {4, 49, 81}

42 Write the set of perfect square integers obtained by multiplying each element of $\{4, {^-10}, {^-17}\}$ by itself. ________________ . {16, 100, 289}

43 Write the set of perfect square integers obtained by multiplying each element of $\{^-3, 5, 24\}$ by itself. ________________.

$\{9, 25, 576\}$

44 Write the set of perfect square integers obtained by multiplying each element of $\{^-9, 15, ^-25\}$ by itself. ________________.

$\{81, 225, 625\}$

45 Write the set of perfect square integers obtained by multiplying each element of $\{^-1, 18, 0, ^-7\}$ by itself. ________________.

$\{1, 324, 0, 49\}$

Self-Quiz # 1

The following questions test the objectives of the preceding section. 100% mastery is desired.

1. Which of the following are perfect square integers? ________ .
(a) 16
(b) 24
(c) $^{-}1$
(d) 64
(e) 35

2. Which of the following are perfect square integers? ________ .
(a) $^{-}49$
(b) 0
(c) 55
(d) 1
(e) 94

3. Which of the following are perfect square integers? ________ .
(a) 100
(b) 18
(c) 56
(d) $^{-}81$
(e) 36

4. Write the set of perfect square integers obtained by multiplying each element of the set $\{^{-}7, 1, 14, 5\}$ by itself. ________ .

5. Write the set of perfect square integers obtained by multiplying each element of the set $\{^{-}3, ^{-}18, 12, 0\}$ by itself. ________ .

46 $\frac{49}{16}$ is a *perfect square* because

$$\frac{7}{4} \cdot \frac{7}{4} = \frac{49}{16}$$

Whenever a rational number is the product of a number multiplied by itself, the product is a *perfect square*. Is $\frac{25}{64}$ a *perfect square*? ________________ .

Yes, $\frac{5}{8} \cdot \frac{5}{8} = \frac{25}{64}$

47 $\frac{4}{9}$ is a perfect square because

$$\frac{2}{3} \cdot \frac{2}{3} = \frac{4}{9}$$

Is $\frac{81}{16}$ a perfect square? ______________ .

Yes, $\frac{9}{4} \cdot \frac{9}{4} = \frac{81}{16}$

48 Is $\frac{36}{25}$ a perfect square? ________ .

Yes

49 Is $\frac{19}{14}$ a perfect square? ________ .

No

50 Is $\frac{16}{25}$ a perfect square? ________ .

Yes

51 Is $\frac{5}{49}$ a perfect square? ________ .

No

52 Is $\frac{10}{9}$ a perfect square? ________ .

No

53 Is $\frac{49}{100}$ a perfect square? ________ .

Yes

54 A rational number is a perfect square whenever both the numerator and denominator are perfect square integers. Is $\frac{36}{49}$ a perfect square? ________ .

Yes

55 A rational number is a perfect square whenever both the numerator and denominator are perfect square integers. Is $\frac{7}{15}$ a perfect square? ________ .

No

56 Is $\frac{^{-}9}{16}$ a perfect square? ________ . No

[*Note: A negative* integer is *not* a perfect square integer.]

57 Is $\frac{6}{25}$ a perfect square? ________ . No

58 Is $\frac{^{-}1}{25}$ a perfect square? ________ . No

59 Is $\frac{1}{25}$ a perfect square? ________ . Yes

60 To determine whether or not a rational number is a *perfect square:*

(a) Simplify the rational number.

(b) If both numerator and denominator, after simplification, are perfect square integers, then the rational number is a perfect square.

$\frac{8}{2}$ is a perfect square because $\frac{8}{2} = \frac{4}{1}$, and both 4 and 1 are perfect square integers. Is $\frac{18}{32}$ a perfect square? ______________ .

Yes, $\frac{18}{32} = \frac{9}{16}$

61 $\frac{12}{50}$ is *not* a perfect square because $\frac{12}{50} = \frac{6}{25}$, and 6 is *not* a perfect square integer. Is $\frac{10}{18}$ a perfect square? ______________ .

No, $\frac{10}{18} = \frac{5}{9}$

62 Is $\frac{18}{8}$ a perfect square? ______________ . Yes, $\frac{18}{8} = \frac{9}{4}$

[*Note:* First simplify $\frac{18}{8}$.]

63 Is $\frac{50}{162}$ a perfect square? ______________ . Yes, $\frac{50}{162} = \frac{25}{81}$

64 Is $\frac{12}{2}$ a perfect square? ______________ .

No, $\frac{12}{2} = \frac{6}{1}$

65 Is $\frac{^-49}{16}$ a perfect square? ________________ .

No, $^-49$ is not a perfect square integer

66 Is $\frac{32}{50}$ a perfect square? ______________ .

Yes, $\frac{32}{50} = \frac{16}{25}$

67 Is $\frac{1}{12}$ a perfect square? _______ .

No

68 $\sqrt{}$ is a symbol called a radical sign.
$\sqrt{4}$ is read as "the square root of 4."
$\sqrt{25}$ is read as "the square root of _______ ."

25

69 $\sqrt{20}$ is read as "the square root of 20."
$\sqrt{52}$ is read as "the square root of _______ ."

52

70 $\sqrt{15}$ is read as "the square root of 15."
$\sqrt{36}$ is read as "the _______ _______
_______ _______ ."

square root
of 36

71 $\sqrt{4} = 2$ because 2 is the positive number that if multiplied by itself gives a product of 4. $\sqrt{9} = 3$ because 3 is the positive number that if multiplied by itself gives a product of _______ .

9

72 $\sqrt{16}$ equals 4 because 4 is the positive number that if multiplied by itself gives a product of 16. $(4 \cdot 4 = 16)$. Fill in the blank below by finding the positive number that if multiplied by itself will give a product of 81.
$\sqrt{81} =$ _______ .

9

73 $\sqrt{4} = 2$ because $2 \cdot 2 = 4$.
$\sqrt{9} = 3$ because $3 \cdot 3 = 9$.
$\sqrt{81} = 9$ because $9 \cdot 9 = 81$.
$\sqrt{36} =$ _______ because $6 \cdot 6 = 36$.

6

74 $\sqrt{9} = 3$ because $3 \cdot 3 = 9$.
$\sqrt{81} = 9$ because $9 \cdot 9 = 81$.
$\sqrt{25} =$ _______ because $5 \cdot 5 = 25$. 5

75 $\sqrt{16} = 4$ because $4 \cdot 4 = 16$.
$\sqrt{49} =$ _______ . 7

76 $\sqrt{49} = 7$ because 7 is the positive number that when multiplied by itself will give a product of 49. $(7 \cdot 7 = 49.)$ $^{-}\sqrt{25} = {}^{-}5$ because $^{-}5$ is the negative number that when multiplied by itself will give a product of _______ . 25

77 $\sqrt{49}$ is a *positive* number, 7.
$^{-}\sqrt{25}$ is a *negative* number, $^{-}5$.
$^{-}\sqrt{36} =$ _______ . $^{-}6$

78 $\sqrt{16} = 4$ because $4 \cdot 4 = 16$.
$\sqrt{81} =$ _______ . 9

79 $^{-}\sqrt{49} = {}^{-}7$.
$^{-}\sqrt{64} =$ _______ . $^{-}8$

80 $\sqrt{100} =$ _______ . 10

81 $\sqrt{0} =$ _______ . 0

82 $^{-}\sqrt{4} =$ _______ . $^{-}2$

83 $^{-}\sqrt{1} =$ _______ . $^{-}1$

84 $\sqrt{25} =$ _______ . 5

85 $^{-}\sqrt{36} =$ _______ . $^{-}6$

86 $\sqrt{81} =$ _______ . 9

87 $^{-}\sqrt{64} =$ _______ . $^{-}8$

88 $^{-}\sqrt{100}$ = ______ . $^{-}10$

89 $\sqrt{16}$ = ______ . 4

90 $^{-}\sqrt{25}$ = ______ . $^{-}5$

91 $\sqrt{49}$ = ______ . 7

92 $^{-}\sqrt{81}$ = ______ . $^{-}9$

93 $\sqrt{1}$ = ______ . 1

94 $\sqrt{9}$ = ______ . 3

Self-Quiz # 2

The following questions test the objectives of the preceding section. 100% mastery is desired.

1. Which of the following are perfect squares? ________ .

(a) $\frac{^-9}{16}$ (b) $\frac{16}{49}$

(c) $\frac{7}{9}$ (d) 8

(e) 100

2. Which of the following are perfect squares? ________ .

(a) 0 (b) $\frac{1}{9}$

(c) 51 (d) $\frac{18}{50}$

(e) $\frac{^-4}{25}$

3. Which of the following are perfect squares? ________ .

(a) $\frac{8}{2}$ (b) $\frac{^-1}{16}$

(c) $\frac{3}{73}$ (d) $\frac{64}{25}$

(e) $\frac{81}{16}$

4. $\sqrt{36}$ = ________ .

5. $^-\sqrt{100}$ = ________ .

6. $^-\sqrt{64}$ = ________ .

7. $\sqrt{4}$ = ________ .

8. $\sqrt{9}$ = ________ .

9. $\sqrt{0}$ = ________ .

10. $^-\sqrt{1}$ = ________ .

95 $\sqrt{\frac{49}{25}} = \frac{7}{5}$ because $\frac{7}{5} \cdot \frac{7}{5} = \frac{49}{25}$.

$\sqrt{\frac{36}{25}} =$ ________ . $\frac{6}{5}$

96 $\sqrt{\frac{1}{9}} = \frac{1}{3}$ because $\frac{1}{3} \cdot \frac{1}{3} = \frac{1}{9}$.

$\sqrt{\frac{25}{64}} =$ ________ . $\frac{5}{8}$

97 $\sqrt{\frac{1}{4}} =$ ________ . $\frac{1}{2}$

98 $^{-}\sqrt{\frac{49}{64}} = \frac{^{-}7}{8}$ because $\frac{^{-}7}{8} \cdot \frac{^{-}7}{8} = \frac{49}{64}$.

$^{-}\sqrt{\frac{25}{81}} =$ ________ . $\frac{^{-}5}{9}$

99 $^{-}\sqrt{\frac{16}{49}} =$ ________ . $\frac{^{-}4}{7}$

100 $^{-}\sqrt{\frac{100}{49}} =$ ________ . $\frac{^{-}10}{7}$

101 $^{-}\sqrt{\frac{9}{49}} =$ ________ . $\frac{^{-}3}{7}$

102 $\sqrt{\frac{1}{36}} =$ ________ . $\frac{1}{6}$

103 $^{-}\sqrt{\frac{1}{81}} =$ ________ . $\frac{^{-}1}{9}$

104 $\sqrt{\frac{25}{1}} = \frac{5}{1} = 5$.

$\sqrt{\frac{49}{1}} =$ ________ . 7

105 $^{-}\sqrt{\frac{81}{1}} = \frac{^{-}9}{1} = {}^{-}9.$

$^{-}\sqrt{\frac{100}{1}} =$ ________ . $\quad$ $^{-}10$

106 $\sqrt{\frac{25}{16}}$ is the *rational number* $\frac{5}{4}$ because both 25 and 16 are perfect square integers. Is $\sqrt{\frac{49}{81}}$ a *rational number*? ________ . $\quad$ Yes, $\frac{7}{9}$

107 $\sqrt{\frac{100}{49}}$ is the *rational number* $\frac{10}{7}$ because both 100 and 49 are perfect square integers. Is $\sqrt{\frac{1}{64}}$ a *rational number*? ________ . $\quad$ Yes, $\frac{1}{8}$

108 $\sqrt{\frac{2}{9}}$ is *not* a rational number because 2 is not a perfect square integer. Is $\sqrt{\frac{5}{16}}$ a rational number? ________ . $\quad$ No

109 $\sqrt{\frac{9}{13}}$ is *not* a rational number because 13 is not a perfect square integer. $\sqrt{\frac{25}{37}}$ is not a rational number because ________ is not a perfect square integer. $\quad$ 37

110 $\sqrt{\frac{7}{81}}$ is not a rational number because ________ is not a perfect square integer. $\quad$ 7

111 $\sqrt{\frac{81}{29}}$ is not a rational number because ________ is not a perfect square integer. $\quad$ 29

112 $\sqrt{\frac{15}{49}}$ is not a rational number because ________ is not a perfect square integer. 15

113 $\sqrt{\frac{16}{21}}$ is not a rational number because ________ is not a perfect square integer. 21

114 $\sqrt{\frac{2}{1}}$ is not a rational number because 2 is not a perfect square integer. Is $\sqrt{\frac{3}{1}}$ a rational number? ________ . No

115 $\sqrt{\frac{5}{1}}$ is *not* a rational number. Is $\sqrt{\frac{7}{1}}$ a rational number? ________ . No

116 $\sqrt{\frac{4}{1}} = \frac{2}{1} = 2$. 2 is a rational number. Is $\sqrt{\frac{25}{1}}$ a rational number? ________ . Yes, 5

117 $\sqrt{9}$ is the rational number 3. Is $\sqrt{16}$ a rational number? ________ . Yes, 4

118 $\sqrt{2}$ is *not* a rational number. $\sqrt{2}$ is an *irrational* number. $\sqrt{3}$ is *not* a rational number. $\sqrt{3}$ is an ________ number. irrational

119 $\sqrt{10}$ is an *irrational* number because there is no rational number that can be multiplied by itself to give a product of 10. Is $\sqrt{13}$ an *irrational* number? ________ . Yes

120 $\sqrt{36} = 6$. 6 is a rational number. $\sqrt{30}$ is an irrational number because there is no rational number that can be multiplied by itself to give a product of 30. Is $\sqrt{51}$ a rational or irrational number? ________ . Irrational

121 Is $\sqrt{21}$ a rational or irrational number?
_______ .

Irrational

122 Is $\sqrt{14}$ a rational or irrational number?
_______ .

Irrational

123 Is $\sqrt{95}$ a rational or irrational number?
_______ .

Irrational

124 Is $\sqrt{100}$ a rational or irrational number?
_________________ .

Rational, 10, because $10 \cdot 10 = 100$

125 Is $\sqrt{56}$ a rational or irrational number?
_______ .

Irrational

126 Is $\sqrt{37}$ a rational or irrational number?
_______ .

Irrational

127 Is $\sqrt{36}$ a rational or irrational number?
_________________ .

Rational, 6, because $6 \cdot 6 = 36$

128 Is $^{-}\sqrt{64}$ a rational or irrational number?
_____________ . [*Note:* $^{-}8 \cdot {}^{-}8 = 64$.]

Rational, $^{-}8$

129 Is $^{-}\sqrt{49}$ a rational or irrational number?
_______ .

Rational

130 Is $^{-}\sqrt{15}$ a rational or irrational number?
_______ .

Irrational

131 Is $^{-}\sqrt{81}$ a rational or irrational number?
_______ .

Rational

132 Is $^{-}\sqrt{73}$ a rational or irrational number?
_______ .

Irrational

133 Which of the following are *rational* numbers?

________ .

(b), (c)

(a) $\sqrt{\frac{49}{65}}$.

(b) $\sqrt{16}$.

(c) $^{-}\sqrt{81}$.

(d) $\sqrt{19}$.

(e) $\sqrt{\frac{3}{25}}$.

134 Which of the following are *irrational* numbers?

________________ .

(c), (d), (e)

(a) $\sqrt{9}$.

(b) $^{-}\sqrt{\frac{4}{25}}$.

(c) $\sqrt{73}$.

(d) $\sqrt{2}$.

(e) $\sqrt{5}$.

Self-Quiz # 3

The following questions test the objectives of the preceding section. 100% mastery is desired.

1. Which of the following are *rational* numbers? ________ .

(a) $\sqrt{\frac{9}{16}}$ (b) $\sqrt{\frac{2}{7}}$

(c) $^{-}\sqrt{\frac{9}{14}}$ (d) $^{-}\sqrt{36}$

(e) $\sqrt{\frac{81}{25}}$

2. Which of the following are *rational* numbers? ________ .

(a) $^{-}\sqrt{49}$ (b) $\sqrt{1}$

(c) $^{-}\sqrt{17}$ (d) $\sqrt{100}$

(e) $\sqrt{63}$

3. Which of the following are *irrational* numbers? ________ .

(a) $^{-}\sqrt{\frac{9}{100}}$ (b) $\sqrt{\frac{17}{16}}$

(c) $^{-}\sqrt{36}$ (d) $\sqrt{65}$

(e) $\sqrt{49}$

4. Which of the following are *irrational* numbers? ________ .

(a) $^{-}\sqrt{63}$ (b) $\sqrt{77}$

(c) $\sqrt{81}$ (d) $^{-}\sqrt{100}$

(e) $\sqrt{\frac{49}{16}}$

Since 13 is not a perfect square integer, $\sqrt{13}$ is an irrational number. $\sqrt{13}$ is greater than 3 because $\sqrt{9} = 3$. $\sqrt{13}$ is less than 4 because $\sqrt{16} = 4$. Therefore, $\sqrt{13}$ is a number between 3 and 4.

$\sqrt{13}$ has a position on the number line between 3 and 4, as shown:

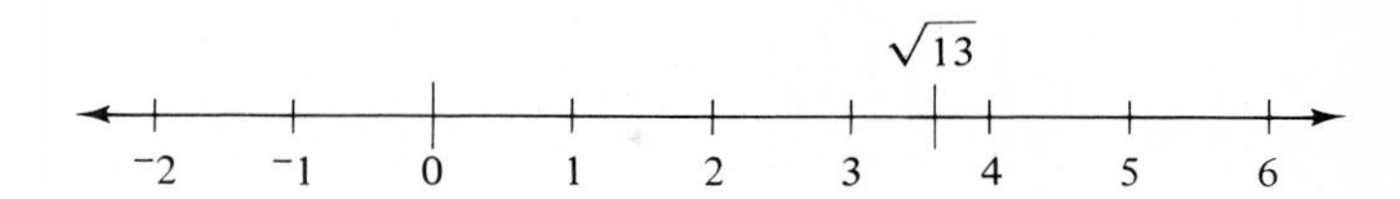

Although $\sqrt{13}$ is not a rational number, it does have a value and a position on the number line like all other real numbers.

---

135 $\sqrt{9} = 3$ and $\sqrt{16} = 4$. Since 11 is between 9 and 16, $\sqrt{11}$ is a number between 3 and 4. $\sqrt{36} = 6$ and $\sqrt{49} = 7$. Since 41 is between 36 and 49, $\sqrt{41}$ is a number between 6 and ________ .

7

136 17 is between 16 and 25. Since $\sqrt{16} = 4$ and $\sqrt{25} = 5$, $\sqrt{17}$ is between 4 and 5. 39 is between 36 and 49. Since $\sqrt{36} = 6$ and $\sqrt{49} = 7$, $\sqrt{39}$ is between 6 and ________ .

7

137 $\sqrt{25} = 5$ and $\sqrt{36} = 6$. $\sqrt{29}$ is between 5 and ________ .

6

138 $\sqrt{1} = 1$ and $\sqrt{4} = 2$. $\sqrt{2}$ is between ________ and ________ .

1, 2

139 $\sqrt{85}$ is between ________ and ________ .

9, 10

140 $\sqrt{31}$ is between ________ and ________ .

5, 6

141 $\sqrt{71}$ is between ________ and ________ .

8, 9

142 $^{-}\sqrt{64} = {}^{-}8$ and $^{-}\sqrt{81} = {}^{-}9$. Since 70 is between 64 and 81, $^{-}\sqrt{70}$ is between $^{-}8$ and $^{-}9$. $^{-}\sqrt{4} = {}^{-}2$ and $^{-}\sqrt{9} = {}^{-}3$. Since 6 is between 4 and 9, $^{-}\sqrt{6}$ is between $^{-}2$ and ________ .

$^{-}3$

143 $^-\sqrt{25} = {}^-5$ and $^-\sqrt{36} = {}^-6$. Since 31 is between 25 and 36, $^-\sqrt{31}$ is between $^-5$ and $^-6$. $^-\sqrt{81} = {}^-9$ and $^-\sqrt{100} = {}^-10$. $^-\sqrt{93}$ is between $^-9$ and ______ .

$^-10$

144 $^-\sqrt{49} = {}^-7$ and $^-\sqrt{64} = {}^-8$. $^-\sqrt{59}$ is between ______ and ______ .

$^-7$, $^-8$

$^-\sqrt{59}$ is a negative *irrational* number. Its value and position can be shown on the number line as follows:

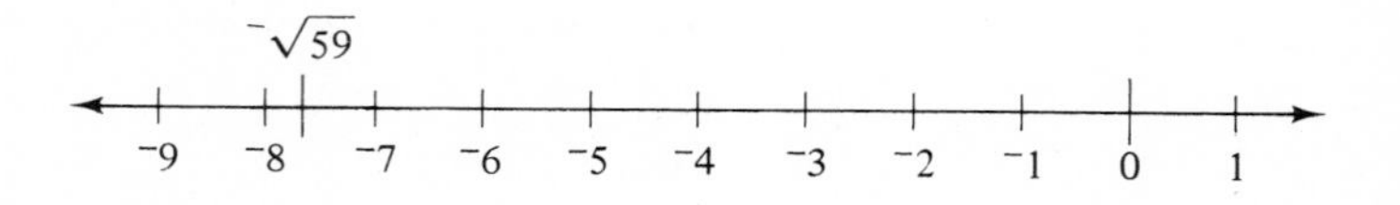

---

145 $^-\sqrt{21}$ is between ______ and ______ .

$^-4$, $^-5$

146 $^-\sqrt{45}$ is between ______ and ______ .

$^-6$, $^-7$

147 $^-\sqrt{87}$ is between ______ and ______ .

$^-9$, $^-10$

148 $\sqrt{47}$ is between ______ and ______ .

6, 7

149 $\sqrt{23}$ is between ______ and ______ .

4, 5

150 $^-\sqrt{27}$ is between ______ and ______ .

$^-5$, $^-6$

151 $\sqrt{33}$ is between ______ and ______ .

5, 6

152 $^-\sqrt{5}$ is between ______ and ______ .

$^-2$, $^-3$

153 $\sqrt{93}$ is between ______ and ______ .

9, 10

154 $\sqrt{6}$ is between ______ and ______ .

2, 3

155 $^-\sqrt{59}$ is between ______ and ______ .

$^-7$, $^-8$

156 $\sqrt{3}$ is between ______ and ______ . 1, 2

157 $\sqrt{84}$ is between ______ and ______ . 9, 10

Self-Quiz # 4

The following questions test the objectives of the preceding section. 100% mastery is desired.

1. $\sqrt{19}$ is between ______ and ______ .
2. $^{-}\sqrt{57}$ is between ______ and ______ .
3. $^{-}\sqrt{87}$ is between ______ and ______ .
4. $\sqrt{61}$ is between ______ and ______ .
5. $\sqrt{90}$ is between ______ and ______ .

$\sqrt{7}$ and $\sqrt{15}$ are *irrational* numbers. In the following section the multiplication of irrational numbers such as $\sqrt{7} \cdot \sqrt{13}$ will be studied.

---

158 $\sqrt{5}$ and $\sqrt{3}$ are both irrational numbers. To multiply $\sqrt{5} \cdot \sqrt{3}$, multiply 5 and 3 and place the product under the radical sign. $\sqrt{5} \cdot \sqrt{3} = \sqrt{15}$. $\sqrt{7} \cdot \sqrt{3} =$ ______ . $\sqrt{21}$

159 $\sqrt{8} \cdot \sqrt{3} = \sqrt{8 \cdot 3} = \sqrt{24}$.
$\sqrt{6} \cdot \sqrt{5} =$ ______ . $\sqrt{30}$

160 $\sqrt{7} \cdot \sqrt{15} = \sqrt{7 \cdot 15} = \sqrt{105}$.
$\sqrt{10} \cdot \sqrt{7} =$ ______ . $\sqrt{70}$

161 $\sqrt{17} \cdot \sqrt{2} = \sqrt{17 \cdot 2} = \sqrt{34}$.
$\sqrt{2} \cdot \sqrt{11} =$ ______ . $\sqrt{22}$

162 $\sqrt{5} \cdot \sqrt{12} =$ ______ . $\sqrt{60}$

163 $\sqrt{7} \cdot \sqrt{13} =$ ______ . $\sqrt{91}$

164 $\sqrt{3} \cdot \sqrt{10} =$ ______ . $\sqrt{30}$

165 $\sqrt{19} \cdot \sqrt{2} =$ ______ . $\sqrt{38}$

166 $\sqrt{2} \cdot \sqrt{13} =$ ______ . $\sqrt{26}$

167 $\sqrt{5} \cdot \sqrt{17} =$ ______ . $\sqrt{85}$

168 $\sqrt{5} \cdot \sqrt{5} = \sqrt{25} = 5.$
$\sqrt{8} \cdot \sqrt{8} = \sqrt{64} =$ ______ . 8

169 $\sqrt{11} \cdot \sqrt{11} = \sqrt{121} = 11.$
$\sqrt{3} \cdot \sqrt{3} = \sqrt{9} =$ ______ . 3

170 $\sqrt{5} \cdot \sqrt{14} =$ ______ . $\sqrt{70}$

171 $\sqrt{917} \cdot \sqrt{917} = 917.$
$\sqrt{1{,}087} \cdot \sqrt{1{,}087} =$ ______ . 1,087

172 $\sqrt{13} \cdot \sqrt{6} =$ ______ . $\sqrt{78}$

173 $\sqrt{97} \cdot \sqrt{97} =$ ______ . 97

174 $\sqrt{17} \cdot \sqrt{6} =$ ______ . $\sqrt{102}$

175 $\sqrt{7} \cdot \sqrt{8} =$ ______ . $\sqrt{56}$

176 $\sqrt{47} \cdot \sqrt{47} =$ ______ . 47

177 $\sqrt{15} \cdot \sqrt{3} =$ ______ . $\sqrt{45}$

178 $\sqrt{9} \cdot \sqrt{12} =$ ______ . $\sqrt{108}$

The radical expression $\sqrt{18}$ can be written as $3\sqrt{2}$ because 9 is a perfect square factor of 18.

$$\begin{aligned}\sqrt{18} &= \sqrt{9} \cdot \sqrt{2} \\ &= 3\sqrt{2}\end{aligned}$$

To simplify a radical expression like $\sqrt{18}$, all perfect square factors of the radicand are removed. The following section explains the simplification of radical expressions.

---

179 $2\sqrt{3}$ means $2 \cdot \sqrt{3}$. Using the multiplication symbol, $(\cdot)$, write $5\sqrt{2}$ ________ . — $5 \cdot \sqrt{2}$

180 $4\sqrt{7}$ means $4 \cdot \sqrt{7}$.
$8\sqrt{3}$ means ________ . — $8 \cdot \sqrt{3}$

181 $5 \cdot \sqrt{7}$ may be written as $5\sqrt{7}$.
$4 \cdot \sqrt{13}$ may be written as ________ . — $4\sqrt{13}$

182 $6 \cdot \sqrt{5}$ may be written as $6\sqrt{5}$.
$9 \cdot \sqrt{7}$ may be written as ________ . — $9\sqrt{7}$

183 $9 \cdot \sqrt{3}$ may be written as ________ . — $9\sqrt{3}$

184 $\sqrt{4}$ can be simplified to 2 because 4 is a perfect square.
$\sqrt{25}$ can be simplified to 5 because ________ is a perfect square. — 25

185 $\sqrt{49}$ can be simplified to 7 because ________ is a perfect square. — 49

186 $^{-}\sqrt{81}$ can be simplified to $^{-}9$ because ________ is a perfect square. — 81

187 $\sqrt{25} = 5$. $\sqrt{81} =$ ________ . — 9

188 $\sqrt{36} = 6$. $\sqrt{16} =$ ________ . — 4

189 $\sqrt{49}$ = _______ . 7

190 $\sqrt{64}$ = _______ . 8

191 $\sqrt{100}$ = _______ . 10

192 $\sqrt{9}$ = _______ . 3

193 $\sqrt{4}$ = _______ . 2

194 To simplify $\sqrt{50}$, it is necessary to find a factor of 50 that is a perfect square integer. $50 = 25 \cdot 2$, and 25 is a perfect square.

$$\sqrt{50} = \sqrt{25 \cdot 2} = \sqrt{25} \cdot \sqrt{2} = \underline{5\sqrt{2}}$$

25 is a perfect square factor of 75. Complete the simplification of $\sqrt{75}$.
$\sqrt{75} = \sqrt{25 \cdot 3} = \sqrt{25} \cdot \sqrt{3} =$ _______ . $5\sqrt{3}$

195 To simplify $\sqrt{12}$, it is necessary to find a perfect square factor of 12. $12 = 4 \cdot 3$, and 4 is a perfect square.

$$\sqrt{12} = \sqrt{4 \cdot 3} = \sqrt{4} \cdot \sqrt{3} = 2\sqrt{3}$$

Using the fact that 9 is a perfect square factor of 18, complete the simplification of $\sqrt{18}$.
$\sqrt{18} = \sqrt{9 \cdot 2} = \sqrt{9} \cdot \sqrt{2} =$ _______ . $3\sqrt{2}$

196 $\sqrt{72}$ is simplified as follows:

$$\sqrt{72} = \sqrt{36} \cdot \sqrt{2} = 6\sqrt{2}$$

Complete the simplification of $\sqrt{32}$.
$\sqrt{32} = \sqrt{16} \cdot \sqrt{2} =$ _______ . $4\sqrt{2}$

197 To simplify $\sqrt{48}$, first find a factor of 48 that is a perfect square. Complete the following simplification:
$\sqrt{48} = \sqrt{16} \cdot \sqrt{3} =$ _______ . $4\sqrt{3}$

198 4 is a perfect square factor of 8. To simplify $\sqrt{8}$, the following steps are used:

$$\sqrt{8} = \sqrt{4} \cdot \sqrt{2} = 2\sqrt{2}$$

Using 4 as a perfect square factor of 24, simplify $\sqrt{24}$. — $2\sqrt{6}$

199 Using 4 as the perfect square factor of 60, complete the simplification of $\sqrt{60}$.
$\sqrt{60} = \sqrt{4} \cdot \sqrt{15} =$ ________ . — $2\sqrt{15}$

200 Using 25 as the perfect square factor of 150, simplify $\sqrt{150}$. ________ . — $5\sqrt{6}$

201 $\sqrt{7}$ *cannot* be simplified because there is no perfect square factor of 7, except 1. Can $\sqrt{13}$ be simplified? ________ . — No

202 $\sqrt{44}$ can be simplified because 4 is a perfect square factor of 44. Complete the simplification of $\sqrt{44}$:
$\sqrt{44} = \sqrt{4} \cdot \sqrt{11} =$ ________ . — $2\sqrt{11}$

203 9 is a perfect square factor of 45. Simplify $\sqrt{45}$. ________ . — $3\sqrt{5}$

204 16 is a perfect square factor of 80. Simplify $\sqrt{80}$. ________ . — $4\sqrt{5}$

205 Simplify $\sqrt{20}$ by first finding a factor of 20 that is a perfect square. ________ . [*Hint:* $\sqrt{20} = \sqrt{4} \cdot \sqrt{5}$] — $2\sqrt{5}$

206 Simplify $\sqrt{27}$ by first finding a factor of 27 that is a perfect square. ________ . — $3\sqrt{3}$

207 To simplify a radical expression, such as $\sqrt{24}$, it is necessary to find a perfect square that is a factor of 24. Simplify $\sqrt{24}$. ________ . — $2\sqrt{6}$

208 To simplify a radical expression, such as $\sqrt{108}$, it is necessary to find a perfect square factor of 108.
$\sqrt{36} \cdot \sqrt{3} = \sqrt{108}$. Simplify $\sqrt{108}$. ________ . — $6\sqrt{3}$

209 Complete the simplification of $\sqrt{28}$:
$\sqrt{28} = \sqrt{4 \cdot 7} = \sqrt{4} \cdot \sqrt{7} =$ ________ .

$2\sqrt{7}$

210 Complete the simplification of $\sqrt{75}$:
$\sqrt{75} = \sqrt{25 \cdot 3} = \sqrt{25} \cdot \sqrt{3} =$ ________ .

$5\sqrt{3}$

211 Complete the simplification of $\sqrt{54}$:
$\sqrt{54} = \sqrt{9} \cdot \sqrt{6} =$ ________ .

$3\sqrt{6}$

212 Complete the simplification of $\sqrt{125}$:
$\sqrt{125} = \sqrt{25} \cdot \sqrt{5} =$ ________ .

$5\sqrt{5}$

213 Simplify $\sqrt{18}$. ________ .

$3\sqrt{2}$

214 Simplify $\sqrt{40}$. ________ .

$2\sqrt{10}$

215 $6 \cdot 25 = 150$. Simplify $\sqrt{150}$. ________ .

$5\sqrt{6}$

216 Simplify $\sqrt{175}$. ________ .

$5\sqrt{7}$

217 Simplify $\sqrt{63}$. ________ .

$3\sqrt{7}$

218 Simplify $\sqrt{36}$. ________ .

6

219 Simplify $\sqrt{90}$. ________ .

$3\sqrt{10}$

220 $2 \cdot 100 = 200$. Simplify $\sqrt{200}$. ________ .

$10\sqrt{2}$

221 Simplify $\sqrt{300}$. ________ .

$10\sqrt{3}$

222 Simplify $\sqrt{50}$. ________ .

$5\sqrt{2}$

223 ${}^{-}\sqrt{36} = {}^{-}6$. Simplify ${}^{-}\sqrt{81}$. ________ .

${}^{-}9$

224 Simplify $\sqrt{162}$. ________ .

$9\sqrt{2}$

225 Simplify $\sqrt{24}$. ________ .

$2\sqrt{6}$

226 Simplify $\sqrt{\frac{4}{9}}$. ________ . $\frac{2}{3}$

227 Simplify $\sqrt{20}$. ________ . $2\sqrt{5}$

228 Simplify $^{-}\sqrt{200}$. ________ . $^{-}10\sqrt{2}$

229 Simplify $\sqrt{500}$. ________ . $10\sqrt{5}$

230 Simplify $\sqrt{\frac{1}{4}}$. ________ . $\frac{1}{2}$

231 Simplify $\sqrt{1,100}$. ________ . $10\sqrt{11}$

232 Simplify $\sqrt{45}$. ________ . $3\sqrt{5}$

233 Simplify $\sqrt{64}$. ________ . 8

234 Simplify $\sqrt{75}$. ________ . $5\sqrt{3}$

235 To simplify $5\sqrt{18}$, the following steps are used:

$$\begin{aligned} 5\sqrt{18} &= 5(\sqrt{9} \cdot \sqrt{2}) \\ &= 5(3 \cdot \sqrt{2}) \\ &= (5 \cdot 3) \cdot \sqrt{2} \\ &= 15\sqrt{2} \end{aligned}$$

Complete the simplification of $2\sqrt{75}$:

$$\begin{aligned} 2\sqrt{75} &= 2(\sqrt{25} \cdot \sqrt{3}) \\ &= 2(5 \cdot \sqrt{3}) \\ &= (2 \cdot 5) \cdot \sqrt{3} = \text{\_\_\_\_\_\_\_\_} . \end{aligned}$$

$10\sqrt{3}$

236 Complete the simplification of $5\sqrt{12}$:

$$\begin{aligned} 5\sqrt{12} &= 5(\sqrt{4} \cdot \sqrt{3}) \\ &= 5(2 \cdot \sqrt{3}) \\ &= (5 \cdot 2)\sqrt{3} = \text{\_\_\_\_\_\_\_\_} . \end{aligned}$$

$10\sqrt{3}$

237 Complete the simplification of $3\sqrt{18}$:
$3\sqrt{18} = 3(\sqrt{9} \cdot \sqrt{2})$
$= 3(3 \cdot \sqrt{2})$
$= (3 \cdot 3)\sqrt{2} =$ ______ . $9\sqrt{2}$

238 Simplify $8\sqrt{50}$. ______ . $40\sqrt{2}$

239 Simplify $3\sqrt{16}$. ______ . 12
[*Note:* $2\sqrt{25} = 2 \cdot 5 = 10$.]

240 Simplify $5\sqrt{49}$. ______ . 35

241 Simplify $7\sqrt{24}$. ______ . $14\sqrt{6}$

242 Simplify $6\sqrt{20}$. ______ . $12\sqrt{5}$

243 Simplify $2\sqrt{300}$. ______ . $20\sqrt{3}$

244 Simplify $5\sqrt{63}$. ______ . $15\sqrt{7}$

245 Simplify $3\sqrt{50}$. ______ . $15\sqrt{2}$

246 To simplify $^{-}3\sqrt{40}$, the following steps are used:

$$
\begin{aligned}
{}^{-}3\sqrt{40} &= {}^{-}3 \cdot (\sqrt{4} \cdot \sqrt{10}) \\
&= {}^{-}3 \cdot (2\sqrt{10}) \\
&= ({}^{-}3 \cdot 2) \cdot \sqrt{10} \\
&= {}^{-}6\sqrt{10}
\end{aligned}
$$

Complete the following simplification:
$^{-}5\sqrt{12} = {}^{-}5 \cdot (\sqrt{4} \cdot \sqrt{3})$
$= {}^{-}5(2\sqrt{3})$
$= ({}^{-}5 \cdot 2)\sqrt{3} =$ ______ . $^{-}10\sqrt{3}$

247 Complete the following simplification:
$^{-}2\sqrt{18} = {}^{-}2(\sqrt{9} \cdot \sqrt{2})$
$= ({}^{-}2 \cdot 3)\sqrt{2} =$ ______ . $^{-}6\sqrt{2}$

248 Simplify $^{-}3\sqrt{50}$. ______ . $^{-}15\sqrt{2}$

249 Simplify $^{-}7\sqrt{18}$. ________ . $^{-}21\sqrt{2}$

250 Simplify $^{-}3\sqrt{200}$. ________ . $^{-}30\sqrt{2}$

251 $^{-}\sqrt{12}$ is equivalent to $^{-}1\sqrt{12}$. To simplify $^{-}\sqrt{12}$, the following steps are used:

$$
\begin{aligned}
^{-}\sqrt{12} &= {}^{-}1\sqrt{12} \\
&= {}^{-}1(\sqrt{4} \cdot \sqrt{3}) \\
&= {}^{-}1(2 \cdot \sqrt{3}) \\
&= ({}^{-}1 \cdot 2)\sqrt{3} \\
&= {}^{-}2\sqrt{3}
\end{aligned}
$$

Complete the simplification of $^{-}\sqrt{20}$:

$$
\begin{aligned}
^{-}\sqrt{20} &= {}^{-}1\sqrt{20} \\
&= {}^{-}1(\sqrt{4} \cdot \sqrt{5}) \\
&= {}^{-}1(2 \cdot \sqrt{5}) \\
&= ({}^{-}1 \cdot 2)\sqrt{5} = \_\_\_\_\_\_\_\_ .
\end{aligned}
$$

$^{-}2\sqrt{5}$

252 Complete the following simplification:

$$
\begin{aligned}
^{-}\sqrt{27} &= {}^{-}1\sqrt{27} \\
&= {}^{-}1(\sqrt{9} \cdot \sqrt{3}) \\
&= {}^{-}1(3 \cdot \sqrt{3}) \\
&= ({}^{-}1 \cdot 3)\sqrt{3} = \_\_\_\_\_\_\_\_ .
\end{aligned}
$$

$^{-}3\sqrt{3}$

253 Simplify $^{-}\sqrt{24}$. ________ . $^{-}2\sqrt{6}$

254 Simplify $^{-}\sqrt{75}$. ________ . $^{-}5\sqrt{3}$

255 Simplify $^{-}\sqrt{12}$. ________ . $^{-}2\sqrt{3}$

256 Simplify $^{-}\sqrt{28}$. ________ . $^{-}2\sqrt{7}$

257 Simplify $^{-}7\sqrt{8}$. ________ . $^{-}14\sqrt{2}$

258 Simplify $\sqrt{64}$. ________ . 8

259 Simplify $2\sqrt{700}$. ________ . $20\sqrt{7}$

260 Simplify $^{-}\sqrt{700}$. ________ . $^{-}10\sqrt{7}$

261 Simplify $\sqrt{54}$. ________ . $3\sqrt{6}$

262 Simplify $\sqrt{36}$. ________ . 6

263 Simplify $\sqrt{98}$. ________ . $7\sqrt{2}$

264 Simplify $^{-}8\sqrt{175}$. ________ . $^{-}40\sqrt{7}$

265 Simplify $^{-}3\sqrt{200}$. ________ . $^{-}30\sqrt{2}$

266 Simplify $5\sqrt{36}$. ________ . 30

267 Simplify $^{-}\sqrt{1}$. ________ . $^{-}1$

268 Simplify $^{-}\sqrt{63}$. ________ . $^{-}3\sqrt{7}$

269 Simplify $^{-}\sqrt{81}$. ________ . $^{-}9$

270 Simplify $^{-}4\sqrt{54}$. ________ . $^{-}12\sqrt{6}$

271 Simplify $3\sqrt{100}$. ________ . 30

272 Simplify $^{-}5\sqrt{20}$. ________ . $^{-}10\sqrt{5}$

273 Simplify $3\sqrt{18}$. ________ . $9\sqrt{2}$

274 Simplify $^{-}\sqrt{75}$. ________ . $^{-}5\sqrt{3}$

275 Simplify $4\sqrt{75}$. ________ . $20\sqrt{3}$

Self-Quiz # 5

The following questions test the objectives of the preceding section. 100% mastery is desired.

Simplify:

1. $\sqrt{16}$. ________ .
2. $^{-}2\sqrt{36}$. ________ .
3. $4\sqrt{18}$. ________ .
4. $^{-}\sqrt{50}$. ________ .
5. $\sqrt{200}$. ________ .
6. $^{-}4\sqrt{8}$. ________ .
7. $2\sqrt{54}$. ________ .
8. $2\sqrt{500}$. ________ .
9. $^{-}\sqrt{81}$. ________ .
10. $5\sqrt{27}$. ________ .

4, 9, and 36 are all perfect square factors of 72. In simplifying $\sqrt{72}$, the best perfect square factor is 36 because it is larger than 4 and 9. The simplification is not complete until the number under the radical sign has *no* perfect square factor except 1.

Below are shown three ways of simplifying $\sqrt{72}$. Note that the first two require repeated simplifications to arrive at the final result, $6\sqrt{2}$.

$$\sqrt{72} = \sqrt{4} \cdot \sqrt{18} = 2\sqrt{18} = 2\sqrt{9}\sqrt{2} = 2 \cdot 3\sqrt{2} = \underline{6\sqrt{2}.}$$

$$\sqrt{72} = \sqrt{9} \cdot \sqrt{8} = 3\sqrt{8} = 3\sqrt{4}\sqrt{2} = 3 \cdot 2\sqrt{2} = \underline{6\sqrt{2}.}$$

$$\sqrt{72} = \sqrt{36} \cdot \sqrt{2} = \underline{6\sqrt{2}.}$$

All three of the above simplifications provide the correct result, but the third simplification is the easiest because 36 is the *largest perfect square factor* of 72.

The simplification of a radical expression is not complete until the number under the radical sign has no perfect square factor except 1.

---

276 $\sqrt{48} = \sqrt{4} \cdot \sqrt{12} = 2\sqrt{12} = 2\sqrt{4} \cdot \sqrt{3} = 4\sqrt{3}$.
$\sqrt{48} = \sqrt{16} \cdot \sqrt{3} =$ ________ .
[*Note:* The second method of simplifying $\sqrt{48}$ is better than the first because 16 is a larger perfect square factor of 48 than 4.]

$4\sqrt{3}$

277 $\sqrt{200} = \sqrt{4}\sqrt{50} = 2\sqrt{50} = 2\sqrt{25}\sqrt{2} = 10\sqrt{2}$.
$\sqrt{200} = \sqrt{100} \cdot \sqrt{2} =$ ________ .
[*Note:* 100 is a larger perfect square factor of 200 than 4.]

$10\sqrt{2}$

278 Simplify $\sqrt{300}$. ________ .

$10\sqrt{3}$

279 Simplify $\sqrt{27}$. ________ .

$3\sqrt{3}$

280 Simplify $\sqrt{125}$. ________ .

$5\sqrt{5}$

281 Simplify $\sqrt{32}$. ________ .

$4\sqrt{2}$

282 Simplify $\sqrt{18}$. ________ .

$3\sqrt{2}$

283 4 and 64 are both perfect square factors of 128. In simplifying $\sqrt{128}$, which of the following is the *best* first step?

$$\sqrt{128} = \sqrt{4} \cdot \sqrt{32} \text{ or } \sqrt{128} = \sqrt{64} \cdot \sqrt{2}$$

________________ .

$\sqrt{64} \cdot \sqrt{2}$

284 Simplify $\sqrt{128}$. ________ .

$8\sqrt{2}$

285 Simplify $\sqrt{84}$. ________ .

$2\sqrt{21}$

286 Simplify $\sqrt{96}$. ________ .
[*Note:* The perfect square factors of 96 are 4 and 16.]

$4\sqrt{6}$

287 Simplify $\sqrt{98}$. ________ .

$7\sqrt{2}$

288 Simplify $\sqrt{99}$. ________ .

$3\sqrt{11}$

289 Simplify $\sqrt{45}$. ________ .

$3\sqrt{5}$

290 To simplify $5\sqrt{28}$, the following steps are used:

$$\begin{aligned} 5 \cdot \sqrt{28} &= 5 \cdot \sqrt{4} \cdot \sqrt{7} \\ &= 5 \cdot 2 \cdot \sqrt{7} \\ &= 10\sqrt{7} \end{aligned}$$

Simplify $3\sqrt{20}$. ________ .

$6\sqrt{5}$

291 Simplify: $^-5\sqrt{72}$. ________ .
[*Note:* First simplify $\sqrt{72}$.]

$^-30\sqrt{2}$

292 Simplify $7\sqrt{300}$. ________ .

$70\sqrt{3}$

293 Simplify $^-6\sqrt{8}$. ________ .

$^-12\sqrt{2}$

294 Simplify $5\sqrt{48}$. ________ .

$20\sqrt{3}$

295 Simplify $^-3\sqrt{200}$. ________ .

$^-30\sqrt{2}$

296 $\frac{14}{21}$ can be simplified to $\frac{2}{3}$ because 7 is a common factor of 14 and 21. Simplify $\frac{10}{15}$. ________ .

$\frac{2}{3}$

297 Simplify $\frac{27}{36}$. ________ .

$\frac{3}{4}$

The rational number $\frac{8}{10}$ is simplified to $\frac{4}{5}$ because 2 is a common factor of 8 and 10.

Similarly, $\frac{\sqrt{15}}{\sqrt{55}}$ can be simplified because the $\sqrt{5}$ is a common factor of $\sqrt{15}$ and $\sqrt{55}$.

The following section explains the simplification of fractions with radical expressions.

---

298 $\frac{7}{7} = 1$, $\frac{15}{15} = 1$. $\frac{109}{109} =$ ______ . — 1

299 $\frac{8}{8} = 1$, $\frac{\sqrt{7}}{\sqrt{7}} = 1$. $\frac{\sqrt{3}}{\sqrt{3}} =$ ______ . — 1

300 $\frac{\sqrt{15}}{\sqrt{15}} = 1$. $\frac{\sqrt{91}}{\sqrt{91}} =$ ______ . — 1

301 $\frac{\sqrt{47}}{\sqrt{47}} =$ ______ . — 1

302 $\frac{\sqrt{113}}{\sqrt{113}} =$ ______ . — 1

303 $\frac{\sqrt{73}}{\sqrt{73}} =$ ______ . — 1

304 $\sqrt{15} = \sqrt{3} \cdot \sqrt{5}$ and $\sqrt{21} = \sqrt{3} \cdot \sqrt{7}$. To simplify $\frac{\sqrt{15}}{\sqrt{21}}$, the following steps are used:

$$\frac{\sqrt{15}}{\sqrt{21}} = \frac{\sqrt{3} \cdot \sqrt{5}}{\sqrt{3} \cdot \sqrt{7}} = 1 \cdot \frac{\sqrt{5}}{\sqrt{7}} = \frac{\sqrt{5}}{\sqrt{7}}$$

Using $\sqrt{10} = \sqrt{5} \cdot \sqrt{2}$ and $\sqrt{15} = \sqrt{5} \cdot \sqrt{3}$, simplify $\frac{\sqrt{10}}{\sqrt{15}}$. ________ .

$\frac{\sqrt{2}}{\sqrt{3}}$

305 $\sqrt{22} = \sqrt{2} \cdot \sqrt{11}$ and $\sqrt{33} = \sqrt{3} \cdot \sqrt{11}$. Simplify $\frac{\sqrt{22}}{\sqrt{33}}$.

________ .

$\frac{\sqrt{2}}{\sqrt{3}}$

306 $\sqrt{15} = \sqrt{3} \cdot \sqrt{5}$ and $\sqrt{6} = \sqrt{3} \cdot \sqrt{2}$. Simplify $\frac{\sqrt{15}}{\sqrt{6}}$.

________ .

$\frac{\sqrt{5}}{\sqrt{2}}$

307 $\sqrt{10} = \sqrt{5} \cdot \sqrt{2}$ and $\sqrt{55} = \sqrt{5} \cdot \sqrt{11}$. Simplify $\frac{\sqrt{10}}{\sqrt{55}}$.

________ .

$\frac{\sqrt{2}}{\sqrt{11}}$

308 $\sqrt{57} = \sqrt{19} \cdot \sqrt{3}$ and $\sqrt{38} = \sqrt{19} \cdot \sqrt{2}$. Simplify $\frac{\sqrt{57}}{\sqrt{38}}$.

________ .

$\frac{\sqrt{3}}{\sqrt{2}}$

309 $\sqrt{39} = \sqrt{13} \cdot \sqrt{3}$ and $\sqrt{65} = \sqrt{13} \cdot \sqrt{5}$. Simplify $\frac{\sqrt{39}}{\sqrt{65}}$.

________ .

$\frac{\sqrt{3}}{\sqrt{5}}$

310 $\sqrt{14} = \sqrt{7} \cdot \sqrt{2}$ and $\sqrt{35} = \sqrt{7} \cdot \sqrt{5}$. Simplify $\frac{\sqrt{14}}{\sqrt{35}}$.

________ .

$\frac{\sqrt{2}}{\sqrt{5}}$

311 $\sqrt{30} = \sqrt{10} \cdot \sqrt{3}$ and $\sqrt{70} = \sqrt{10} \cdot \sqrt{7}$. Simplify $\dfrac{\sqrt{30}}{\sqrt{70}}$.

________ .

$\dfrac{\sqrt{3}}{\sqrt{7}}$

312 Simplify $\dfrac{\sqrt{15}}{\sqrt{35}}$. ________ .

$\dfrac{\sqrt{3}}{\sqrt{7}}$

313 Simplify $\dfrac{\sqrt{14}}{\sqrt{21}}$. ________ .

$\dfrac{\sqrt{2}}{\sqrt{3}}$

314 To simplify $\dfrac{\sqrt{19}}{\sqrt{38}}$, the following steps are used:

$$\frac{\sqrt{19}}{\sqrt{38}} = \frac{\sqrt{19} \cdot \sqrt{1}}{\sqrt{19} \cdot \sqrt{2}} = \frac{\sqrt{1}}{\sqrt{2}} = \frac{1}{\sqrt{2}}$$

Simplify $\dfrac{\sqrt{5}}{\sqrt{10}}$. ________ .

$\dfrac{1}{\sqrt{2}}$

315 Simplify $\dfrac{\sqrt{10}}{\sqrt{30}}$. ________ . [*Note:* $\sqrt{1}$ simplifies to 1.]

$\dfrac{1}{\sqrt{3}}$

316 Simplify $\dfrac{\sqrt{42}}{\sqrt{35}}$. ________ .

$\dfrac{\sqrt{6}}{\sqrt{5}}$

317 Simplify $\dfrac{\sqrt{26}}{\sqrt{6}}$. ________ .

$\dfrac{\sqrt{13}}{\sqrt{3}}$

318 Simplify $\dfrac{\sqrt{6}}{\sqrt{30}}$. ________ .

$\dfrac{1}{\sqrt{5}}$

319 Simplify $\dfrac{\sqrt{22}}{\sqrt{55}}$. ________ .

$\dfrac{\sqrt{2}}{\sqrt{5}}$

320 To simplify $\frac{\sqrt{14}}{\sqrt{2}}$, the following steps are used:

$$\frac{\sqrt{14}}{\sqrt{2}} = \frac{\sqrt{2} \cdot \sqrt{7}}{\sqrt{2} \cdot \sqrt{1}} = \frac{\sqrt{7}}{\sqrt{1}} = \sqrt{7}$$

Simplify $\frac{\sqrt{22}}{\sqrt{2}}$. ________ .

$\sqrt{11}$

321 $\frac{\sqrt{35}}{\sqrt{5}} = \frac{\sqrt{5} \cdot \sqrt{7}}{\sqrt{5} \cdot \sqrt{1}} = \frac{\sqrt{7}}{\sqrt{1}} = \frac{\sqrt{7}}{1} = \sqrt{7}.$

Simplify $\frac{\sqrt{30}}{\sqrt{6}}$. ________ .

$\sqrt{5}$

322 Simplify $\frac{\sqrt{38}}{\sqrt{19}}$. ________ .

$\sqrt{2}$

323 Simplify $\frac{\sqrt{55}}{\sqrt{11}}$. ________ .

$\sqrt{5}$

324 Simplify $\frac{\sqrt{26}}{\sqrt{22}}$. ________ .

$\frac{\sqrt{13}}{\sqrt{11}}$

325 Simplify $\frac{\sqrt{11}}{\sqrt{77}}$. ________ .

$\frac{1}{\sqrt{7}}$

326 Simplify $\frac{\sqrt{66}}{\sqrt{11}}$. ________ .

$\sqrt{6}$

327 Simplify $\frac{\sqrt{20}}{\sqrt{14}}$. ________ .

$\frac{\sqrt{10}}{\sqrt{7}}$

328 Simplify $\frac{\sqrt{26}}{\sqrt{2}}$. ________ .

$\sqrt{13}$

329 $\sqrt{2}$ is not equal to 2. Therefore, $\frac{\sqrt{2}}{2}$ is not equal to 1.

$\sqrt{3}$ is not equal to 3. Is $\frac{\sqrt{3}}{3}$ equal to 1?

________ . — No

330 $\sqrt{5}$ is not equal to 5. Therefore, $\frac{\sqrt{5}}{5}$ is not equal to 1.

$\sqrt{7}$ is not equal to 7. Is $\frac{\sqrt{7}}{7}$ equal to 1? ________ . — No

331 $\frac{\sqrt{11}}{11}$ is not equal to 1 because $\sqrt{11}$ is not equal to 11. Is

$\frac{\sqrt{19}}{19}$ equal to 1? ________ . — No

332 $\frac{17}{\sqrt{17}}$ is not equal to 1. Is $\frac{46}{\sqrt{46}}$ equal to 1? ________ . — No

333 $\frac{\sqrt{73}}{73}$ is not equal to 1. Is $\frac{\sqrt{29}}{29}$ equal to 1? ________ . — No

334 $\frac{\sqrt{10}}{2}$ *cannot* be simplified because $\frac{\sqrt{2}}{2}$ is not equal to 1.

$\frac{\sqrt{15}}{3} = \frac{\sqrt{3} \cdot \sqrt{5}}{3}$. Can $\frac{\sqrt{15}}{3}$ be simplified? ________ . — No

335 $\frac{\sqrt{22}}{4} = \frac{\sqrt{2} \cdot \sqrt{11}}{2 \cdot 2}$. $\frac{\sqrt{22}}{4}$ *cannot* be simplified because $\frac{\sqrt{2}}{2}$ is *not* equal to 1. $\frac{\sqrt{26}}{6} = \frac{\sqrt{2} \cdot \sqrt{13}}{2 \cdot 3}$. Can $\frac{\sqrt{26}}{4}$ be simplified? ________ . — No

336 $\frac{7}{\sqrt{14}}$ *cannot* be simplified because $\frac{7}{\sqrt{7}}$ is *not* equal to 1.

Can $\frac{13}{\sqrt{26}}$ be simplified? ________ . | No

337 $\frac{\sqrt{7}}{35}$ *cannot* be simplified because $\frac{\sqrt{7}}{7}$ is *not* equal to 1.

Can $\frac{\sqrt{5}}{45}$ be simplified? ________ . | No

338 $\frac{\sqrt{15}}{10} = \frac{\sqrt{5} \cdot \sqrt{3}}{5 \cdot 2} \cdot \frac{\sqrt{15}}{10}$ *cannot* be simplified because $\frac{\sqrt{5}}{5}$ is *not* equal to 1. $\frac{\sqrt{21}}{14} = \frac{\sqrt{7} \cdot \sqrt{3}}{7 \cdot 2} \cdot$ Can $\frac{\sqrt{21}}{14}$ be simplified? ________ . | No

339 $\frac{\sqrt{10}}{\sqrt{2}}$ can be simplified to $\sqrt{5}$ because both 10 and 2 are under radical signs.

$\frac{\sqrt{10}}{\sqrt{2}} = \frac{\sqrt{2} \cdot \sqrt{5}}{\sqrt{2}} = 1 \cdot \sqrt{5} = \sqrt{5}$. $\frac{\sqrt{10}}{2}$ *cannot* be simplified because 2 ________ (is, is not) under a radical sign. | is not

340 $\frac{\sqrt{21}}{\sqrt{3}}$ can be simplified to $\sqrt{7}$ because *both* 21 and 3 are under radical signs. $\frac{\sqrt{21}}{3}$ *cannot* be simplified, because 3 ________ (is, is not) under a radical sign. | is not

341 $\frac{\sqrt{30}}{6}$ *cannot* be simplified. $\frac{\sqrt{35}}{\sqrt{7}}$ can be simplified. Can $\frac{\sqrt{45}}{5}$ be simplified? ________ . | No

342 Can $\frac{\sqrt{30}}{\sqrt{6}}$ be simplified? ________ .

Yes, $\sqrt{5}$

343 Which of the following *cannot* be simplified?

________ .

(b), (d)

(a) $\frac{\sqrt{10}}{\sqrt{30}}$.

(b) $\frac{\sqrt{7}}{14}$.

(c) $\frac{\sqrt{17}}{\sqrt{34}}$.

(d) $\frac{\sqrt{5}}{10}$.

344 Which of the following *cannot* be simplified?

________ .

(d)

(a) $\frac{\sqrt{14}}{\sqrt{35}}$.

(b) $\frac{\sqrt{15}}{\sqrt{21}}$.

(c) $\frac{\sqrt{33}}{\sqrt{22}}$.

(d) $\frac{\sqrt{7}}{7}$.

345 Which of the following *cannot* be simplified?

________ .

(b), (d)

(a) $\frac{\sqrt{5}}{\sqrt{15}}$.

(b) $\frac{\sqrt{14}}{2}$.

(c) $\frac{\sqrt{38}}{\sqrt{19}}$.

(d) $\frac{\sqrt{15}}{3}$.

Self-Quiz # 6

The following questions test the objectives of the preceding section. 100% mastery is desired.

Simplify:

1. $\sqrt{45}$ = ________ .

2. $\dfrac{\sqrt{35}}{\sqrt{14}}$ = ________ .

3. $\dfrac{\sqrt{7}}{\sqrt{21}}$ = ________ .

4. $\sqrt{48}$ = ________ .

5. $^{-}\sqrt{81}$ = ________ .

6. $\dfrac{\sqrt{5}}{\sqrt{35}}$ = ________ .

7. $\dfrac{\sqrt{20}}{\sqrt{30}}$ = ________ .

8. $\dfrac{\sqrt{45}}{\sqrt{9}}$ = ________ .

9. Which of the following fractions *cannot* be simplified? ________ .

(a) $\dfrac{\sqrt{15}}{\sqrt{65}}$ (b) $\dfrac{\sqrt{23}}{\sqrt{23}}$ (c) $\dfrac{\sqrt{5}}{10}$ (d) $\dfrac{\sqrt{6}}{\sqrt{10}}$

(e) $\dfrac{\sqrt{11}}{\sqrt{55}}$ (f) $\dfrac{\sqrt{19}}{19}$ (g) $\dfrac{\sqrt{26}}{\sqrt{30}}$ (h) $\dfrac{\sqrt{6}}{\sqrt{42}}$

(i) $\dfrac{\sqrt{55}}{\sqrt{15}}$ (j) $\dfrac{\sqrt{22}}{4}$

In simplifying radical fractions, no radical should appear in the denominator of the fraction.

In the following section a method is shown for eliminating the radical from the denominator.

---

346 $\frac{\sqrt{3}}{\sqrt{3}} = 1.\quad \frac{\sqrt{5}}{\sqrt{5}} =$ ________ .

1

347 $\frac{\sqrt{13}}{\sqrt{13}} = 1.\quad \frac{\sqrt{29}}{\sqrt{29}} =$ ________ .

1

348 $\frac{\sqrt{47}}{\sqrt{47}} =$ ________ .

1

349 $\frac{\sqrt{913}}{\sqrt{913}} =$ ________ .

1

350 $\frac{\sqrt{2}}{\sqrt{2}} =$ ________ .

1

351 $\frac{\sqrt{3}}{\sqrt{3}} = 1$ and $\sqrt{9} = 3$. In eliminating the radical from the denominator of $\frac{\sqrt{7}}{\sqrt{3}}$ the fraction is multiplied by $\frac{\sqrt{3}}{\sqrt{3}}$ as follows:

$$\frac{\sqrt{7}}{\sqrt{3}} = \frac{\sqrt{7}}{\sqrt{3}} \cdot \frac{\sqrt{3}}{\sqrt{3}} = \frac{\sqrt{21}}{\sqrt{9}} = \frac{\sqrt{21}}{3}$$

Simplify $\frac{\sqrt{10}}{\sqrt{3}}$ by multiplying both the numerator and denominator by $\sqrt{3}$. ________ .

$\frac{\sqrt{30}}{3}$

352 In simplifying $\frac{\sqrt{3}}{\sqrt{5}}$ the following steps are used:

$$\frac{\sqrt{3}}{\sqrt{5}} = \frac{\sqrt{3}}{\sqrt{5}} \cdot \frac{\sqrt{5}}{\sqrt{5}} = \frac{\sqrt{15}}{\sqrt{25}} = \frac{\sqrt{15}}{5}$$

Simplify $\frac{\sqrt{7}}{\sqrt{2}}$ by multiplying both the numerator and denominator by $\sqrt{2}$. ________ .

$\frac{\sqrt{14}}{2}$

353 $\frac{\sqrt{15}}{\sqrt{2}}$ is simplified as follows:

$$\frac{\sqrt{15}}{\sqrt{2}} = \frac{\sqrt{15}}{\sqrt{2}} \cdot \frac{\sqrt{2}}{\sqrt{2}} = \frac{\sqrt{30}}{\sqrt{4}} = \frac{\sqrt{30}}{2}$$

Simplify $\frac{\sqrt{11}}{\sqrt{2}}$ by first multiplying both the numerator and denominator by $\sqrt{2}$. ________ .

$\frac{\sqrt{22}}{2}$

354 Simplify $\frac{\sqrt{7}}{\sqrt{5}}$ by first multiplying both the numerator and denominator by $\sqrt{5}$. ________ .

$\frac{\sqrt{35}}{5}$

355 Simplify $\frac{\sqrt{3}}{\sqrt{10}}$ by multiplying both the numerator and denominator by $\sqrt{10}$. ________ .

$\frac{\sqrt{30}}{10}$

356 Simplify $\frac{\sqrt{7}}{\sqrt{11}}$. ________ .

$\frac{\sqrt{77}}{11}$

357 Simplify $\frac{\sqrt{13}}{\sqrt{2}}$. ________ .

$\frac{\sqrt{26}}{2}$

358 Simplify $\frac{\sqrt{21}}{\sqrt{2}}$. ________ .

$\frac{\sqrt{42}}{2}$

359 Simplify $\frac{\sqrt{11}}{\sqrt{10}}$. ________ .

$\frac{\sqrt{110}}{10}$

360 Simplify $\frac{\sqrt{23}}{\sqrt{3}}$. ________ .

$\frac{\sqrt{69}}{3}$

361 Simplify $\frac{\sqrt{5}}{\sqrt{11}}$. ________ .

$\frac{\sqrt{55}}{11}$

362 The first step in simplifying $\frac{\sqrt{15}}{\sqrt{10}}$ is to write $\frac{\sqrt{15}}{\sqrt{10}}$ as $\frac{\sqrt{3}}{\sqrt{2}}$. The simplification is then completed by multiplying both the numerator and denominator of $\frac{\sqrt{3}}{\sqrt{2}}$ by $\sqrt{2}$.

$\frac{\sqrt{15}}{\sqrt{10}} = \frac{\sqrt{3}}{\sqrt{2}} = \frac{\sqrt{3}}{\sqrt{2}} \cdot \frac{\sqrt{2}}{\sqrt{2}} = \frac{\sqrt{6}}{\sqrt{4}} =$ ________ .

$\frac{\sqrt{6}}{2}$

363 Complete the following simplification of $\frac{\sqrt{6}}{\sqrt{21}}$:

$\frac{\sqrt{6}}{\sqrt{21}} = \frac{\sqrt{2}}{\sqrt{7}} =$ ________ .

$\frac{\sqrt{14}}{7}$

[*Note:* Multiply both the numerator and denominator of $\frac{\sqrt{2}}{\sqrt{7}}$ by $\sqrt{7}$.]

364 Complete the following simplification of $\frac{\sqrt{35}}{\sqrt{15}}$:

$\frac{\sqrt{35}}{\sqrt{15}} = \frac{\sqrt{7}}{\sqrt{3}} =$ ________ .

$\frac{\sqrt{21}}{3}$

365 Simplify $\frac{\sqrt{55}}{\sqrt{30}}$ by first dividing out the common factor, $\sqrt{5}$. ______ .

$\frac{\sqrt{66}}{6}$

366 Simplify $\frac{\sqrt{21}}{\sqrt{33}}$ by first dividing out the common factor, $\sqrt{3}$. ______ .

$\frac{\sqrt{77}}{11}$

367 Simplify $\frac{\sqrt{6}}{\sqrt{10}}$. ______ .

$\frac{\sqrt{15}}{5}$

368 Simplify $\frac{\sqrt{6}}{\sqrt{21}}$. ______ .

$\frac{\sqrt{14}}{7}$

369 Simplify $\frac{\sqrt{10}}{\sqrt{15}}$. ______ .

$\frac{\sqrt{6}}{3}$

370 Simplify $\frac{\sqrt{30}}{\sqrt{35}}$. ______ .

$\frac{\sqrt{42}}{7}$

371 Simplify $\frac{\sqrt{14}}{\sqrt{10}}$. ______ .

$\frac{\sqrt{35}}{5}$

372 Simplify $\frac{\sqrt{22}}{\sqrt{77}}$. ______ .

$\frac{\sqrt{14}}{7}$

373 Simplify $\frac{\sqrt{15}}{\sqrt{65}}$. ______ .

$\frac{\sqrt{39}}{13}$

374 To simplify $\frac{2\sqrt{3}}{\sqrt{5}}$, multiply both numerator and denominator by $\sqrt{5}$, as follows:

$$\frac{2\sqrt{3}}{\sqrt{5}} = \frac{2\sqrt{3}}{\sqrt{5}} \cdot \frac{\sqrt{5}}{\sqrt{5}} = \frac{2\sqrt{15}}{\sqrt{25}} = \frac{2\sqrt{15}}{5}$$

Complete the following simplification:

$\frac{5\sqrt{7}}{\sqrt{3}} = \frac{5\sqrt{7}}{\sqrt{3}} \cdot \frac{\sqrt{3}}{\sqrt{3}} =$ ________ .

$\frac{5\sqrt{21}}{3}$

375 Complete the simplification:

$\frac{4\sqrt{7}}{\sqrt{3}} = \frac{4\sqrt{7}}{\sqrt{3}} \cdot \frac{\sqrt{3}}{\sqrt{3}} = \frac{4\sqrt{21}}{\sqrt{9}} =$ ________ .

$\frac{4\sqrt{21}}{3}$

376 Simplify $\frac{3\sqrt{10}}{\sqrt{7}}$ by first multiplying both the numerator and denominator by $\sqrt{7}$. ________ .

$\frac{3\sqrt{70}}{7}$

377 Simplify $\frac{5\sqrt{3}}{\sqrt{2}}$. ________ .

$\frac{5\sqrt{6}}{2}$

378 Simplify $\frac{5\sqrt{3}}{\sqrt{11}}$. ________ .

$\frac{5\sqrt{33}}{11}$

379 To simplify $\frac{\sqrt{6}}{2\sqrt{5}}$, multiply both the numerator and denominator by $\sqrt{5}$, as follows:

$$\frac{\sqrt{6}}{2\sqrt{5}} = \frac{\sqrt{6}}{2\sqrt{5}} \cdot \frac{\sqrt{5}}{\sqrt{5}} = \frac{\sqrt{30}}{2\sqrt{25}} = \frac{\sqrt{30}}{2 \cdot 5} = \frac{\sqrt{30}}{10}$$

Complete the following simplification:

$\frac{\sqrt{10}}{7\sqrt{3}} = \frac{\sqrt{10}}{7\sqrt{3}} \cdot \frac{\sqrt{3}}{\sqrt{3}} = \frac{\sqrt{30}}{7\sqrt{9}} =$ ________ .

$\frac{\sqrt{30}}{21}$

380 Complete the simplification:

$$\frac{\sqrt{11}}{5\sqrt{3}} = \frac{\sqrt{11}}{5\sqrt{3}} \cdot \frac{\sqrt{3}}{\sqrt{3}} = \frac{\sqrt{33}}{5\sqrt{9}} = \_\_\_\_\_\_ .$$

$$\frac{\sqrt{33}}{15}$$

381 Complete the simplification:

$$\frac{\sqrt{5}}{3\sqrt{7}} = \frac{\sqrt{5}}{3\sqrt{7}} \cdot \frac{\sqrt{7}}{\sqrt{7}} = \frac{\sqrt{35}}{3\sqrt{49}} = \_\_\_\_\_\_ .$$

$$\frac{\sqrt{35}}{21}$$

382 Simplify $\frac{\sqrt{13}}{2\sqrt{5}}$ by first multiplying both the numerator and denominator by $\sqrt{5}$. ______ .

$$\frac{\sqrt{65}}{10}$$

383 Simplify $\frac{\sqrt{7}}{2\sqrt{11}}$. ______ .

$$\frac{\sqrt{77}}{22}$$

384 Simplify $\frac{\sqrt{14}}{5\sqrt{3}}$. ______ .

$$\frac{\sqrt{42}}{15}$$

385 Simplify $\frac{3\sqrt{6}}{\sqrt{11}}$. ______ .

$$\frac{3\sqrt{66}}{11}$$

386 Simplify $\frac{\sqrt{7}}{5\sqrt{3}}$. ______ .

$$\frac{\sqrt{21}}{15}$$

387 Simplify $\frac{5\sqrt{2}}{\sqrt{3}}$. ______ .

$$\frac{5\sqrt{6}}{3}$$

388 Complete the simplification:

$$\frac{8\sqrt{3}}{5\sqrt{7}} = \frac{8\sqrt{3}}{5\sqrt{7}} \cdot \frac{\sqrt{7}}{\sqrt{7}} = \frac{8\sqrt{21}}{5\sqrt{49}} = \_\_\_\_\_\_ .$$

$$\frac{8\sqrt{21}}{35}$$

389 Complete the simplification:

$$\frac{3\sqrt{6}}{7\sqrt{5}} = \frac{3\sqrt{6}}{7\sqrt{5}} \cdot \frac{\sqrt{5}}{\sqrt{5}} = \frac{3\sqrt{30}}{7\sqrt{25}} = \_\_\_\_\_\_ .$$

$\frac{3\sqrt{30}}{35}$

390 Simplify $\frac{5\sqrt{10}}{2\sqrt{7}}$. ______ .

$\frac{5\sqrt{70}}{14}$

391 Simplify $\frac{7\sqrt{3}}{2\sqrt{5}}$. ______ .

$\frac{7\sqrt{15}}{10}$

392 $\frac{4\sqrt{5}}{6}$ can be simplified to $\frac{2\sqrt{5}}{3}$ because 2 is a common factor of 4 and 6.

$$\frac{4\sqrt{5}}{6} = \frac{2 \cdot 2\sqrt{5}}{2 \cdot 3} = \frac{2\sqrt{5}}{3}$$

Simplify $\frac{8\sqrt{17}}{10}$. ______ .

$\frac{4\sqrt{17}}{5}$

393 Simplify $\frac{9\sqrt{7}}{6}$. ______ .

$\frac{3\sqrt{7}}{2}$

394 Simplify $\frac{5\sqrt{17}}{15}$. ______ .

$\frac{\sqrt{17}}{3}$

395 Simplify $\frac{10\sqrt{3}}{5}$. ______ .

$2\sqrt{3}$

396 Simplify $\frac{14\sqrt{3}}{21}$. ______ .

$\frac{2\sqrt{3}}{3}$

397 To simplify $\frac{4\sqrt{5}}{\sqrt{6}}$, the following steps are used:

$$\frac{4\sqrt{5}}{\sqrt{6}} = \frac{4\sqrt{5}}{\sqrt{6}} \frac{\sqrt{6}}{\sqrt{6}} = \frac{4\sqrt{30}}{6} = \frac{2\sqrt{30}}{3}$$

Simplify $\frac{6\sqrt{3}}{\sqrt{10}}$ by first multiplying both the numerator and denominator by $\sqrt{10}$. _______ .

$\frac{3\sqrt{30}}{5}$

398 Simplify $\frac{9\sqrt{5}}{\sqrt{6}}$ · _______ .

$\frac{3\sqrt{30}}{2}$

399 Simplify $\frac{4\sqrt{3}}{\sqrt{14}}$ · _______ .

$\frac{2\sqrt{42}}{7}$

400 Simplify $\frac{6\sqrt{7}}{\sqrt{3}}$ · _______ .

$2\sqrt{21}$

401 Simplify $\frac{5\sqrt{3}}{\sqrt{5}}$ · _______ .

$\sqrt{15}$

402 Simplify $\frac{15\sqrt{7}}{2\sqrt{5}}$ · _______ .

$\frac{3\sqrt{35}}{2}$

403 Simplify $\frac{12\sqrt{5}}{\sqrt{3}}$ · _______ .

$4\sqrt{15}$

404 To simplify $\frac{\sqrt{11}}{\sqrt{12}}$, first simplify $\sqrt{12}$.

$$\frac{\sqrt{11}}{\sqrt{12}} = \frac{\sqrt{11}}{\sqrt{4} \cdot \sqrt{3}} = \frac{\sqrt{11}}{2\sqrt{3}} = \frac{\sqrt{11} \cdot \sqrt{3}}{2\sqrt{3} \cdot \sqrt{3}} = \frac{\sqrt{33}}{6}$$

Simplify $\frac{\sqrt{5}}{\sqrt{18}}$ by first simplifying $\sqrt{18}$. _______ .

$\frac{\sqrt{10}}{6}$

405 Simplify $\frac{\sqrt{3}}{\sqrt{50}}$ by first simplifying $\sqrt{50}$. ________ .

$\frac{\sqrt{6}}{10}$

406 Simplify $\frac{\sqrt{6}}{\sqrt{125}}$ by first simplifying $\sqrt{125}$. ________ .

$\frac{\sqrt{30}}{25}$

407 Simplify $\frac{2\sqrt{7}}{\sqrt{8}}$ by first simplifying $\sqrt{8}$. ________ .

$\frac{\sqrt{14}}{2}$

408 Simplify $\frac{5}{\sqrt{18}}$. ________ .

$\frac{5\sqrt{2}}{6}$

409 Simplify $\frac{12\sqrt{5}}{\sqrt{6}}$. ________ .

$2\sqrt{30}$

410 Simplify $\frac{10\sqrt{11}}{\sqrt{32}}$. ________ .

$\frac{5\sqrt{22}}{4}$

411 Simplify $\frac{14\sqrt{7}}{\sqrt{18}}$. ________ .

$\frac{7\sqrt{14}}{3}$

412 Simplify $\frac{18\sqrt{5}}{\sqrt{72}}$. ________ .

$\frac{3\sqrt{10}}{2}$

413 To simplify $\frac{\sqrt{27}}{\sqrt{20}}$, first simplify both $\sqrt{27}$ and $\sqrt{20}$.

$$\frac{\sqrt{27}}{\sqrt{20}} = \frac{\sqrt{9} \cdot \sqrt{3}}{\sqrt{4} \cdot \sqrt{5}} = \frac{3\sqrt{3}}{2\sqrt{5}} = \frac{3\sqrt{3}}{2\sqrt{5}} \cdot \frac{\sqrt{5}}{\sqrt{5}} = \frac{3\sqrt{15}}{10}$$

Simplify $\frac{\sqrt{32}}{\sqrt{75}}$ by first simplifying both $\sqrt{32}$ and $\sqrt{75}$.

________ .

$\frac{4\sqrt{6}}{15}$

414 Simplify $\frac{\sqrt{8}}{\sqrt{27}}$ by first simplifying both $\sqrt{8}$ and $\sqrt{27}$.

_______ .

$\frac{2\sqrt{6}}{9}$

415 Simplify $\frac{\sqrt{18}}{\sqrt{125}}$ by first simplifying both $\sqrt{18}$ and $\sqrt{125}$.

_______ .

$\frac{3\sqrt{10}}{25}$

416 Simplify $\frac{\sqrt{28}}{\sqrt{75}}$. _______ .

$\frac{2\sqrt{21}}{15}$

417 Simplify $\frac{\sqrt{20}}{\sqrt{63}}$. _______ .

$\frac{2\sqrt{35}}{21}$

418 Simplify $\frac{3\sqrt{32}}{\sqrt{45}}$ by first simplifying both the numerator and denominator. _______ .

$\frac{4\sqrt{10}}{5}$

419 Simplify $\frac{5\sqrt{27}}{\sqrt{72}}$. _______ .

$\frac{5\sqrt{6}}{4}$

420 Simplify $\frac{3\sqrt{5}}{\sqrt{12}}$. _______ .

$\frac{\sqrt{15}}{2}$

421 Simplify $\frac{8\sqrt{5}}{\sqrt{32}}$. _______ .

$\sqrt{10}$

422 $\sqrt{\frac{3}{5}}$ can be written $\frac{\sqrt{3}}{\sqrt{5}}$.

$\sqrt{\frac{15}{22}}$ can be written as _______ .

$\frac{\sqrt{15}}{\sqrt{22}}$

423 $\sqrt{\frac{2}{3}}$ can be written as $\frac{\sqrt{2}}{\sqrt{3}}$.

$\sqrt{\frac{5}{7}}$ can be written as ______ .

$\frac{\sqrt{5}}{\sqrt{7}}$

424 $\sqrt{\frac{10}{11}}$ can be written as ______ .

$\frac{\sqrt{10}}{\sqrt{11}}$

425 $\sqrt{\frac{5}{13}}$ can be written as ______ .

$\frac{\sqrt{5}}{\sqrt{13}}$

426 $\sqrt{\frac{7}{19}}$ can be written as ______ .

$\frac{\sqrt{7}}{\sqrt{19}}$

427 To simplify $\sqrt{\frac{3}{5}}$, first write the fraction as $\frac{\sqrt{3}}{\sqrt{5}}$. Simplify $\frac{\sqrt{3}}{\sqrt{5}}$ by multiplying both numerator and denominator by $\sqrt{5}$. ______ .

$\frac{\sqrt{15}}{5}$

428 To simplify $\sqrt{\frac{3}{4}}$, first write the fraction $\frac{\sqrt{3}}{\sqrt{4}}$. Simplify $\frac{\sqrt{3}}{\sqrt{4}}$. ______ .

$\frac{\sqrt{3}}{2}$

429 To simplify $\sqrt{\frac{9}{10}}$, first write $\frac{\sqrt{9}}{\sqrt{10}}$. Simplify $\frac{\sqrt{9}}{\sqrt{10}}$. ______ .

$\frac{3\sqrt{10}}{10}$

430 To simplify $\sqrt{\frac{7}{12}}$, first write $\frac{\sqrt{7}}{\sqrt{12}}$. Simplify $\sqrt{\frac{7}{12}}$. ______ .

$\frac{\sqrt{21}}{6}$

431 Simplify $\sqrt{\frac{2}{5}}$. ______ .

$\frac{\sqrt{10}}{5}$

In simplifying a radical fraction the following four-step process should be used:

1. Reduce the fraction by dividing out any common factors.

$$\frac{8\sqrt{35}}{12\sqrt{90}} = \frac{2\sqrt{7}}{3\sqrt{18}}$$

2. Simplify separately the radical expressions in the numerator and denominator.

$$\frac{2\sqrt{7}}{3\sqrt{18}} = \frac{2\sqrt{7}}{3 \cdot 3\sqrt{2}} = \frac{2\sqrt{7}}{9\sqrt{2}}$$

3. Multiply the numerator and denominator by the radical expression that will make the radicand in the denominator a perfect square.

$$\frac{2\sqrt{7}}{9\sqrt{2}} = \frac{2\sqrt{7}}{9\sqrt{2}} \cdot \frac{\sqrt{2}}{\sqrt{2}} = \frac{2\sqrt{14}}{9\sqrt{4}} = \frac{2\sqrt{14}}{18}$$

4. Reduce the common fraction again, if possible.

$$\frac{2\sqrt{14}}{18} = \frac{\sqrt{14}}{9}$$

---

432 Simplify $\sqrt{\frac{2}{7}}$. ________ . $\frac{\sqrt{14}}{7}$

433 Simplify $\frac{10\sqrt{3}}{\sqrt{5}}$. ________ . $2\sqrt{15}$

434 Simplify $\frac{\sqrt{20}}{\sqrt{4}}$. ________ . $\sqrt{5}$

435 Simplify $^{-}\sqrt{64}$. ________ . $^{-}8$

436 Simplify $\frac{3\sqrt{32}}{8}$. ________ . $\frac{3\sqrt{2}}{2}$

437 Simplify $\frac{\sqrt{75}}{\sqrt{25}}$. ________ . $\sqrt{3}$

438 Simplify $\frac{6}{5\sqrt{18}}$. ________ . $\frac{\sqrt{2}}{5}$

439 Simplify $\frac{2}{\sqrt{5}}$. ________ . $\frac{2\sqrt{5}}{5}$

440 Simplify $\sqrt{\frac{5}{3}}$. ________ . $\frac{\sqrt{15}}{3}$

441 Simplify $\sqrt{54}$. ________ . $3\sqrt{6}$

442 Simplify $\frac{3\sqrt{200}}{\sqrt{48}}$. ________ . $\frac{5\sqrt{6}}{2}$

443 Simplify $\frac{3\sqrt{32}}{\sqrt{45}}$. ________ . $\frac{4\sqrt{10}}{5}$

444 Simplify $\frac{8}{\sqrt{6}}$. ________ . $\frac{4\sqrt{6}}{3}$

445 Simplify $\frac{3}{\sqrt{3}}$. ________ . $\sqrt{3}$

446 Simplify $\frac{15\sqrt{7}}{2\sqrt{3}}$. ________ . $\frac{5\sqrt{21}}{2}$

Self-Quiz # 7

The following questions test the objectives of the preceding section. 100% mastery is desired.

Simplify each of the following:

1. $\frac{5\sqrt{2}}{\sqrt{7}}$. ________ .

2. $\frac{\sqrt{30}}{\sqrt{5}}$. ________ .

3. $^{-}\sqrt{81}$. ________ .

4. $\frac{2\sqrt{50}}{15}$. ________ .

5. $\sqrt{\frac{4}{7}}$. ________ .

6. $\frac{3}{2\sqrt{5}}$. ________ .

7. $\sqrt{\frac{2}{3}}$. ________ .

8. $\frac{7}{\sqrt{7}}$. ________ .

9. $\frac{3\sqrt{5}}{4\sqrt{3}}$. ________ .

10. $\frac{\sqrt{32}}{2}$. ________ .

The addition of $2\sqrt{7} + 3\sqrt{7}$ is accomplished using the Distributive Law of Multiplication over Addition in a manner similar to the addition of $2x + 3x$.

$2x + 3x$	$2\sqrt{7} + 3\sqrt{7}$
$(2 + 3)x$	$(2 + 3) \cdot \sqrt{7}$
$5x$	$5\sqrt{7}$

In the following section the addition of radical expressions is explained.

---

447 To add $4x + 3x$, the Distributive Law of Multiplication over Addition is used as follows:

$$4x + 3x$$
$$(4 + 3)x$$
$$7x$$

To add $5\sqrt{2} + 6\sqrt{2}$, the Distributive Law of Multiplication over Addition is used as follows:
$5\sqrt{2} + 6\sqrt{2}$
$(5 + 6) \cdot \sqrt{2}$
________ .

$11\sqrt{2}$

448 $2x$ and $7x$ are called *like terms* because each has the same letter, x. $2x + 7x$ can be simplified to $9x$ by adding the coefficients of x. Can $5y + 7z$ be simplified? ________ .

No

449 $4\sqrt{3}$ and $7\sqrt{3}$ are called *like terms* because each has the same radical expression, $\sqrt{3}$. Are $8\sqrt{5}$ and $5\sqrt{8}$ like terms? ________________ .

No, $\sqrt{5} \neq \sqrt{8}$

450 To simplify $4\sqrt{3} + 7\sqrt{3}$, the like terms are combined.

$$4\sqrt{3} + 7\sqrt{3} = 11\sqrt{3}$$

Are $8\sqrt{2}$ and $9\sqrt{2}$ like terms? ________________ .

Yes, $\sqrt{2} = \sqrt{2}$

451 To simplify $8\sqrt{2} + 9\sqrt{2}$, the like terms are combined.

$$8\sqrt{2} + 9\sqrt{2} = 17\sqrt{2}$$

Simplify $6\sqrt{13} + 15\sqrt{13}$. _______ . — $21\sqrt{13}$

452 $9\sqrt{7} - 4\sqrt{7} = 5\sqrt{7}$.
$6\sqrt{5} - 2\sqrt{5} =$ _______ . — $4\sqrt{5}$

453 $9\sqrt{11} - 12\sqrt{11} =$ _______ . — $^{-}3\sqrt{11}$

454 $^{-}4\sqrt{17} + 2\sqrt{17} =$ _______ . — $^{-}2\sqrt{17}$

455 $3\sqrt{17} + \sqrt{17} =$ _______ . — $4\sqrt{17}$
[*Note:* $\sqrt{17}$ is equivalent to $1\sqrt{17}$.]

456 $\sqrt{5} + 8\sqrt{5} =$ _______ . — $9\sqrt{5}$

457 $\sqrt{7} - 3\sqrt{7} =$ _______ . — $^{-}2\sqrt{7}$

458 $4\sqrt{11} - \sqrt{11} =$ _______ . — $3\sqrt{11}$

459 To add $3 + 2\sqrt{17} + 5 + 9\sqrt{17}$, the like terms are combined.

$$3 + 2\sqrt{17} + 5 + 9\sqrt{17}$$
$$3 + 5 + 2\sqrt{17} + 9\sqrt{17}$$
$$8 + 11\sqrt{17}$$

Complete the following by combining like terms:
$6 + 3\sqrt{7} + 5 + 9\sqrt{7}$
$6 + 5 + 3\sqrt{7} + 9\sqrt{7}$
____________ . — $11 + 12\sqrt{7}$

460 $5 + 7\sqrt{3} + 9 + \sqrt{3} =$ ____________ . — $14 + 8\sqrt{3}$

461 Complete the following:
$8 - 3\sqrt{21} + 7 + 5\sqrt{21}$
$8 + 7 - 3\sqrt{21} + 5\sqrt{21}$
____________ . — $15 + 2\sqrt{21}$

462 Complete the following:
$7 + 2\sqrt{5} + 2 - 6\sqrt{5}$
$7 + 2 + 2\sqrt{5} - 6\sqrt{5}$
________ .

$9 - 4\sqrt{5}$

463 $9 - 6\sqrt{10} - 6 - 4\sqrt{10} =$ ________________ .

$3 - 10\sqrt{10}$

464 $6 - 7\sqrt{3} + 4 + 8\sqrt{3} =$ ________ .

$10 + \sqrt{3}$

465 $6 + 17\sqrt{5} + 4 - 8\sqrt{5} =$ ________________ .

$10 + 9\sqrt{5}$

466 $3 - \sqrt{7} - 8 + 4\sqrt{7} =$ ________________ .

$^{-}5 + 3\sqrt{7}$

467 The expression $3x + 5y$ *cannot* be simplified because $3x$ and $5y$ are not like terms. $3\sqrt{2} + 5\sqrt{7}$ cannot be simplified because $3\sqrt{2}$ and $5\sqrt{7}$ have different radicands. Can $7\sqrt{3} + 5\sqrt{11}$ be simplified? ________ .

No

468 $4\sqrt{5} + 9\sqrt{7}$ cannot be simplified because $4\sqrt{5}$ and $9\sqrt{7}$ have different radicands. Can $5\sqrt{17} + 6\sqrt{11}$ be simplified? ________ .

No

469 Can $7\sqrt{10} - 6\sqrt{2}$ be simplified? ________ .

No

470 $4\sqrt{7} + 5\sqrt{7}$ can be simplified because $4\sqrt{7}$ and $5\sqrt{7}$ have the common radical factor $\sqrt{7}$ and are like terms. Can $8\sqrt{3} + 17\sqrt{3}$ be simplified? ________________ .

Yes, $25\sqrt{3}$

471 Can $5\sqrt{2} + 2\sqrt{15}$ be simplified? ________ .

No

472 $2\sqrt{3} + 4\sqrt{7} + 5\sqrt{3} - \sqrt{7}$ is simplified by combining like terms.

$$2\sqrt{3} + 4\sqrt{7} + 5\sqrt{3} - \sqrt{7}$$
$$2\sqrt{3} + 5\sqrt{3} + 4\sqrt{7} - \sqrt{7}$$
$$7\sqrt{3} + 3\sqrt{7}$$

Simplify $5\sqrt{10} - 3\sqrt{6} + 4\sqrt{10} + 8\sqrt{6}$.
________________ .

$9\sqrt{10} + 5\sqrt{6}$

473 Simplify $2\sqrt{11} + 4\sqrt{13} - \sqrt{11} + 9\sqrt{13}$.

________________ .

$\sqrt{11} + 13\sqrt{13}$

474 Simplify $4\sqrt{6} + 2\sqrt{6} - 3\sqrt{5} + 5\sqrt{5}$.

________________ .

$6\sqrt{6} + 2\sqrt{5}$

475 Even though the numbers under the radical signs in $6\sqrt{5} + \sqrt{20}$ are not the same, the addition is possible because $\sqrt{20}$ can be simplified to $2\sqrt{5}$. $6\sqrt{5} + \sqrt{20}$ is simplified as follows:

$$6\sqrt{5} + \sqrt{20}$$
$$6\sqrt{5} + \sqrt{4} \cdot \sqrt{5}$$
$$6\sqrt{5} + 2\sqrt{5}$$
$$8\sqrt{5}$$

Add $\sqrt{18} + 10\sqrt{2}$ by first simplifying $\sqrt{18}$.

________ .

$13\sqrt{2}$

476 $2\sqrt{3} + 5\sqrt{12}$ can be simplified as follows:

$$2\sqrt{3} + 5\sqrt{12}$$
$$2\sqrt{3} + 5\sqrt{4} \cdot \sqrt{3}$$
$$2\sqrt{3} + 10\sqrt{3}$$
$$12\sqrt{3}$$

Add $6\sqrt{2} + 3\sqrt{50}$ by first simplifying $3\sqrt{50}$.

________ .

$21\sqrt{2}$

477 $8\sqrt{5} + \sqrt{45}$ is simplified as follows:

$$8\sqrt{5} + \sqrt{45}$$
$$8\sqrt{5} + \sqrt{9} \cdot \sqrt{5}$$
$$8\sqrt{5} + 3\sqrt{5}$$
$$11\sqrt{5}$$

Add $5\sqrt{6} + \sqrt{24}$ by first simplifying $\sqrt{24}$.

________ .

$7\sqrt{6}$

478 Add $2\sqrt{6} + \sqrt{150}$ by first simplifying $\sqrt{150}$.

________ .

$7\sqrt{6}$

479 Add $6\sqrt{2} + 3\sqrt{8}$ by first simplifying $3\sqrt{8}$.
________ . $12\sqrt{2}$

480 Add $4\sqrt{7} + \sqrt{63}$ by first simplifying $\sqrt{63}$. ________ . $7\sqrt{7}$

481 Add $\sqrt{27} + \sqrt{12}$ by first simplifying both $\sqrt{27}$ and $\sqrt{12}$ as follows:

$$\sqrt{27} + \sqrt{12}$$
$$3\sqrt{3} + 2\sqrt{3}$$
$$5\sqrt{3}$$

Add $\sqrt{20} + \sqrt{45}$ by first simplifying both $\sqrt{20}$ and $\sqrt{45}$.
________ . $5\sqrt{5}$

482 Add $\sqrt{8} + \sqrt{18}$ by first simplifying both $\sqrt{8}$ and $\sqrt{18}$.
________ . $5\sqrt{2}$

483 $6\sqrt{3} + 5\sqrt{27} =$ ________ . $21\sqrt{3}$

484 $5\sqrt{18} - 6\sqrt{2} =$ ________ . $9\sqrt{2}$

485 $5\sqrt{2} + \sqrt{32} =$ ________ . $9\sqrt{2}$

486 $6\sqrt{8} - 3\sqrt{50} =$ ________ . $^{-}3\sqrt{2}$

487 $5\sqrt{40} - \sqrt{90} =$ ________ . $7\sqrt{10}$

488 $\sqrt{54} + \sqrt{6} =$ ________ . $4\sqrt{6}$

489 $2\sqrt{6} + \sqrt{54} =$ ________ . $5\sqrt{6}$

Self-Quiz # 8

The following questions test the objectives of the preceding section. 100% mastery is desired.

Simplify:

1. $5\sqrt{2} + 6\sqrt{2} =$ ________ .

2. $8\sqrt{17} - 6\sqrt{17} =$ ________ .

3. $3\sqrt{5} - 2\sqrt{7} + 5\sqrt{7} - \sqrt{5} =$ ________________ .

4. $6\sqrt{13} + 4 + 5\sqrt{13} - 9 =$ ________________ .

5. $4\sqrt{12} + 5\sqrt{3} =$ ________ .

6. $9\sqrt{2} - \sqrt{50} =$ ________ .

7. $3\sqrt{20} + \sqrt{45} =$ ________ .

8. $6\sqrt{63} - 5\sqrt{28} =$ ________ .

Chapter Summary

Rational numbers such as 4, 0, $^{-}17$, $\frac{3}{5}$, $\frac{^{-}7}{2}$, $\sqrt{16}$, and $^{-}\sqrt{\frac{4}{9}}$ were studied in this chapter. Also studied were irrational numbers such as $\sqrt{3}$, $^{-}\sqrt{17}$, $\sqrt{\frac{3}{4}}$, and $\sqrt{\frac{2}{5}}$.

The arithmetic of radical expressions was explained:

1. Like terms (terms containing the same radical expression) can be added.

$$5\sqrt{3} + 7\sqrt{3} = 12\sqrt{3}$$

2. Radical expressions can be multiplied by multiplying their radicands.

$$\sqrt{7} \cdot \sqrt{5} = \sqrt{35}$$

Expressions involving radicals were simplified:

1. $\sqrt{18} = \sqrt{9} \cdot \sqrt{2} = 3\sqrt{2}$.
2. $\dfrac{\sqrt{15}}{\sqrt{55}} = \dfrac{\sqrt{3} \cdot \sqrt{5}}{\sqrt{11} \cdot \sqrt{5}} = \dfrac{\sqrt{3}}{\sqrt{11}} \cdot \dfrac{\sqrt{11}}{\sqrt{11}} = \dfrac{\sqrt{33}}{11}$.
3. $\dfrac{6\sqrt{7}}{5\sqrt{2}} = \dfrac{6\sqrt{7}}{5\sqrt{2}} \cdot \dfrac{\sqrt{2}}{\sqrt{2}} = \dfrac{6\sqrt{14}}{10} = \dfrac{3\sqrt{14}}{5}$.

CHAPTER POST-TEST

The following questions test the objectives of this chapter. A score of 90% indicates sufficient mastery, and the student may proceed to the next chapter.

I Which of the following are perfect square integers?

(a) 64 (b) 0 (c) $^{-}9$ (d) 18 (e) 49

II Which of the following are perfect squares?

(a) $\frac{25}{16}$ (b) $\frac{^{-}9}{64}$ (c) $\frac{18}{50}$ (d) $\frac{8}{2}$ (e) $\frac{17}{36}$

III Which of the following are rational numbers?

(a) $\sqrt{\frac{16}{25}}$ (b) $\sqrt{\frac{3}{4}}$ (c) $^{-}\sqrt{49}$ (d) $\sqrt{75}$ (e) $^{-}\sqrt{\frac{1}{9}}$

IV Simplify:

1. $4\sqrt{72}$

2. $^{-}5\sqrt{54}$

3. $^{-}\sqrt{49}$

4. $\frac{\sqrt{70}}{\sqrt{10}}$

5. $\frac{3}{\sqrt{13}}$

6. $\frac{\sqrt{18}}{\sqrt{6}}$

7. $\frac{4\sqrt{6}}{5\sqrt{10}}$

8. $\sqrt{\frac{2}{3}}$

9. $15 - 6\sqrt{11} + 4 + \sqrt{11}$

10. $2\sqrt{5} - 3\sqrt{20}$

11. $\sqrt{63} - 4\sqrt{7}$

12. $4\sqrt{12} + \sqrt{27}$

13. $9\sqrt{32} - 2\sqrt{8}$

14. $8\sqrt{28} + 3\sqrt{7}$

15. $\sqrt{18} - \sqrt{200}$

Solving Quadratic Equations with Irrational Solutions

1
2
3
4
5
6
7
8
9
10

CHAPTER PRE-TEST

The following questions indicate the objectives of this chapter. A score of 90% indicates sufficient mastery, and the student may immediately take the Chapter Post-Test.

Solve:

1. $x^2 = 9$

2. $x^2 = 31$

3. $x^2 = {}^{-}25$

4. $(x - 2)^2 = 49$

5. $(x + 8)^2 = 19$

6. $x^2 + 7x + 4 = 0$

7. $x^2 + x - 5 = 0$

8. $x^2 - 10x + 9 = 0$

9. $3x^2 - 5x - 1 = 0$

10. $2x^2 - 8x + 7 = 0$

11. $5x^2 - 2x - 4 = 0$

12. $x^2 + 7x = 30$

$x^2 + 5x + 6 = 0$ is a quadratic equation. In Chapter 7 the truth sets for some quadratic equations were found by factoring as follows:

$$x^2 + 5x + 6 = 0$$

$$(x + 3)(x + 2) = 0$$

Since the product of $(x + 3)$ and $(x + 2)$ is zero, then one of the factors must be zero. Hence, the truth set of $(x + 3)(x + 2) = 0$ is $\{^-2, ^-3\}$.

Notice that $x^2 + 5x + 3 = 0$ cannot be solved by factoring because the trinomial $x^2 + 5x + 3$ cannot be factored.

In this chapter truth sets for quadratic equations such as $x^2 + 5x + 3 = 0$ will be found.

---

1 $x^2 = 16$ is a quadratic equation. There are two numbers that can replace x to obtain a true statement. The truth set of $x^2 = 16$ is $\{4, ^-4\}$ because $4 \cdot 4 = 16$ and $^-4 \cdot {^-4} = 16$. What is the truth set of $x^2 = 25$? ________.

$\{5, ^-5\}$

2 The truth set of $x^2 = 81$ is $\{9, ^-9\}$ because $9 \cdot 9 = 81$ and $^-9 \cdot {^-9} = 81$. What is the truth set of $x^2 = 49$? ________.

$\{7, ^-7\}$

3 The truth set of $x^2 = 9$ is $\{3, ^-3\}$ because $3 \cdot 3 = 9$ and $^-3 \cdot {^-3} = 9$. What is the truth set of $x^2 = 1$? ________.

$\{1, ^-1\}$

4 The truth set of $x^2 = \frac{25}{16}$ is $\left\{\frac{5}{4}, \frac{^-5}{4}\right\}$ because $\frac{5}{4} \cdot \frac{5}{4} = \frac{25}{16}$ and $\frac{^-5}{4} \cdot \frac{^-5}{4} = \frac{25}{16}$. What is the truth set of $x^2 = \frac{36}{49}$? ________.

$\left\{\frac{6}{7}, \frac{^-6}{7}\right\}$

5 What is the truth set of $x^2 = \frac{16}{81}$? ________.

$\left\{\frac{4}{9}, \frac{^-4}{9}\right\}$

6 What is the truth set of $x^2 = \frac{1}{25}$? ________.

$\left\{\frac{1}{5}, \frac{^-1}{5}\right\}$

7 What is the truth set of $x^2 = \frac{1}{81}$? ________ . $\left\{\frac{1}{9}, \frac{^-1}{9}\right\}$

8 The truth set of $x^2 = 36$ is $\{6, {}^-6\}$. Could the truth set also be written as $\{\sqrt{36}, {}^-\sqrt{36}\}$? ________ . Yes
[*Note*: $(\sqrt{36})^2 = 36$ and $({}^-\sqrt{36})^2 = 36$.]

9 The truth set of $x^2 = 4$ is $\{2, {}^-2\}$. Could the truth set also be written as $\{\sqrt{4}, {}^-\sqrt{4}\}$? ________ . Yes

10 The truth set of $x^2 = 100$ is $\{10, {}^-10\}$. Could the truth set also be written as $\{\sqrt{100}, {}^-\sqrt{100}\}$? ________ . Yes

11 The truth set of $x^2 = 2$ is $\{\sqrt{2}, {}^-\sqrt{2}\}$ because $(\sqrt{2})^2 = 2$ and $({}^-\sqrt{2})^2 = 2$ are true statements. What is the truth set of $x^2 = 3$? ________ . $\{\sqrt{3}, {}^-\sqrt{3}\}$
[*Note*: $(\sqrt{3})^2 = 3$ and $({}^-\sqrt{3})^2 = 3$.]

12 The truth set of $x^2 = 5$ is $\{\sqrt{5}, {}^-\sqrt{5}\}$ because $(\sqrt{5})^2 = 5$ and $({}^-\sqrt{5})^2 = 5$ are true statements. What is the truth set of $x^2 = 6$? ________ . $\{\sqrt{6}, {}^-\sqrt{6}\}$

13 What is the truth set of $x^2 = 7$?________ . $\{\sqrt{7}, {}^-\sqrt{7}\}$
[*Note*: $(\sqrt{7})^2 = 7$ and $({}^-\sqrt{7})^2 = 7$.]

14 What is the truth set of $x^2 = 13$? ________________ . $\{\sqrt{13}, {}^-\sqrt{13}\}$

15 What is the truth set of $x^2 = 16$?
____________________ . $\{\sqrt{16}, {}^-\sqrt{16}\}$ or $\{4, {}^-4\}$, but the simplified answer is preferred

16 What is the truth set of $x^2 = 23$? ________________ . $\{\sqrt{23}, {}^-\sqrt{23}\}$

17 What is the truth set of $x^2 = 47$? ________________ . $\{\sqrt{47}, {}^-\sqrt{47}\}$

18 What is the truth set of $x^2 = 50$? ________________ . $\{5\sqrt{2}, {}^-5\sqrt{2}\}$
[*Note*: $\sqrt{50} = 5\sqrt{2}$.]

19 What is the truth set of $x^2 = 20$? ______________ . $\{2\sqrt{5}, {}^-2\sqrt{5}\}$

20 { } is the truth set of $x^2 = {}^-35$ because there is no rational or irrational number that can be multiplied by itself to give a negative product. What is the truth set of $x^2 = {}^-16$? ________ . { }

21 What is the truth set of $x^2 = {}^-49$? ________ . { }

22 What is the truth set of $x^2 = 19$? ______________ . $\{\sqrt{19}, {}^-\sqrt{19}\}$

23 What is the truth set of $x^2 = 81$? ________ . $\{9, {}^-9\}$

24 What is the truth set of $x^2 = 93$? ______________ . $\{\sqrt{93}, {}^-\sqrt{93}\}$

25 What is the truth set of $x^2 = {}^-5$? ________ . { }

26 What is the truth set of $x^2 = {}^-12$? ________ . { }

27 What is the truth set of $x^2 = 49$? ________ . $\{7, {}^-7\}$

28 What is the truth set of $x^2 = 53$? ______________ . $\{\sqrt{53}, {}^-\sqrt{53}\}$

29 $(5)^2 = 25$ and $({}^-5)^2 = 25$. What two numbers can be placed in the blank to obtain a true statement? $(\_\_\_\_\_\_)^2 = 36$. 6 or ${}^-6$

30 $(\sqrt{7})^2 = 7$ and $({}^-\sqrt{7})^2 = 7$. What two numbers can be placed in the blank to obtain a true statement? $(\_\_\_\_\_\_)^2 = 15$. $\sqrt{15}$ or ${}^-\sqrt{15}$

31 What two numbers can be placed in the blank to obtain a true statement? $(\_\_\_\_\_\_)^2 = 100$. 10 or ${}^-10$

32 What two numbers can be placed in the blank to obtain a true statement? $(\_\_\_\_\_\_)^2 = 37$. $\sqrt{37}$ or ${}^-\sqrt{37}$

33 What two numbers can be placed in the blank to obtain a true statement? $(\_\_\_\_\_\_)^2 = 19$.

$\sqrt{19}$ or $^-\sqrt{19}$

34 What two numbers can be placed in the blank to obtain a true statement? $(\_\_\_\_\_\_)^2 = 81$.

9 or $^-9$

35 What two numbers can be placed in the blank to obtain a true statement? $(\_\_\_\_\_\_)^2 = 1$

1 or $^-1$

36 What two numbers can be placed in the blank to obtain a true statement? $(\_\_\_\_\_\_)^2 = 9$.

3 or $^-3$

37 For the equation $(\underline{x + 2})^2 = 25$, if $(\underline{x + 2})$ is replaced by 5 or $^-5$, a true statement is obtained. What two numbers would replace $(\underline{x - 3})$ in the equation $(\underline{x - 3})^2 = 49$ to obtain a true statement? ______ .

7 or $^-7$

38 For the equation $(\underline{x + 4})^2 = 49$, if $(\underline{x + 4})$ is replaced by 7 or $^-7$, a true statement is obtained. What two numbers could replace $(\underline{x - 5})$ in the equation $(\underline{x - 5})^2 = 4$ to obtain a true statement? ______ .

2 or $^-2$

39 For the equation $(\underline{x - 7})^2 = 81$, if $(\underline{x - 7})$ is replaced by 9 or $^-9$, a true statement is obtained. What two numbers could replace $(\underline{x + 8})$ in the equation $(\underline{x + 8})^2 = 16$ to obtain a true statement? ______ .

4 or $^-4$

40 To obtain a true statement from $(\underline{x + 2})^2 = 64$, $(\underline{x + 2})$ must be replaced by 8 or ______ .

$^-8$

41 To obtain a true statement from $(\underline{x - 5})^2 = 81$, $(\underline{x - 5})$ must be replaced by $^-9$ or ______ .

9

42 To obtain a true statement from $(\underline{x - 13})^2 = 100$, $(\underline{x - 13})$ must be replaced by 10 or ______ .

$^-10$

43 To obtain a true statement from $(\underline{x + 12})^2 = 36$, $(\underline{x + 12})$ must be replaced by ______ or ______ .

$6, ^-6$

44 To obtain a true statement from $(\underline{x - 5})^2 = 16$, $(\underline{x - 5})$ must be replaced by ______ or ______ .

$4, {}^-4$

45 To obtain a true statement from $(\underline{x + 9})^2 = 1$, $(\underline{x + 9})$ must be replaced by ______ or ______ .

$1, {}^-1$

46 To obtain a true statement from $(\underline{x - 1})^2 = 25$, $(\underline{x - 1})$ must be replaced by ______ or ______ .

$5, {}^-5$

47 To obtain a true statement from $(\underline{x - 3})^2 = 71$, $(\underline{x - 3})$ must be replaced by $\sqrt{71}$ or ${}^-\sqrt{71}$ because $(\sqrt{71})^2 = 71$ and $({}^-\sqrt{71})^2 = 71$. To obtain a true statement from $(\underline{x + 8})^2 = 53$, $(\underline{x + 8})$ must be replaced by $\sqrt{53}$ or ______ .

${}^-\sqrt{53}$

48 To obtain a true statement from $(\underline{x + 7})^2 = 17$, $(\underline{x + 7})$ must be replaced by $\sqrt{17}$ or ______ .

${}^-\sqrt{17}$

49 To obtain a true statement from $(\underline{x + 1})^2 = 77$, $(\underline{x + 1})$ must be replaced by ${}^-\sqrt{77}$ or ______ .

$\sqrt{77}$

50 To obtain a true statement from $(\underline{x - 10})^2 = 91$, $(\underline{x - 10})$ must be replaced by $\sqrt{91}$ or ______ .

${}^-\sqrt{91}$

51 What *two* numbers can replace $(x + 5)$ to make $(x + 5)^2 = 29$ a true statement? ____________ .

$\sqrt{29}, {}^-\sqrt{29}$

52 What two numbers can replace $(x + 1)$ to make $(x + 1)^2 = 23$ a true statement? ____________ .

$\sqrt{23}, {}^-\sqrt{23}$

53 What two numbers can replace $(x - 7)$ to make $(x - 7)^2 = 64$ a true statement? ____________ .

$8, {}^-8$

54 What two numbers can replace $(x + 3)$ to make $(x + 3)^2 = 49$ a true statement? ______ .

$7, {}^-7$

55 What two numbers can replace $(x - 3)$ to make $(x - 3)^2 = 19$ a true statement? ____________ .

$\sqrt{19}, {}^-\sqrt{19}$

56 What two numbers can replace $(x + 5)$ to make $(x + 5)^2 = 13$ a true statement? ________ .

$\sqrt{13}, {}^{-}\sqrt{13}$

57 What two numbers can replace $(x + 8)$ to make $(x + 8)^2 = 81$ a true statement? ________ .

$9, {}^{-}9$

Self-Quiz # 1

The following questions test the objectives of the preceding section. 100% mastery is desired.

1. Solve $x^2 = 16$. ________ .
2. Solve $x^2 = 100$. ________ .
3. Solve $x^2 = {}^{-}9$. ________ .
4. Solve $x^2 = 11$. ________ .
5. Solve $x^2 = 8$. ________ .
6. What two numbers can replace $(x + 5)$ to make $(x + 5)^2 = 49$ a true statement? ________ .
7. What two numbers can replace $(x - 8)$ to make $(x - 8)^2 = 17$ a true statement? ________ .

58 To solve $(x + 3)^2 = 64$, $(x + 3)$ must be either 8 or ${}^{-}8$. Two linear equations, $\underline{x + 3 = 8}$ or $\underline{x + 3 = {}^{-}8}$, are obtained. The two linear equations obtained from $(x - 2)^2 = 25$ are $x - 2 = 5$ or ________ .

$x - 2 = {}^{-}5$

59 To solve $(x + 13)^2 = 49$, $(x + 13)$ must be either 7 or ${}^{-}7$. Two linear equations $\underline{x + 13 = 7}$ or $\underline{x + 13 = {}^{-}7}$ are obtained. The two linear equations obtained from $(x - 7)^2 = 100$ are $x - 7 = 10$ or ________ .

$x - 7 = {}^{-}10$

60 To solve $(x + 7)^2 = 81$, $(x + 7)$ must be replaced by either 9 or $^-9$. Two linear equations, $\underline{x + 7 = 9}$ or $\underline{x + 7 = {}^-9}$, are obtained. The two linear equations obtained from $(x + 3)^2 = 16$ are $x + 3 = {}^-4$ or ________________.

$x + 3 = 4$

61 To solve $(x + 3)^2 = 16$, $(x + 3)$ must be 4 or $^-4$. Two linear equations, $\underline{x + 3 = 4}$ or $\underline{x + 3 = {}^-4}$, are obtained. For the equation $(x - 9)^2 = 25$, what two linear equations are obtained?

________________ or ________________.

$x - 9 = 5$,
$x - 9 = {}^-5$

62 For the equation $(x - 12)^2 = 4$, what two linear equations are obtained?

________________ or ________________.

$x - 12 = 2$,
$x - 12 = {}^-2$

63 For the equation $(x - 1)^2 = 100$, what two linear equations are obtained?

________________ or ________________.

$x - 1 = 10$,
$x - 1 = {}^-10$

64 For the equation $(x + 8)^2 = 1$, what two linear equations are obtained?

________________ or ________________.

$x + 8 = 1$,
$x + 8 = {}^-1$

65 For the equation $(x + 7)^2 = 9$, what two linear equations are obtained?

________________ or ________________.

$x + 7 = 3$,
$x + 7 = {}^-3$

66 For the equation $(x + 7)^2 = 19$, the two linear equations $x + 7 = \sqrt{19}$ or $x + 7 = {}^-\sqrt{19}$ are obtained. For the equation $(x + 5)^2 = 13$, what two linear equations are obtained?

________________ or ________________.

$x + 5 = \sqrt{13}$,
$x + 5 = {}^-\sqrt{13}$

67 For the equation $(x + 5)^2 = 13$, the two linear equations are $x + 5 = \sqrt{13}$ or $x + 5 = {}^{-}\sqrt{13}$. For the equation $(x + 7)^2 = 21$, what two linear equations are obtained? ______ or ______.

$x + 7 = \sqrt{21}$,
$x + 7 = {}^{-}\sqrt{21}$

68 For the equation $(x + 11)^2 = 97$, what two linear equations are obtained?

______ or ______.

$x + 11 = \sqrt{97}$,
$x + 11 = {}^{-}\sqrt{97}$

69 For the equation $(x - 6)^2 = 85$, what two linear equations are obtained?

______ or ______.

$x - 6 = \sqrt{85}$,
$x - 6 = {}^{-}\sqrt{85}$

70 For the equation $(x + 4)^2 = 81$, what two linear equations are obtained?

______ or ______.

$x + 4 = 9$,
$x + 4 = {}^{-}9$

71 For the equation $(x - 2)^2 = 37$, what two linear equations are obtained?

______ or ______.

$x - 2 = \sqrt{37}$,
$x - 2 = {}^{-}\sqrt{37}$

72 For the equation $(x + 9)^2 = 35$, what two linear equations are obtained?

______ or ______.

$x + 9 = \sqrt{35}$,
$x + 9 = {}^{-}\sqrt{35}$

73 For the equation $(x + 4)^2 = 36$, what two linear equations are obtained?

______ or ______.

$x + 4 = 6$,
$x + 4 = {}^{-}6$

74 For the equation $(x - 14)^2 = 15$, what two linear equations are obtained?

______ or ______.

$x - 14 = \sqrt{15}$,
$x - 14 = {}^{-}\sqrt{15}$

75 To solve $(x - 5)^2 = 36$, the following steps are used:

$$(x - 5)^2 = 36$$

$$\begin{array}{rcl} x - 5 = 6 & \text{or} & x - 5 = {}^-6 \\ +5 \quad {}^+5 & & +5 \quad {}^+5 \\ \hline x \quad = 11 & & x \quad = {}^-1 \end{array}$$

The truth set for $(x - 5)^2 = 36$ is $\{11, {}^-1\}$. Solve $(x - 3)^2 = 49$. ________ .

$\{10, {}^-4\}$

76 Solve $(x - 7)^2 = 1$ by solving $x - 7 = 1$ or $x - 7 = {}^-1$. ________ .

$\{8, 6\}$

77 Solve $(x - 2)^2 = 4$ by solving $x - 2 = 2$ or $x - 2 = {}^-2$. ________ .

$\{4, 0\}$

78 Solve $(x + 5)^2 = 81$ by solving $x + 5 = 9$ or $x + 5 = {}^-9$. ________ .

$\{4, {}^-14\}$

79 Solve $(x + 4)^2 = 49$ by solving $x + 4 = 7$ or $x + 4 = {}^-7$. ________ .

$\{3, {}^-11\}$

80 Solve $(x + 3)^2 = 100$ by solving $x + 3 = 10$ or $x + 3 = {}^-10$. ________ .

$\{7, {}^-13\}$

81 Solve $(x + 3)^2 = 81$. ________ .

$\{6, {}^-12\}$

82 Solve $(x - 7)^2 = 1$. ________ .

$\{8, 6\}$

83 Solve $(x + 5)^2 = 64$. ________ .

$\{3, {}^-13\}$

84 Solve $(x - 7)^2 = 49$. ________ .

$\{14, 0\}$

85 Solve $(x - 3)^2 = 36$. ________ .

$\{9, {}^-3\}$

86 Solve $(x - 4)^2 = 4$. ________ .

$\{6, 2\}$

87 To solve $x + 3 = \sqrt{17}$, the following step is used:

$$\begin{array}{lcl} x + 3 & = & \sqrt{17} \\ \quad - 3 & & {}^{-}3 \\ \hline x & = & {}^{-}3 + \sqrt{17} \end{array}$$

${}^{-}3 + \sqrt{17}$ cannot be simplified because ${}^{-}3$ and $\sqrt{17}$ are not like terms. Solve $x + 5 = \sqrt{21}$.

________________ .

$\{{}^{-}5 + \sqrt{21}\}$

88 To solve $x - 5 = \sqrt{29}$, add 5 to both sides of the equation. $5 + \sqrt{29}$ cannot be simplified. $\{5 + \sqrt{29}\}$ is the truth set of $x - 5 = \sqrt{29}$. Find the truth set of $x - 8 = \sqrt{34}$. ________________ .

$\{8 + \sqrt{34}\}$

89 To solve $x + 9 = {}^{-}\sqrt{19}$, add ${}^{-}9$ to both sides of the equation. Since ${}^{-}9 - \sqrt{19}$ cannot be simplified, $\{{}^{-}9 - \sqrt{19}\}$ is the truth set of $x + 9 = {}^{-}\sqrt{19}$. Find the truth set of $x + 11 = {}^{-}\sqrt{22}$. ________________ .

$\{{}^{-}11 - \sqrt{22}\}$

90 To solve $x - 3 = {}^{-}\sqrt{13}$, add 3 to both sides of the equation. Since $3 - \sqrt{13}$ cannot be simplified $\{3 - \sqrt{13}\}$ is the truth set of $x - 3 = {}^{-}\sqrt{13}$. Find the truth set of $x - 5 = {}^{-}\sqrt{17}$. ________________ .

$\{5 - \sqrt{17}\}$

91 To solve $(x + 4)^2 = 11$, the following steps are used:

$$(x + 4)^2 = 11$$

$$\begin{array}{lcl} x + 4 & = & \sqrt{11} \\ \quad - 4 & & {}^{-}4 \\ \hline x & = & {}^{-}4 + \sqrt{11} \end{array} \quad \text{or} \quad \begin{array}{lcl} x + 4 & = & {}^{-}\sqrt{11} \\ \quad - 4 & & {}^{-}4 \\ \hline x & = & {}^{-}4 - \sqrt{11} \end{array}$$

The truth set of $(x + 4)^2 = 11$ is $\{{}^{-}4 + \sqrt{11}, {}^{-}4 - \sqrt{11}\}$. Solve $(x + 5)^2 = 15$.

________________ .

$\{{}^{-}5 + \sqrt{15}, {}^{-}5 - \sqrt{15}\}$

92 Solve $(x + 5)^2 = 7$ by solving $x + 5 = \sqrt{7}$ or $x + 5 = {}^{-}\sqrt{7}$. ________________ .

$\{{}^{-}5 + \sqrt{7}, {}^{-}5 - \sqrt{7}\}$

93 Solve $(x - 2)^2 = 13$ by solving $x - 2 = \sqrt{13}$ or $x - 2 = {}^{-}\sqrt{13}$. ________________ .

$\{2 + \sqrt{13}, 2 - \sqrt{13}\}$

94 Solve $(x + 2)^2 = 23$ by solving $x + 2 = \sqrt{23}$ or $x + 2 = {}^{-}\sqrt{23}$. ________________ .

$\{{}^{-}2 + \sqrt{23}, {}^{-}2 - \sqrt{23}\}$

95 Solve $(x - 3)^2 = 38$ by solving $x - 3 = \sqrt{38}$ or $x - 3 = {}^{-}\sqrt{38}$. ________________ .

$\{3 + \sqrt{38}, 3 - \sqrt{38}\}$

96 Solve $(x + 5)^2 = 26$. ________________ .

$\{{}^{-}5 + \sqrt{26}, {}^{-}5 - \sqrt{26}\}$

97 Solve $(x - 6)^2 = 47$. ________________ .

$\{6 + \sqrt{47}, 6 - \sqrt{47}\}$

98 Solve $(x + 7)^2 = 61$. ________________ .

$\{{}^{-}7 + \sqrt{61}, {}^{-}7 - \sqrt{61}\}$

99 Solve $(x - 1)^2 = 100$. ________________ .

$\{11, {}^{-}9\}$

100 Solve $(x - 2)^2 = 25$. ________________ .

$\{7, {}^{-}3\}$

101 To solve $x - \frac{7}{2} = \sqrt{17}$, add $\frac{7}{2}$ to both sides of the equation to obtain the truth set $\left\{\frac{7}{2} + \sqrt{17}\right\}$. Solve $x - \frac{5}{2} = \sqrt{37}$ by adding $\frac{5}{2}$ to both sides of the equation. ________________ .

$$\begin{array}{rcl} x - \frac{5}{2} & = & \sqrt{37} \\ +\frac{5}{2} & & +\frac{5}{2} \\ \hline x & = & \frac{5}{2} + \sqrt{37} \end{array}$$

$\left\{\frac{5}{2} + \sqrt{37}\right\}$

102 Solve $x + \frac{7}{2} = \sqrt{5}$ by adding $\frac{{}^{-}7}{2}$ to both sides of the equation. ________________ .

$\left\{\frac{{}^{-}7}{2} + \sqrt{5}\right\}$

103 To solve $\left(x - \frac{5}{2}\right)^2 = 19$, use the equations $x - \frac{5}{2} = \sqrt{19}$ or $x - \frac{5}{2} = {}^{-}\sqrt{19}$. The truth set is $\left\{\frac{5}{2} + \sqrt{19}, \frac{5}{2} - \sqrt{19}\right\}$. Using the equations $x - \frac{3}{2} = \sqrt{23}$ or $x - \frac{3}{2} = {}^{-}\sqrt{23}$, solve $\left(x - \frac{3}{2}\right)^2 = 23$ ________________ .

$\left\{\frac{3}{2} + \sqrt{23}, \frac{3}{2} - \sqrt{23}\right\}$

104 Using the equations $x + \frac{15}{2} = \sqrt{7}$ or $x + \frac{15}{2} = {}^{-}\sqrt{7}$, solve $\left(x + \frac{15}{2}\right)^2 = 7$. ________________ .

$\left\{\frac{{}^{-}15}{2} + \sqrt{7}, \frac{{}^{-}15}{2} - \sqrt{7}\right\}$

105 Solve $\left(x - \frac{1}{2}\right)^2 = 3$. ________________ .

$\left\{\frac{1}{2} + \sqrt{3}, \frac{1}{2} - \sqrt{3}\right\}$

106 Solve $\left(x + \frac{21}{2}\right)^2 = 29$. ________________ .

$\left\{\frac{{}^{-}21}{2} + \sqrt{29}, \frac{{}^{-}21}{2} - \sqrt{29}\right\}$

107 Solve $(x + 1)^2 = 3$. ________________ .

$\{{}^{-}1 + \sqrt{3}, {}^{-}1 - \sqrt{3}\}$

108 Solve $(x + 6)^2 = 1$. ________ .

$\{{}^{-}5, {}^{-}7\}$

109 Solve $\left(x + \frac{5}{2}\right)^2 = 11$. ________________ .

$\left\{\frac{{}^{-}5}{2} + \sqrt{11}, \frac{{}^{-}5}{2} - \sqrt{11}\right\}$

110 Solve $(x - 7)^2 = 43$. ________________ .

$\{7 + \sqrt{43}, 7 - \sqrt{43}\}$

111 Solve $(x + 3)^2 = 16$. ________ .

$\{1, {}^{-}7\}$

Self-Quiz # 2

The following questions test the objectives of the preceding section. 100% mastery is desired.

Solve:

1. $(x + 2)^2 = 25$. ______.

2. $(x - 3)^2 = 49$. ______.

3. $(x + 5)^2 = 17$. ______.

4. $(x - 7)^2 = 29$. ______.

5. $\left(x - \frac{5}{2}\right)^2 = 17$. ______.

6. $(x + 9)^2 = 1$. ______.

7. $(x + 16)^2 = 13$. ______.

8. $(x - 4)^2 = 81$. ______.

9. $(x + 15)^2 = 100$. ______.

10. $\left(x + \frac{11}{2}\right)^2 = 19$. ______.

Solving quadratic equations such as $x^2 + 6x = 11$ is dependent upon an ability to form a perfect square trinomial from $x^2 + 6x$.

A perfect square trinomial is obtained by multiplying a binomial by itself.

$$\begin{aligned}(x + 6)^2 &= (x + 6)(x + 6)\\ &= x^2 + 12x + 36\end{aligned}$$

$x^2 + 12x + 36$ is a perfect square trinomial because it is obtained by multiplying $(x + 6)$ by itself.

---

112 $(x - 4)^2 = (x - 4)(x - 4) = x^2 - 8x + 16$

$x^2 - 8x + 16$ is a perfect square trinomial because it is obtained by multiplying the binomial, $(x - 4)$, by itself. What perfect square trinomial is obtained from $(x - 6)^2$? ________________ .

$x^2 - 12x + 36$

113 $(x + 7)^2 = (x + 7)(x + 7) = x^2 + 14x + 49$

$x^2 + 14x + 49$ is a perfect square trinomial because it is obtained by multiplying the binomial, $(x + 7)$, by itself. What perfect square trinomial is obtained from $(x + 3)^2$? ________________ .

$x^2 + 6x + 9$

114 $(x + 2)^2 = (x + 2)(x + 2) = x^2 + 4x + 4$

$x^2 + 4x + 4$ is a perfect square trinomial because it is obtained by multiplying the binomial, $(x + 2)$, by itself. What perfect square trinomial is obtained from $(x + 5)^2$? ________________ .

$x^2 + 10x + 25$

115 $(x - 3)^2 = (x - 3)(x - 3) = x^2 - 6x + 9$

$x^2 - 6x + 9$ is a perfect square trinomial because it is obtained by multiplying the bionomial, $(x - 3)$, by itself. What perfect square trinomial is obtained from $(x - 8)^2$? ________________ .

$x^2 - 16x + 64$

116 What perfect square trinomial is obtained by multiplying $(x - 5)$ by itself? ________________ .

$x^2 - 10x + 25$

117 What perfect square trinomial is obtained by multiplying $(x + 1)$ by itself? ________________ .

$x^2 + 2x + 1$

118 What perfect square trinomial is obtained by multiplying $(x + 9)$ by itself? ________________ .

$x^2 + 18x + 81$

119 $x^2 + 6x + 9$ is a perfect square trinomial because $(x + 3)^2 = x^2 + 6x + 9$. Is $x^2 + 10x + 25$ a perfect square trinomial? ________ .
[*Note*: Factor $x^2 + 10x + 25$.]

Yes

120 $x^2 - 14x + 49$ is a perfect square trinomial because $(x - 7)^2 = x^2 - 14x + 49$. Is $x^2 - 20x + 100$ a perfect square trinomial? ________ .
[*Note*: Factor $x^2 - 20x + 100$.]

Yes

121 Is $x^2 + 14x + 49$ a perfect square trinomial? ________________ .

Yes, $(x + 7)^2$

122 Is $x^2 - 12x + 36$ a perfect square trinomial? ________________ .

Yes, $(x - 6)^2$

123 The factors of $x^2 + 7x + 10$ are $(x + 5)(x + 2)$. Is $x^2 + 7x + 10$ a perfect square trinomial? ________ .

No

124 Is $x^2 - 2x - 8$ a perfect square trinomial? ________ .

No

125 Is $x^2 + 2x + 1$ a perfect square trinomial? ________________ .

Yes, $(x + 1)^2$

126 Is $x^2 - 10x + 25$ a perfect square trinomial? ________________ .

Yes, $(x - 5)^2$

127 Is $x^2 + 7x + 12$ a perfect square trinomial? ________ .

No

128 Is $x^2 + 20x + 100$ a perfect square trinomial? ____________________ .

Yes, $(x + 10)^2$

129 $x^2 + 10x + 25$ is a perfect square trinomial. The terms $\underline{10x}$ and $\underline{25}$ are related in the following manner:

$$\left(\frac{1}{2} \cdot 10\right)^2 = (5)^2 = 25$$

$x^2 + \underline{6}x + \underline{9}$ is a perfect square trinomial. Is $\left(\frac{1}{2} \cdot 6\right)^2$ equal to the third term of $x^2 + 6x + 9$? ________ .

Yes

130 $x^2 + 20x + 100$ is a perfect square trinomial. The terms $\underline{20}x$ and $\underline{100}$ are related in the following manner:

$$\left(\frac{1}{2} \cdot 20\right)^2 = (10)^2 = 100$$

$x^2 + \underline{8}x + \underline{16}$ is a perfect square trinomial. Is $\left(\frac{1}{2} \cdot 8\right)^2$ equal to the third term of $x^2 + 8x + \underline{16}$? ________ .

Yes

131 $x^2 - 2x + 1$ is a perfect square trinomial. The second term, $^-\underline{2}x$, and the third term, $\underline{1}$, are related. Is $\left(\frac{1}{2} \cdot {}^-\underline{2}\right)^2$ equal to the third term of $x^2 - 2x + \underline{1}$? ________ .

Yes

132 $x^2 - 10x + 25$ is a perfect square trinomial. The second term, $^-\underline{10}x$, and the third term, $\underline{25}$, are related. Is $\left(\frac{1}{2} \cdot {}^-\underline{10}\right)^2$ equal to $\underline{25}$? ________

Yes

133 $x^2 + \underline{16}x + \underline{64}$ is a perfect square trinomial. Is $\left(\frac{1}{2} \cdot \underline{16}\right)^2$ equal to $\underline{64}$? ________ .

Yes

134 $x^2 - \underline{4}x + \underline{4}$ is a perfect square trinomial. Is $\left(\frac{1}{2} \cdot {}^-\underline{4}\right)^2$ equal to $\underline{4}$? ________ .

Yes

135 $x^2 - \underline{6}x + \underline{9}$ is a perfect square trinomial. Is $\left(\frac{1}{2} \cdot {}^-\underline{6}\right)^2$ equal to $\underline{9}$? ________ .

Yes

136 $x^2 + \underline{14}x + \underline{49}$ is a perfect square trinomial. Is $\left(\frac{1}{2} \cdot \underline{14}\right)^2$ equal to $\underline{49}$? ________ .

Yes

137 $x^2 + \underline{18}x + \underline{81}$ is a perfect square trinomial. Is $\left(\frac{1}{2} \cdot \underline{18}\right)^2$ equal to $\underline{81}$? ________ .

Yes

138 The second and third terms of a perfect square trinomial are related. $x^2 + 8x + 16$, $x^2 + 12x + 36$, and $x^2 + 18x + 81$ are all perfect square trinomials.

For $x^2 + 8x + 16$, $\frac{1}{2} \cdot 8 = 4$ and $(4)^2 = 16$.

For $x^2 + 12x + 36$, $\frac{1}{2} \cdot 12 = 6$ and $(6)^2 = 36$.

For $x^2 + 18x + 81$, $\frac{1}{2} \cdot 18 = 9$. Is $(9)^2$ the third term of $x^2 + 18x + 81$? ________ .

Yes

139 To form a perfect square trinomial from $x^2 + 8x$, the following steps are used:

(a) The second term of $x^2 + 8x$ is $8x$. Multiply $\frac{1}{2} \cdot 8$.

(b) $\frac{1}{2} \cdot 8 = 4$. Multiply $4 \cdot 4$.

(c) $4 \cdot 4 = 16$. Using 16 as the third term, the trinomial $x^2 + 8x + 16$ is obtained.

Is $x^2 + 8x + 16$ a perfect square trinomial? ________ .

Yes

140 To form a perfect square trinomial from $x^2 - 16x$, the following steps are used:

(a) The second term of $x^2 - 16x$ is ^-16x. Multiply $\frac{1}{2} \cdot {}^-16$.

(b) $\frac{1}{2} \cdot {}^-16 = {}^-8$. Multiply $^-8 \cdot {}^-8$.

(c) $^-8 \cdot {}^-8 = 64$. Using 64 as the third term, the trinomial $x^2 - 16x + 64$ is obtained.

Is $x^2 - 16x + 64$ a perfect square trinomial?

________ .

Yes

141 To form a perfect square trinomial from $x^2 + 12x$, multiply $\frac{1}{2} \cdot 12$ and multiply the product, 6, by itself.

$$\frac{1}{2} \cdot 12 = 6$$

$$(6)^2 = 36$$

Is $x^2 + 12x + 36$ a perfect square trinomial?

________.

Yes

142 To form a perfect square trinomial from $x^2 - 18x$, multiply $\frac{1}{2} \cdot {}^{-}18$ and multiply the product, $^{-}9$, by itself.

$$\frac{1}{2} \cdot {}^{-}18 = {}^{-}9$$

$$({}^{-}9)^2 = 81$$

Is $x^2 - 18x + 81$ a perfect square trinomial?

________.

Yes

143 Form a perfect square trinomial from $x^2 - 10x$ by multiplying $\frac{1}{2} \cdot {}^{-}10$. Then multiply the product, $^{-}5$, by itself to obtain the third term. ________________.

$x^2 - 10x + 25$

144 Form a perfect square trinomial from $x^2 + 2x$ by multiplying $\frac{1}{2} \cdot 2$. Then multiply the product, 1, by itself to obtain the third term. ________________.

$x^2 + 2x + 1$

145 Fill the blank of $x^2 + 6x +$ ________ to obtain a perfect square trinomial.

$\left(\frac{1}{2} \cdot 6\right)^2 =$ ________________.

$x^2 + 6x + \underline{9}$

146 Fill the blank of $x^2 - 14x +$ ________ to obtain a perfect square trinomial.

$\left(\frac{1}{2} \cdot {}^{-}14\right)^2 =$ ________________.

$x^2 - 14x + \underline{49}$

147 $x^2 + 8x +$ ________ is a perfect square trinomial. | 16

148 $x^2 - 16x +$ ________ is a perfect square trinomial. | 64

149 $x^2 + 18x +$ ________ is a perfect square trinomial. | 81

150 $x^2 - 10x +$ ________ is a perfect square trinomial. | 25

151 To obtain a perfect square trinomial from $x^2 + 5x$, the following steps are used:

(a) $\frac{1}{2} \cdot 5 = \frac{5}{2}$.

(b) $\frac{5}{2} \cdot \frac{5}{2} = \frac{25}{4}$.

The perfect square trinomial is

$$x^2 + 5x + \frac{25}{4}$$

$x^2 + 7x +$ ________ is a perfect square trinomial. | $\frac{49}{4}$

152 To obtain a perfect square trinomial from $x^2 - 3x$, the following steps are used:

(a) $\frac{1}{2} \cdot {}^{-}3 = \frac{{}^{-}3}{2}$.

(b) $\frac{{}^{-}3}{2} \cdot \frac{{}^{-}3}{2} = \frac{9}{4}$.

The perfect square trinomial is

$$x^2 - 3x + \frac{9}{4}$$

$x^2 - 9x +$ ________ is a perfect square trinomial. | $\frac{81}{4}$

153 $x^2 - 7x +$ ________ is a perfect square trinomial.

$\left(\frac{1}{2} \cdot {}^{-}7\right)^2 =$ ________. | $\frac{49}{4}$

154 $x^2 + 15x +$ ________ is a perfect square trinomial.

$\left(\frac{1}{2} \cdot 15\right)^2 =$ ________. | $\frac{225}{4}$

155 $x^2 + 9x +$ ________ is a perfect square trinomial.

$\left(\frac{1}{2} \cdot 9\right)^2 =$ ________ .

$\frac{81}{4}$

156 $x^2 + 13x +$ ________ is a perfect square trinomial.

$\frac{169}{4}$

157 $x^2 - 19x +$ ________ is a perfect square trinomial.

$\frac{361}{4}$

158 $x^2 + x +$ ________ is a perfect square trinomial.

[*Note*: x is equivalent to $1x$.]

$\frac{1}{4}$

159 $x^2 - x +$ ________ is a perfect square trinomial.

$\frac{1}{4}$

160 $x^2 - 6x +$ ________ is a perfect square trinomial.

9

161 $x^2 + 3x +$ ________ is a perfect square trinomial.

$\frac{9}{4}$

162 $x^2 + 11x +$ ________ is a perfect square trinomial.

$\frac{121}{4}$

163 $x^2 - 10x +$ ________ is a perfect square trinomial.

25

Self-Quiz # 3

The following questions test the objectives of the preceding section. 100% mastery is desired.

Find the third term for each expression to obtain a perfect square trinomial.

1. $x^2 + 6x +$ ______.

2. $x^2 - 10x +$ ______.

3. $x^2 + 7x +$ ______.

4. $x^2 - 8x +$ ______.

5. $x^2 - 9x +$ ______.

6. $x^2 + 15x +$ ______.

The factors of a perfect square trinomial can always be written as the square of a binomial expression. $x^2 - 10x + 25$ is a perfect square trinomial that factors as $(x - 5)(x - 5)$ or $(x - 5)^2$.

---

164 $x^2 + 8x + 16$ is factored as $(x + 4)(x + 4)$ or $(x + 4)^2$ because the factors of 16 that have a sum of 8 are 4 and 4. What are the factors of $x^2 + 12x + 36$? ______.

$(x + 6)^2$

165 $x^2 + 10x + 25$ is a perfect square trinomial. Factor $x^2 + 10x + 25$ by finding factors of 25 that have a sum of 10. ______.

$(x + 5)^2$

166 $x^2 + 4x + 4$ is a perfect square trinomial. Factor $x^2 + 4x + 4$. ______.

$(x + 2)^2$

167 $x^2 - 6x + 9$ is factored as $(x - 3)(x - 3)$ or $(x - 3)^2$ because the factors of 9 that have a sum of $^-6$ are $^-3$ and $^-3$. What are the factors of $x^2 - 10x + 25$? ________.

$(x - 5)^2$

168 $x^2 - 14x + 49$ is a perfect square trinomial and is factored as $(x - 7)^2$. Factor $x^2 - 16x + 64$. ________.

$(x - 8)^2$

169 $x^2 - 2x + 1$ is a perfect square trinomial. Factor $x^2 - 2x + 1$ by finding factors of 1 that have a sum of $^-2$. ________.

$(x - 1)^2$

170 $x^2 - 12x + 36$ is a perfect square trinomial. Factor $x^2 - 12x + 36$. ________.

$(x - 6)^2$

171 $x^2 + 7x + \frac{49}{4}$ is a perfect square trinomial. To factor $x^2 + 7x + \frac{49}{4}$, it is necessary to find factors of $\frac{49}{4}$ that have a sum of 7.

$$\frac{7}{2} \cdot \frac{7}{2} = \frac{49}{4} \quad \text{and} \quad \frac{7}{2} + \frac{7}{2} = 7$$

Therefore,

$$x^2 + 7x + \frac{49}{4} = \left(x + \frac{7}{2}\right)^2$$

Factor the perfect square trinomial $x^2 + 9x + \frac{81}{4}$ by finding the factors of $\frac{81}{4}$ that have a sum of 9. ________.

$\left(x + \frac{9}{2}\right)^2$

172 $x^2 + 5x + \frac{25}{4}$ is factored as $\left(x + \frac{5}{2}\right)^2$ because $\frac{5}{2} \cdot \frac{5}{2} = \frac{25}{4}$ and $\frac{5}{2} + \frac{5}{2} = 5$. Factor the perfect square trinomial $x^2 + 3x + \frac{9}{4}$. ________.

$\left(x + \frac{3}{2}\right)^2$

173 $x^2 - 5x + \frac{25}{4}$ can be factored as $\left(x - \frac{5}{2}\right)^2$ because $\frac{^-5}{2} \cdot \frac{^-5}{2} = \frac{25}{4}$ and $\frac{^-5}{2} + \frac{^-5}{2} = {}^-5$. Factor the perfect square trinomial $x^2 - 9x + \frac{81}{4}$. ________ .

$\left(x - \frac{9}{2}\right)^2$

174 $x^2 - x + \frac{1}{4}$ is factored as $\left(x - \frac{1}{2}\right)^2$ because $\frac{^-1}{2} \cdot \frac{^-1}{2} = \frac{1}{4}$ and $\frac{^-1}{2} + \frac{^-1}{2} = {}^-1$. Factor the perfect square trinomial $x^2 - 7x + \frac{49}{4}$. ________ .

$\left(x - \frac{7}{2}\right)^2$

175 Factor the perfect square trinomial $x^2 + 5x + \frac{25}{4}$.

________ .

$\left(x + \frac{5}{2}\right)^2$

176 Factor the perfect square trinomial $x^2 - x + \frac{1}{4}$.

________ .

$\left(x - \frac{1}{2}\right)^2$

177 Factor the perfect square trinomial $x^2 - 16x + 64$.

________ .

$(x - 8)^2$

178 Factor the perfect square trinomial $x^2 - 7x + \frac{49}{4}$.

________ .

$\left(x - \frac{7}{2}\right)^2$

179 To obtain a perfect square trinomial from $x^2 + 6x$, the following steps are used:

(a) $\frac{1}{2} \cdot 6 = 3$.

(b) $(3)^2 = 9$.

The perfect square trinomial is $x^2 + 6x + 9$. Factor $x^2 + 6x + 9$. ________ .

$(x + 3)^2$

180 To obtain a perfect square trinomial from $x^2 + 7x$, the following steps are used:

(a) $\frac{1}{2} \cdot 7 = \frac{7}{2}$.

(b) $\frac{7}{2} \cdot \frac{7}{2} = \frac{49}{4}$.

The perfect square trinomial is $x^2 + 7x + \frac{49}{4}$.

Factor $x^2 + 7x + \frac{49}{4}$. ________ .

$\left(x + \frac{7}{2}\right)^2$

181 $x^2 + 8x +$ ________ is a perfect square trinomial.

16

182 Factor the perfect square trinomial $x^2 + 8x + 16$.

________ .

$(x + 4)^2$

183 $x^2 + 9x +$ ________ is a perfect square trinomial.

$\frac{81}{4}$

184 Factor the perfect square trinomial $x^2 + 9x + \frac{81}{4}$.

________ .

$\left(x + \frac{9}{2}\right)^2$

185 $x^2 - 12x +$ ________ is a perfect square trinomial.

36

186 Factor the perfect square trinomial $x^2 - 12x + 36$.

________ .

$(x - 6)^2$

187 $x^2 - 3x +$ ________ is a perfect square trinomial.

$\frac{9}{4}$

188 Factor the perfect square trinomial $x^2 - 3x + \frac{9}{4}$.

________ .

$\left(x - \frac{3}{2}\right)^2$

189 $x^2 + 10x +$ ________ is a perfect square trinomial.

25

190 Factor the perfect square trinomial $x^2 + 10x + 25$.

________ .

$(x + 5)^2$

191 $x^2 + 7x +$ ________ is a perfect square trinomial.

$\frac{49}{4}$

192 Factor the perfect square trinomial $x^2 + 7x + \frac{49}{4}$.

________ .

$\left(x + \frac{7}{2}\right)^2$

193 Add a third term to $x^2 + 16x$ to form a perfect square trinomial. ____________ .

$x^2 + 16x \underline{+ 64}$

194 Factor the perfect square trinomial $x^2 + 16x + 64$.

________ .

$(x + 8)^2$

195 Add a third term to $x^2 - 6x$ to form a perfect square trinomial. ____________ .

$x^2 - 6x \underline{+ 9}$

196 Factor the perfect square trinomial $x^2 - 6x + 9$.

________ .

$(x - 3)^2$

197 Add a third term to $x^2 + 5x$ to form a perfect square trinomial. ____________ .

$x^2 + 5x \underline{+ \frac{25}{4}}$

198 Add a third term to $x^2 + 10x$ to obtain a perfect square trinomial. What are the factors of the perfect square trinomial? ________ .

$(x + 5)^2$

199 Add a third term to $x^2 - 8x$ to obtain a perfect square trinomial. What are the factors of the perfect square trinomial? ________ .

$(x - 4)^2$

200 Add a third term to $x^2 + 3x$ to obtain a perfect square trinomial. What are the factors of the perfect square trinomial? ____________ .

$x^2 + 3x + \frac{9}{4}$

$\left(x + \frac{3}{2}\right)^2$

201 Add a third term to $x^2 + 12x$ to obtain a perfect square trinomial. What are the factors of the perfect square trinomial? ________ .

$(x + 6)^2$

202 Add a third term to $x^2 + 7x$ to obtain a perfect square trinomial. What are the factors of the perfect square trinomial? ________ .

$\left(x + \frac{7}{2}\right)^2$

Self-Quiz # 4

The following questions test the objectives of the preceding section. 100% mastery is desired.

Add a third term to each binomial to obtain a perfect square trinomial, and write the factors of the perfect square trinomial.

1. $x^2 + 4x +$ ________ .
2. $x^2 - 18x +$ ________ .
3. $x^2 + x +$ ________ .
4. $x^2 - 14x +$ ________ .
5. $x^2 - 5x +$ ________ .
6. $x^2 + 3x +$ ________ .

To solve $(x + 5)^2 = 17$, the equation is used to write two linear equations.

$$(x + 5)^2 = 17$$

$$x + 5 = \sqrt{17} \quad \text{or} \quad x + 5 = {}^{-}\sqrt{17}$$

Each of the linear equations is solved by adding $^{-}5$ to both sides of the equations.

$$\begin{array}{rcl} x + 5 & = & \sqrt{17} \\ -5 & & {}^{-}5 \\ \hline x & = & {}^{-}5 + \sqrt{17} \end{array} \qquad \begin{array}{rcl} x + 5 & = & {}^{-}\sqrt{17} \\ -5 & & {}^{-}5 \\ \hline x & = & {}^{-}5 - \sqrt{17} \end{array}$$

The truth set of $(x + 5)^2 = 17$ is $\{{}^{-}5 + \sqrt{17}, {}^{-}5 - \sqrt{17}\}$.

---

203 To solve $(x - 6)^2 = 25$, the two linear equations $x - 6 = 5$ or $x - 6 = {}^{-}5$ are used. The truth set of $(x - 6)^2 = 25$ is $\{11, 1\}$. Solve $(x - 9)^2 = 4$. ______ .

$\{11, 7\}$

204 To solve $(x - 3)^2 = 17$, the two linear equations $x - 3 = \sqrt{17}$ or $x - 3 = {}^{-}\sqrt{17}$ are used. If 3 is added to both sides of the equations, the truth set of $(x - 3)^2 = 17$ is $\{3 + \sqrt{17}, 3 - \sqrt{17}\}$. Solve $(x - 5)^2 = 21$. ______ .

$\{5 + \sqrt{21}, 5 - \sqrt{21}\}$

205 To solve $(x + 6)^2 = 11$, the two linear equations $x + 6 = \sqrt{11}$ or $x + 6 = {}^{-}\sqrt{11}$ are used. If $^{-}6$ is added to both sides of the equations, the truth set of $(x + 6)^2 = 11$ is $\{{}^{-}6 + \sqrt{11}, {}^{-}6 - \sqrt{11}\}$. Solve $(x + 2)^2 = 29$. ______ .

$\{{}^{-}2 + \sqrt{29}, {}^{-}2 - \sqrt{29}\}$

206 Solve $(x + 5)^2 = 7$. ______ .

$\{{}^{-}5 + \sqrt{7}, {}^{-}5 - \sqrt{7}\}$

207 Solve $(x - 8)^2 = 23$. ______ .

$\{8 + \sqrt{23}, 8 - \sqrt{23}\}$

208 Solve $(x - 5)^2 = {}^-6$. ______ .
[*Note*: There is no rational or irrational number that can be multiplied by itself to obtain a negative number.]

{ }

209 Solve $(x + 3)^2 = 16$. ______ .
[*Note*: $\sqrt{16} = 4$.]

$\{1, {}^-7\}$

210 Solve $(x - 4)^2 = 5$. ______ .

$\{4 + \sqrt{5}, 4 - \sqrt{5}\}$

211 Solve $(x + 1)^2 = {}^-15$. ______ .

{ }

212 Solve $(x - 5)^2 = 9$. ______ .

{8, 2}

213 To solve $x^2 + 10x + 25 = 6$, first factor $x^2 + 10x + 25$ as follows:

$$x^2 + 10x + 25 = 6$$
$$(x + 5)^2 = 6$$

Complete the solution of $x^2 + 10x + 25 = 6$ by solving $(x + 5)^2 = 6$. ______ .

$\{{}^-5 + \sqrt{6}, {}^-5 - \sqrt{6}\}$

214 Solve $x^2 - 6x + 9 = 7$ by first factoring $x^2 - 6x + 9$.
______ .

$\{3 + \sqrt{7}, 3 - \sqrt{7}\}$

215 Solve $x^2 - 16x + 64 = 10$ by first factoring $x^2 - 16x + 64$. ______ .

$\{8 + \sqrt{10}, 8 - \sqrt{10}\}$

216 To solve $x^2 - 3x + \frac{9}{4} = 7$, the following steps are used:

$$x^2 - 3x + \frac{9}{4} = 7$$

$$\left(x - \frac{3}{2}\right)^2 = 7$$

$$x - \frac{3}{2} = \sqrt{7} \quad \text{or} \quad x - \frac{3}{2} = {}^-\sqrt{7}$$

The truth set is $\left\{\frac{3}{2} + \sqrt{7}, \frac{3}{2} - \sqrt{7}\right\}$. Complete the following solution:

$$x^2 - 5x + \frac{25}{4} = 13$$

$$\left(x - \frac{5}{2}\right)^2 = 13$$

$$x - \frac{5}{2} = \sqrt{13} \quad \text{or} \quad x - \frac{5}{2} = {}^-\sqrt{13}$$

____________________.

$\left\{\frac{5}{2} + \sqrt{13}, \frac{5}{2} - \sqrt{13}\right\}$

217 Complete the following solution:

$$x^2 + 7x + \frac{49}{4} = 11$$

$$\left(x + \frac{7}{2}\right)^2 = 11$$

$$x + \frac{7}{2} = \sqrt{11} \quad \text{or} \quad x + \frac{7}{2} = {}^-\sqrt{11}$$

____________________.

$\left\{\frac{^-7}{2} + \sqrt{11}, \frac{^-7}{2} - \sqrt{11}\right\}$

218 Solve $x^2 + 5x + \frac{25}{4} = 7$ by first factoring $x^2 + 5x + \frac{25}{4}$. ____________________.

$\left\{\frac{^-5}{2} + \sqrt{7}, \frac{^-5}{2} - \sqrt{7}\right\}$

219 Solve $x^2 - x + \frac{1}{4} = 6$ by first factoring $x^2 - x + \frac{1}{4}$.

____________________ .

$\left\{\frac{1}{2} + \sqrt{6}, \frac{1}{2} - \sqrt{6}\right\}$

220 Solve $x^2 + 9x + \frac{81}{4} = 37$ by first factoring $x^2 + 9x + \frac{81}{4}$. ____________________ .

$\left\{\frac{^-9}{2} + \sqrt{37}, \frac{^-9}{2} - \sqrt{37}\right\}$

221 Solve $x^2 - 8x + 16 = 23$ by first factoring $x^2 - 8x + 16$. ____________________ .

$\{4 + \sqrt{23}, 4 - \sqrt{23}\}$

222 Solve $x^2 - 10x + 25 = 3$. ____________________ .

$\{5 + \sqrt{3}, 5 - \sqrt{3}\}$

223 To solve $x^2 - 6x = 5$, the first step is to add a number to both sides of the equation so that $x^2 - 6x$ will become a perfect square trinomial.
$x^2 - 6x +$ ________ is a perfect square trinomial.

9

224 To solve $x^2 - 6x = 5$, the first step is to add 9 to both sides of the equation because $x^2 - 6x + 9$ is a perfect square trinomial.

$$x^2 - 6x = 5$$
$$x^2 - 6x + 9 = 5 + 9$$
$$(x - 3)^2 = 14$$

Complete the solution of $x^2 - 6x = 5$ by solving $(x - 3)^2 = 14$. ____________________ .

$\{3 + \sqrt{14}, 3 - \sqrt{14}\}$

225 To solve $x^2 + 10x = 4$, the first step is to add a number to both sides of the equation to form a perfect square trinomial from $x^2 + 10x$. $x^2 + 10x +$ ________ is a perfect square trinomial.

25

226 To solve $x^2 + 10x = 4$, first add 25 to both sides of the equation because $x^2 + 10x + 25$ is a perfect square trinomial.

$$x^2 + 10x = 4$$
$$x^2 + 10x + 25 = 4 + 25$$
$$(x + 5)^2 = 29$$

Complete the solution of $x^2 + 10x = 4$ by solving $(x + 5)^2 = 29$. ________________ .

$\{^{-}5 + \sqrt{29}, ^{-}5 - \sqrt{29}\}$

227 To solve $x^2 - 8x = 3$, the first step is to add a number to both sides of the equation to form a perfect square trinomial from $x^2 - 8x$. $x^2 - 8x +$ ________ is a perfect square trinomial.

16

228 To solve $x^2 - 8x = 3$, first add 16 to both sides of the equation because $x^2 - 8x + 16$ is a perfect square trinomial.

$$x^2 - 8x = 3$$
$$x^2 - 8x + 16 = 3 + 16$$

Complete the solution of $x^2 - 8x = 3$.

________________ .

$\{4 + \sqrt{19}, 4 - \sqrt{19}\}$

229 What number can be added to both sides of $x^2 - 18x = 12$ to form a perfect square trinomial from $x^2 - 18x$? ________ .

81

230 To solve $x^2 - 18x = 12$, first add 81 to both sides of the equation.

$$x^2 - 18x = 12$$
$$x^2 - 18x + 81 = 12 + 81$$

Complete the solution of $x^2 - 18x = 12$.

________________ .

$\{9 + \sqrt{93}, 9 - \sqrt{93}\}$

231 Solve $x^2 - 6x = 2$ by first adding 9 to both sides of the equation. ________________ .

[*Note*: $x^2 - 6x + 9$ is a perfect square trinomial.]

$\{3 + \sqrt{11}, 3 - \sqrt{11}\}$

232 Solve $x^2 - 12x = 1$ by first adding 36 to both sides of the equation. ________________ .
[*Note*: $x^2 - 12x + 36$ is a perfect square trinomial.]

$\{6 + \sqrt{37}, 6 - \sqrt{37}\}$

233 Solve $x^2 - 16x = {}^{-}62$ by first adding 64 to both sides of the equation. ________________ .

$\{8 + \sqrt{2}, 8 - \sqrt{2}\}$

234 Solve $x^2 - 8x = {}^{-}3$ by first adding a number to both sides of the equation to form a perfect square trinomial from $x^2 - 8x$. ________________ .

$\{4 + \sqrt{13}, 4 - \sqrt{13}\}$

235 Solve $x^2 + 2x = 6$ by first adding a number to both sides of the equation to form a perfect square trinomial from $x^2 + 2x$. ________________ .

$\{{}^{-}1 + \sqrt{7}, {}^{-}1 - \sqrt{7}\}$

236 Solve $x^2 - 6x = {}^{-}4$. ________________ .

$\{3 + \sqrt{5}, 3 - \sqrt{5}\}$

237 Solve $x^2 - 8x = 1$. ________________ .

$\{4 + \sqrt{17}, 4 - \sqrt{17}\}$

238 To solve $x^2 - 6x = 5$, the following steps are used:

$$x^2 - 6x = 5$$
$$x^2 - 6x + 9 = 5 + 9$$
$$(x - 3)^2 = 14$$
$$x - 3 = \sqrt{14} \quad \text{or} \quad x - 3 = {}^{-}\sqrt{14}$$
$$\{3 + \sqrt{14}, 3 - \sqrt{14}\}$$

Solve $x^2 - 8x = 3$. ________________ .

$\{4 + \sqrt{19}, 4 - \sqrt{19}\}$

239 Solve $x^2 + 10x = 21$. ________________ .

$\{{}^{-}5 + \sqrt{46}, {}^{-}5 - \sqrt{46}\}$

240 Solve $x^2 - 12x = 13$. ________ .

$\{13, {}^{-}1\}$

241 Solve $x^2 + 6x = 1$. ________________ .

$\{{}^{-}3 + \sqrt{10}, {}^{-}3 - \sqrt{10}\}$

242 Solve $x^2 - 14x = 10$. ________________ .

$\{7 + \sqrt{59}, 7 - \sqrt{59}\}$

243 Solve $x^2 - 2x = 3$. ________ .

$\{3, {}^{-}1\}$

244 Solve $x^2 + 12x = {}^{-}19$. ______________ .

$\{{}^{-}6 + \sqrt{17}, {}^{-}6 - \sqrt{17}\}$

245 To solve $x^2 - 3x = \frac{3}{4}$, first add $\frac{9}{4}$ to both sides of the equation because $x^2 - 3x + \frac{9}{4}$ is a perfect square trinomial.

$$x^2 - 3x = \frac{3}{4}$$

$$x^2 - 3x + \frac{9}{4} = \frac{3}{4} + \frac{9}{4}$$

$$\left(x - \frac{3}{2}\right)^2 = 3$$

Complete the solution of $x^2 - 3x = \frac{3}{4}$ by solving $\left(x - \frac{3}{2}\right)^2 = 3$. ______________ .

$\left\{\frac{3}{2} + \sqrt{3}, \frac{3}{2} - \sqrt{3}\right\}$

246 To solve $x^2 + 7x = \frac{3}{4}$, first add $\frac{49}{4}$ to both sides of the equation because $x^2 + 7x + \frac{49}{4}$ is a perfect square trinomial.

$$x^2 + 7x = \frac{3}{4}$$

$$x^2 + 7x + \frac{49}{4} = \frac{3}{4} + \frac{49}{4}$$

$$\left(x + \frac{7}{2}\right)^2 = 13$$

Complete the solution of $x^2 + 7x = \frac{3}{4}$ by solving $\left(x + \frac{7}{2}\right)^2 = 13$. ______________ .

$\left\{\frac{{}^{-}7}{2} + \sqrt{13}, \frac{{}^{-}7}{2} - \sqrt{13}\right\}$

247 Solve $x^2 - 5x = \frac{3}{4}$ by first adding $\frac{25}{4}$ to both sides of the equation. ______________ .

$\left\{\frac{5}{2} + \sqrt{7}, \frac{5}{2} - \sqrt{7}\right\}$

248 Solve $x^2 + 9x = \frac{3}{4}$ by first adding $\frac{81}{4}$ to both sides of the equation. ______________ .

$\left\{\frac{^-9}{2} + \sqrt{21}, \frac{^-9}{2} - \sqrt{21}\right\}$

249 Solve $x^2 - 7x = \frac{^-5}{4}$ by first adding $\frac{49}{4}$ to both sides of the equation. ______________ .

$\left\{\frac{7}{2} + \sqrt{11}, \frac{7}{2} - \sqrt{11}\right\}$

250 Solve $x^2 - 3x = \frac{11}{4}$. ______________ .

$\left\{\frac{3}{2} + \sqrt{5}, \frac{3}{2} - \sqrt{5}\right\}$

251 To solve $x^2 - 3x = \frac{^-1}{4}$, the following steps are used:

$$x^2 - 3x = \frac{^-1}{4}$$

$$x^2 - 3x + \frac{9}{4} = \frac{^-1}{4} + \frac{9}{4}$$

$$\left(x - \frac{3}{2}\right)^2 = 2$$

The truth set is $\left\{\frac{3}{2} + \sqrt{2}, \frac{3}{2} - \sqrt{2}\right\}$. Solve $x^2 - 5x = \frac{^-1}{4}$. ______________ .

$\left\{\frac{5}{2} + \sqrt{6}, \frac{5}{2} - \sqrt{6}\right\}$

252 Solve $x^2 - x = \frac{11}{4}$. ______________ .

$\left\{\frac{1}{2} + \sqrt{3}, \frac{1}{2} - \sqrt{3}\right\}$

253 Solve $x^2 + 5x = \frac{^-17}{4}$. ______________ .

$\left\{\frac{^-5}{2} + \sqrt{2}, \frac{^-5}{2} - \sqrt{2}\right\}$

254 Solve $x^2 + x = \frac{23}{4}$. ______________ .

$\left\{\frac{^-1}{2} + \sqrt{6}, \frac{^-1}{2} - \sqrt{6}\right\}$

255 Solve $x^2 - 6x = 6$. ______________ .

$\{3 + \sqrt{15}, 3 - \sqrt{15}\}$

256 Solve $x^2 + 10x = {}^-12$. ______________ .

$\{^-5 + \sqrt{13}, {}^-5 - \sqrt{13}\}$

257 To solve $(x - 2)^2 = \frac{25}{4}$, the following steps are used:

$$(x - 2)^2 = \frac{25}{4}$$

$$\begin{array}{lll} x - 2 = \frac{5}{2} & \text{or} & x - 2 = \frac{^-5}{2} \\ +2 \quad {}^+2 & & +2 \quad {}^+2 \\ \hline x \quad = 2 + \frac{5}{2} & & x \quad = 2 - \frac{5}{2} \\ x \quad = \frac{9}{2} & & x \quad = \frac{^-1}{2} \end{array}$$

Solve $x^2 - 5x = 6$ by first adding $\frac{25}{4}$ to both sides of the equation. ____________________ .

$\{6, {}^-1\}$

258 Solve $x^2 + 3x = 4$. ____________________ .

$\{{}^-4, 1\}$

259 Solve $x^2 + 7x = \frac{^-41}{4}$. ____________________ .

$\left\{\frac{^-7}{2} + \sqrt{2}, \frac{^-7}{2} - \sqrt{2}\right\}$

To multiply $\left(\frac{5+\sqrt{41}}{4}\right)^2$, the FOIL multiplication method is used:

$$\left(\frac{5+\sqrt{41}}{4}\right)^2 = \left(\frac{5+\sqrt{41}}{4}\right)\left(\frac{5+\sqrt{41}}{4}\right)$$

$$= \frac{\overset{\text{F}}{25} + \overset{\text{O}}{5\sqrt{41}} + \overset{\text{I}}{5\sqrt{41}} + \overset{\text{L}}{41}}{16}$$

$$= \frac{66 + 10\sqrt{41}}{16}$$

To check $\frac{5+\sqrt{41}}{4}$ as a solution of $2x^2 - 5x - 2 = 0$, the following steps are used:

$$2x^2 - 5x - 2 = 0 \qquad \text{becomes}$$

$$2\left(\frac{5+\sqrt{41}}{4}\right)^2 - 5\left(\frac{5+\sqrt{41}}{4}\right) - 2 = 0 \qquad \text{when } x = \frac{5+\sqrt{41}}{4}$$

$$2\left(\frac{66+10\sqrt{41}}{16}\right) - 5\left(\frac{5+\sqrt{41}}{4}\right) - 2 = 0$$

$$\frac{66+10\sqrt{41}}{8} - \frac{25+5\sqrt{41}}{4} - 2 = 0$$

$$\frac{66+10\sqrt{41}}{8} - \frac{50+10\sqrt{41}}{8} - 2 = 0$$

$$\frac{66+10\sqrt{41} - 50 - 10\sqrt{41}}{8} - 2 = 0$$

$$\frac{16}{8} - 2 = 0 \qquad \text{is true}$$

The other solution for $2x^2 - 5x - 2 = 0$ is $\frac{5-\sqrt{41}}{4}$, and it should be checked in the same manner.

260 Solve and check $x^2 + 3x = \frac{19}{4}$.

____________________.

$\left\{\frac{^-3}{2} + \sqrt{7}, \frac{^-3}{2} - \sqrt{7}\right\}$

261 Solve and check $x^2 - 7x = {}^-10$. ________.

$\{2, 5\}$

262 Solve and check $x^2 - 18x = {}^-78$.

____________________.

$\{9 + \sqrt{3}, 9 - \sqrt{3}\}$

263 To solve $x^2 - 6x = {}^-12$, the following steps are used:

$$x^2 - 6x = {}^-12$$
$$x^2 - 6x + 9 = {}^-12 + 9$$
$$(x - 3)^2 = {}^-3$$

The truth set is { } because no rational or irrational number can be multiplied by itself to give a negative number. Solve $x^2 - 10x = {}^-30$. ________.

{ }

264 Solve $x^2 - 2x = {}^-3$. ________.

{ }

265 Solve $x^2 - 10x = 12$. ____________________.

$\{5 + \sqrt{37}, 5 - \sqrt{37}\}$

266 Solve $x^2 - 10x = 4$. ____________________.

$\{5 + \sqrt{29}, 5 - \sqrt{29}\}$

267 The truth set of $\left(x - \frac{3}{2}\right)^2 = {}^-5$ is { } because there is no number that can be multiplied by itself to obtain a negative number. Solve $x^2 - 9x = \frac{^-85}{4}$. ________.

{ }

268 Solve $x^2 + 5x = \frac{^-5}{4}$. ____________________.

$\left\{\frac{^-5}{2} + \sqrt{5}, \frac{^-5}{2} - \sqrt{5}\right\}$

269 Solve $x^2 - 12x = {}^-29$. ____________________.

$\{6 + \sqrt{7}, 6 - \sqrt{7}\}$

270 Solve $x^2 - 6x = {}^-15$. ________.

{ }

271 Solve $x^2 - 7x = \frac{3}{4}$. ____________________.

$\left\{\frac{7}{2} + \sqrt{13}, \frac{7}{2} - \sqrt{13}\right\}$

Self-Quiz # 5

The following questions test the objectives of the preceding section. 100% mastery is desired.

Solve:

1. $x^2 - 8x = 9$. ________ .
2. $x^2 + 6x = 4$. ________ .
3. $x^2 + 2x = {}^{-}3$. ________ .
4. $x^2 - 5x = \frac{^{-}13}{4}$. ________ .
5. $x^2 + 7x = {}^{-}10$. ________ .

272 To solve $x^2 + 8x - 6 = 0$, the first step is to add 6 to both sides of the equation:

$$x^2 + 8x - 6 = 0$$
$$x^2 + 8x = 6$$

Complete the solution of $x^2 + 8x = 6$ by adding 16 to both sides of the equation and factoring the perfect square trinomial, $x^2 + 8x + 16$.

________________ .

$\{^{-}4 + \sqrt{22}, {}^{-}4 - \sqrt{22}\}$

273 To solve $x^2 - 6x - 1 = 0$, first add 1 to both sides of the equation. Complete the following solution:

$x^2 - 6x - 1 = 0$
$x^2 - 6x = 1$
$x^2 - 6x + 9 = 1 + 9$
$(x - 3)^2 = 10$

________________ .

$\{3 + \sqrt{10}, 3 - \sqrt{10}\}$

274 Solve $x^2 - 4x + 1 = 0$ by first adding $^{-}1$ to both sides of the equation. Complete the following solution:

$x^2 - 4x + 1 = 0$
$x^2 - 4x = {}^{-}1$

________________ .

$\{2 + \sqrt{3}, 2 - \sqrt{3}\}$

275 Solve $x^2 - 10x + 6 = 0$ by first adding $^{-}6$ to both sides of the equation. ________________.

$\{5 + \sqrt{19}, 5 - \sqrt{19}\}$

276 Solve $x^2 - 6x - 4 = 0$ by first adding 4 to both sides of the equation. ________________.

$\{3 + \sqrt{13}, 3 - \sqrt{13}\}$

277 Solve $x^2 - 2x - 2 = 0$. ________________.

$\{1 + \sqrt{3}, 1 - \sqrt{3}\}$

278 Solve $x^2 - 12x - 3 = 0$. ________________.

$\{6 + \sqrt{39}, 6 - \sqrt{39}\}$

279 Solve $x^2 + 4x + 1 = 0$. ________________.

$\{^{-}2 + \sqrt{3}, ^{-}2 - \sqrt{3}\}$

280 Solve $x^2 - 10x - 1 = 0$. ________________.

$\{5 + \sqrt{26}, 5 - \sqrt{26}\}$

281 Solve $x^2 + 8x + 5 = 0$. ________________.

$\{^{-}4 + \sqrt{11}, ^{-}4 - \sqrt{11}\}$

282 To solve $3x^2 - 12x = 6$, first multiply both sides of the equation by $\frac{1}{3}$.

$$\frac{1}{3}(3x^2 - 12x) = \frac{1}{3} \cdot 6$$

$$x^2 - 4x = 2$$

Complete the solution of $3x^2 - 12x = 6$ by solving $x^2 - 4x = 2$. ________________.

$\{2 + \sqrt{6}, 2 - \sqrt{6}\}$

283 To solve $5x^2 + 30x = 20$, first multiply both sides of the equation by $\frac{1}{5}$.

$$5x^2 + 30x = 20$$

$$\frac{1}{5}(5x^2 + 30x) = \frac{1}{5} \cdot 20$$

$$x^2 + 6x = 4$$

Complete the solution of $5x^2 + 30x = 20$ by solving $x^2 + 6x = 4$. ________________.

$\{^{-}3 + \sqrt{13}, ^{-}3 - \sqrt{13}\}$

284 Solve $2x^2 - 8x = 20$. ______________________. $\{2 + \sqrt{14}, 2 - \sqrt{14}\}$

[*Note*: First multiply both sides of the equation by $\frac{1}{2}$.]

285 Solve $4x^2 - 16x = 8$. ______________________. $\{2 + \sqrt{6}, 2 - \sqrt{6}\}$

[*Note*: First multiply both sides of the equation by $\frac{1}{4}$.]

286 Solve $3x^2 - 18x = 6$. ______________________. $\{3 + \sqrt{11}, 3 - \sqrt{11}\}$

[*Note:* First multiply both sides of the equation by $\frac{1}{3}$.]

287 Solve $5x^2 + 30x = 20$. ______________________. $\{^-3 + \sqrt{13}, {^-3} - \sqrt{13}\}$

288 Solve $2x^2 + 12x = {^-4}$. ______________________. $\{^-3 + \sqrt{7}, {^-3} - \sqrt{7}\}$

289 Solve $3x^2 + 6x = 12$. ______________________. $\{^-1 + \sqrt{5}, {^-1} - \sqrt{5}\}$

290 Solve $4x^2 + 16x = {^-4}$. ______________________. $\{^-2 + \sqrt{3}, {^-2} - \sqrt{3}\}$

291 Solve $x^2 - 8x + 7 = 0$. __________. $\{1, 7\}$

292 Solve $2x^2 - 8x = 2$. ______________________. $\{2 + \sqrt{5}, 2 - \sqrt{5}\}$

293 Solve $x^2 - 3x + \frac{1}{4} = 0$. ______________________. $\left\{\frac{3}{2} + \sqrt{2}, \frac{3}{2} - \sqrt{2}\right\}$

294 Solve $3x^2 + 12x = 3$. ______________________. $\{^-2 + \sqrt{5}, {^-2} - \sqrt{5}\}$

295 Solve $x^2 - 3x = \frac{3}{4}$. ______________________. $\left\{\frac{3}{2} + \sqrt{3}, \frac{3}{2} - \sqrt{3}\right\}$

296 Solve $7x^2 - 14x = 42$. ______________________. $\{1 + \sqrt{7}, 1 - \sqrt{7}\}$

297 Solve $5x^2 - 20x = 45$. ______________________. $\{2 + \sqrt{13}, 2 - \sqrt{13}\}$

298 Solve $x^2 - 5x = 6$. __________. $\{^-1, 6\}$

Self-Quiz # 6

The following questions test the objectives of the preceding section. 100% mastery is desired.

Solve:

1. $2x^2 + 8x + 2 = 0$. ________ .

2. $x^2 - 6x - 1 = 0$. ________ .

3. $x^2 - 4x = 1$. ________ .

4. $5x^2 - 20x = 10$. ________ .

5. $4x^2 + 16x - 12 = 0$. ________ .

The equations in Self-Quiz # 6 were solved by the completing the square method. The same method may be applied to the general quadratic equation, $ax^2 + bx + c = 0$, to derive a formula for solving any quadratic equation.

The derivation of the "quadratic formula" is shown in Appendix B of this text. The use of the "quadratic formula" is explained in the following frames.

The truth set of $ax^2 + bx + c = 0$ is

$$\left\{\frac{^-b + \sqrt{b^2 - 4ac}}{2a}, \frac{^-b - \sqrt{b^2 - 4ac}}{2a}\right\}$$

This is a statement of the *quadratic formula*, which is more commonly written as

$$x = \frac{^-b \pm \sqrt{b^2 - 4ac}}{2a}$$

---

299 For any quadratic equation in the form $ax^2 + bx + c = 0$, the truth set is

$$\left\{\frac{^-b + \sqrt{b^2 - 4ac}}{2a}, \frac{^-b - \sqrt{b^2 - 4ac}}{2a}\right\}$$

To use the *quadratic formula*, the specific values for a, b, and c must be obtained from the equation to be solved. For the equation $3x^2 + 5x + 7 = 0$, $a = 3$, $b = 5$, and $c = 7$. For the equation $2x^2 + 4x \underline{+ 9} = 0$, $a = 2$, $b = 4$, and $c =$ ______ .

9

300 $ax^2 + bx + c = 0$. To use the *quadratic formula*, the values of a, b, and c must be determined from the equation to be solved. For the equation $7x^2 + 3x - 5 = 0$, $a = 7$, $b = 3$, and $c = {}^-5$. For the equation $5x^2 \underline{- 3}x + 8 = 0$, $a = 5$, $b =$ ______, and $c = 8$.

$^-3$

301 $ax^2 + bx + c = 0$. The correct values for a, b, and c must be used in solving a quadratic equation by using the quadratic formula. For the equation $4x^2 - 3x + 5 = 0$, $a = \underline{4}$, $b = {}^-3$, and $c = 5$. For the equation $\underline{6}x^2 + 7x - 8 = 0$, $a =$ ______, $b = 7$, and $c = {}^-8$.

6

302 The *quadratic formula* can be used to solve any quadratic equation in the form $ax^2 + bx + c = 0$. To solve $5x^2 - 9x + 2 = 0$, $a = 5$, $b = {}^-9$, and $c = 2$. What values for a, b, and c must be used to solve $8x^2 + 4x - 1 = 0$?
$a =$ ______, $b =$ ______, $c =$ ______ .

$a = 8, b = 4, c = {}^-1$

303 To use the *quadratic formula* to solve $3x^2 - 7x + 11 = 0$, $a = 3$, $b = {}^-7$, and $c = 11$. What values for a, b, and c must be used to solve the equation $10x^2 - 7x + 9 = 0$?
$a =$ ______, $b =$ ______, $c =$ ______ .

$a = 10, b = {}^-7, c = 9$

304 $ax^2 + bx + c = 0$. For the equation
$3x^2 + 4x + 7 = 0$,
$a =$ ______, $b =$ ______, $c =$ ______.

$a = 3, b = 4, c = 7$

The quadratic formula, $x = \dfrac{^{-}b \pm \sqrt{b^2 - 4ac}}{2a}$, has three parts, ^{-}b, $\sqrt{b^2 - 4ac}$, and $2a$. In the following frames these three parts are evaluated separately as preparation for combining them to find the truth sets for quadratic equations.

---

305 The quadratic formula is $x = \dfrac{^{-}b \pm \sqrt{b^2 - 4ac}}{2a}$. To use the formula for solving the equation

$3x^2 + 7x - 1 = 0$, $a = 3$, $b = 7$, and $c = {^{-}1}$.

$$^{-}b = {^{-}1} \cdot b = {^{-}1} \cdot 7 = {^{-}7}$$

$$\sqrt{b^2 - 4ac} = \sqrt{7^2 - 4 \cdot 3 \cdot {^{-}1}} = \sqrt{49 + 12} = \sqrt{61}$$

$2a = 2 \cdot 3 =$ ______.

6

306 The quadratic formula, $x = \dfrac{^{-}b \pm \sqrt{b^2 - 4ac}}{2a}$, can be used to solve any quadratic equation. To use the formula to solve $6x^2 - 3x - 2 = 0$, $a = 6$, $b = {^{-}3}$, and $c = {^{-}2}$.

$$^{-}b = {^{-}1} \cdot b = {^{-}1} \cdot {^{-}3} = 3$$

$$\sqrt{b^2 - 4ac} = \sqrt{({^{-}3})^2 - 4 \cdot 6 \cdot {^{-}2}} = \sqrt{9 + 48} = \sqrt{57}$$

$2a =$ ______.

12

307 The quadratic formula is $x = \dfrac{{}^{-}b \pm \sqrt{b^2 - 4ac}}{2a}$. To use the formula for solving the equation

$5x^2 - 7x + 1 = 0$, $a = 5$, $b = {}^{-}7$, and $c = 1$.

$\sqrt{b^2 - 4ac} = \sqrt{({}^{-}7)^2 - 4 \cdot 5 \cdot 1} = \sqrt{49 - 20} = \sqrt{29}$

$2a = 2 \cdot 5 = 10$

${}^{-}b = {}^{-}1 \cdot b = {}^{-}1 \cdot {}^{-}7 =$ ________ .

7

308 The quadratic formula is $x = \dfrac{{}^{-}b \pm \sqrt{b^2 - 4ac}}{2a}$ · To use the formula for solving the equation

$3x^2 + 5x - 2 = 0$, $a = 3$, $b = 5$, and $c = {}^{-}2$,

$\sqrt{b^2 - 4ac} = \sqrt{5^2 - 4 \cdot 3 \cdot {}^{-}2} = \sqrt{25 + 24} = \sqrt{49} = 7$

$2a = 2 \cdot 3 = 6$

${}^{-}b =$ ________ .

${}^{-}5$

309 The quadratic formula is $x = \dfrac{{}^{-}b \pm \sqrt{b^2 - 4ac}}{2a}$.

To use the formula for solving

$x^2 + 5x + 2 = 0$, $a = 1$, $b = 5$, and $c = 2$.

${}^{-}b = {}^{-}1 \cdot b = {}^{-}1 \cdot 5 = {}^{-}5$

$2a = 2 \cdot 1 = 2$

$\sqrt{b^2 - 4ac} = \sqrt{5^2 - 4 \cdot 1 \cdot 2} = \sqrt{25 - 8}$

$=$ ________ .

$\sqrt{17}$

310 The quadratic formula is $x = \dfrac{{}^{-}b \pm \sqrt{b^2 - 4ac}}{2a}$.

To use the formula for solving

$x^2 - 3x - 5 = 0$, $a = 1$, $b = {}^{-}3$, and $c = {}^{-}5$.

$${}^{-}b = {}^{-}1 \cdot {}^{-}3 = 3$$

$$2a = 2 \cdot 1 = 2$$

$$\sqrt{b^2 - 4ac} = \sqrt{({}^{-}3)^2 - 4 \cdot 1 \cdot {}^{-}5} = \sqrt{9 + 20}$$

$=$ ________ .

$\sqrt{29}$

311 $x = \dfrac{{}^{-}b \pm \sqrt{b^2 - 4ac}}{2a}$. To use the formula for solving $2x^2 - 5x + 1 = 0$, $a = 2$, $b = {}^{-}5$, $c = 1$.

${}^{-}b =$ ________

$2a =$ ________

$\sqrt{b^2 - 4ac} =$ ________ .

${}^{-}b = 5$
$2a = 4$
$\sqrt{b^2 - 4ac} = \sqrt{17}$

312 $x = \dfrac{{}^{-}b \pm \sqrt{b^2 - 4ac}}{2a}$. To use the formula for solving $x^2 - 7x + 3 = 0$, $a = 1$, $b = {}^{-}7$, $c = 3$.

${}^{-}b =$ ________

$2a =$ ________

$\sqrt{b^2 - 4ac} =$ ________.

${}^{-}b = 7$
$2a = 2$
$\sqrt{b^2 - 4ac} = \sqrt{37}$

313 $x = \dfrac{{}^{-}b \pm \sqrt{b^2 - 4ac}}{2a}$. To use the formula for solving $2x^2 - 7x + 1 = 0$, $a = 2$, $b = {}^{-}7$, $c = 1$.

${}^{-}b =$ ________

$2a =$ ________

$\sqrt{b^2 - 4ac} =$ ________ .

${}^{-}b = 7$
$2a = 4$
$\sqrt{b^2 - 4ac} = \sqrt{41}$

314 $x = \dfrac{^{-}b \pm \sqrt{b^2 - 4ac}}{2a}$. For the equation $2x^2 - 7x + 3 = 0$.

$^{-}b =$ ________

$2a =$ ________

$\sqrt{b^2 - 4ac} =$ ________.

$^{-}b = 7$
$2a = 4$
$\sqrt{b^2 - 4ac} = \sqrt{25} = 5$

Self-Quiz # 7

The following questions test the objectives of the preceding section. 100% mastery is desired.

For each of the following equations find:

$^{-}b =$ ________,
$2a =$ ________,
$\sqrt{b^2 - 4ac} =$ ________.

1. $2x^2 - 3x - 7 = 0$. ________, ________, ________.
2. $7x^2 + 5x - 1 = 0$. ________, ________, ________.
3. $x^2 + 3x - 9 = 0$. ________, ________, ________.
4. $2x^2 - x - 5 = 0$. ________, ________, ________.
5. $x^2 + x - 6 = 0$. ________, ________, ________.
6. $2x^2 - 10x + 5 = 0$. ________, ________, ________.
7. $x^2 + 7x + 3 = 0$. ________, ________, ________.

The quadratic formula, $x = \frac{^{-}b \pm \sqrt{b^2 - 4ac}}{2a}$, is a common way of writing the truth set $\left\{\frac{^{-}b + \sqrt{b^2 - 4ac}}{2a}, \frac{^{-}b - \sqrt{b^2 - 4ac}}{2a}\right\}$ for any quadratic equation in the form $ax^2 + bx + c = 0$.

To solve $2x^2 + 7x + 1 = 0$, the quadratic formula is used as follows:

$$x = \frac{^{-}b \pm \sqrt{b^2 - 4ac}}{2a}$$

$$x = \frac{^{-}7 \pm \sqrt{49 - 8}}{2 \cdot 2}$$

$$x = \frac{^{-}7 \pm \sqrt{41}}{4}$$

The $\pm$ symbol in the formula is used to provide two solutions,

$$\frac{^{-}7 + \sqrt{41}}{4} \quad \text{or} \quad \frac{^{-}7 - \sqrt{41}}{4}$$

---

315 Complete the solution *for* $3x^2 - x - 5 = 0$.

$$x = \frac{^{-}b \pm \sqrt{b^2 - 4ac}}{2a}$$

$$x = \frac{1 \pm \sqrt{1 + 60}}{6}$$

Using the plus symbol of $\pm$, the solution is

$$x = \frac{1 + \sqrt{61}}{6}$$

Using the minus symbol of $\pm$, the solution is $x =$ ______________ .

$$\frac{1 - \sqrt{61}}{6}$$

316 Complete the solution for $x^2 - 5x + 3 = 0$.

$$x = \frac{^-b \pm \sqrt{b^2 - 4ac}}{2a}$$

$$x = \frac{5 \pm \sqrt{25 - 12}}{2}$$

$x = \dfrac{5 + \sqrt{13}}{2}$ or $x =$ ______________ .

$\dfrac{5 - \sqrt{13}}{2}$

317 Complete the solution for $x^2 - 9x + 14 = 0$.

$$x = \frac{^-b \pm \sqrt{b^2 - 4ac}}{2a}$$

$$x = \frac{9 \pm \sqrt{81 - 56}}{2}$$

$$x = \frac{9 \pm \sqrt{25}}{2}$$

$x = 7$ or $x =$ ________ .

2

318 Complete the solution for $3x^2 - 5x + 1 = 0$,

$$x = \frac{^-b \pm \sqrt{b^2 - 4ac}}{2a}$$

$$x = \frac{5 \pm \sqrt{25 - 12}}{6}$$

$x =$ ______________ or $x = \dfrac{5 - \sqrt{13}}{6}$.

$\dfrac{5 + \sqrt{13}}{6}$

For some equations the radical expression $\sqrt{b^2 - 4ac}$ can be simplified as in the following example:

$$2x^2 - 7x + 3 = 0$$

$$x = \frac{7 \pm \sqrt{49 - 24}}{4}$$

$$= \frac{7 \pm \sqrt{25}}{4}$$

$$= \frac{7 \pm 5}{4}$$

Using the plus symbol, $x = \frac{7 + 5}{4} = \frac{12}{4} = 3$.

Using the minus symbol, $x = \frac{7 - 5}{4} = \frac{2}{4} = \frac{1}{2}$.

---

319 Complete the solution for $3x^2 - 5x + 2 = 0$.

$$x = \frac{{}^{-}b \pm \sqrt{b^2 - 4ac}}{2a}$$

$$x = \frac{5 \pm \sqrt{25 - 24}}{6}$$

$$x = \frac{5 \pm 1}{6}$$

$x = 1$ or $x =$ ________.

$\frac{2}{3}$

320 Complete the solution for $4x^2 - x - 2 = 0$.

$$x = \frac{{}^{-}b \pm \sqrt{b^2 - 4ac}}{2a}$$

$$x = \frac{1 \pm \sqrt{1 + 32}}{8}$$

$x = \frac{1 + \sqrt{33}}{8}$ or $x =$ ________________.

$\frac{1 - \sqrt{33}}{8}$

321 Solve $x^2 - 5x + 2 = 0$.

$$x = \frac{{}^{-}b \pm \sqrt{b^2 - 4ac}}{2a}$$

$x =$ ________ or $x =$ ________.

$\frac{5 + \sqrt{17}}{2}, \frac{5 - \sqrt{17}}{2}$

322 Solve $3x^2 - 3x - 5 = 0$.

$$x = \frac{{}^{-}b \pm \sqrt{b^2 - 4ac}}{2a}$$

$x =$ ________ or $x =$ ________.

$\frac{3 + \sqrt{69}}{6}, \frac{3 - \sqrt{69}}{6}$

323 Solve $x^2 - 8x + 7 = 0$.

$$x = \frac{{}^{-}b \pm \sqrt{b^2 - 4ac}}{2a}$$

$x =$ ________ or $x =$ ________.

1, 7

324 Solve $3x^2 + x - 1 = 0$.

$$x = \frac{{}^{-}b \pm \sqrt{b^2 - 4ac}}{2a}$$

$x =$ ________ or $x =$ ________.

$\frac{{}^{-}1 + \sqrt{13}}{6}, \frac{{}^{-}1 - \sqrt{13}}{6}$

325 Solve $2x^2 - 5x - 1 = 0$.

$$x = \frac{{}^{-}b \pm \sqrt{b^2 - 4ac}}{2a}$$

$x =$ ________ or $x =$ ________.

$\frac{5 + \sqrt{33}}{4}, \frac{5 - \sqrt{33}}{4}$

326 Solve $3x^2 + 5x + 1 = 0$.
$x =$ ________ or $x =$ ________.

$\frac{{}^{-}5 + \sqrt{13}}{6}, \frac{{}^{-}5 - \sqrt{13}}{6}$

327 Solve $x^2 - 3x - 1 = 0$.
$x =$ ________ or $x =$ ________.

$\frac{3 + \sqrt{13}}{2}, \frac{3 - \sqrt{13}}{2}$

328 Solve $x^2 - 7x + 12 = 0$.
$x =$ ________ or $x =$ ________.

3, 4

329 Solve $3x^2 - 9x + 5 = 0$.
$x =$ ________ or $x =$ ________.

$\frac{9 + \sqrt{21}}{6}, \frac{9 - \sqrt{21}}{6}$

330 Solve $5x^2 + 3x - 1 = 0$.
$x =$ ________ or $x =$ ________.

$\frac{^-3 + \sqrt{29}}{10}, \frac{^-3 - \sqrt{29}}{10}$

331 For the equation $5x^2 + 3x - 1 = 0$,

$x = \frac{^-3 + \sqrt{29}}{10}$ or $x = \frac{^-3 - \sqrt{29}}{10}$

The truth set of $5x^2 + 3x - 1 = 0$ is $\left\{\frac{^-3 + \sqrt{29}}{10}, \frac{^-3 - \sqrt{29}}{10}\right\}$. Find the truth set of $2x^2 + 7x - 1 = 0$. ________________.

$\left\{\frac{^-7 + \sqrt{57}}{4}, \frac{^-7 - \sqrt{57}}{4}\right\}$

332 Find the truth set of $x^2 - 5x + 1 = 0$.
________________.

$\left\{\frac{5 + \sqrt{21}}{2}, \frac{5 - \sqrt{21}}{2}\right\}$

333 Find the truth set of $x^2 + 3x - 2 = 0$.
________________.

$\left\{\frac{^-3 + \sqrt{17}}{2}, \frac{^-3 - \sqrt{17}}{2}\right\}$

334 Find the truth set of $x^2 - 5x + 6 = 0$. ________.

$\{3, 2\}$

335 Find the truth set of $4x^2 - 3x - 1 = 0$.
________.

$\left\{1, \frac{^-1}{4}\right\}$

336 Solve $x^2 - 3x - 7 = 0$. ________________.

$\left\{\frac{3 + \sqrt{37}}{2}, \frac{3 - \sqrt{37}}{2}\right\}$

[*Note*: "Solve" means to find the truth set.]

337 Solve $2x^2 - 5x + 1 = 0$. ________________.

$\left\{\frac{5 + \sqrt{17}}{4}, \frac{5 - \sqrt{17}}{4}\right\}$

The radical expression $\sqrt{125}$ is simplified in the following way:

$$\sqrt{125} = \sqrt{25} \cdot \sqrt{5} = 5\sqrt{5}$$

This simplification is needed in solving $x^2 + 11x - 1 = 0$, as shown below:

$$x^2 + 11x - 1 = 0$$

$$x = \frac{^-11 \pm \sqrt{121 + 4}}{2}$$

$$= \frac{^-11 \pm \sqrt{125}}{2}$$

$$= \frac{^-11 \pm 5\sqrt{5}}{2}$$

The truth set of $x^2 + 11x - 1 = 0$ should be shown as

$$\left\{\frac{^-11 + 5\sqrt{5}}{2}, \frac{^-11 - 5\sqrt{5}}{2}\right\}$$

---

338 To solve $x^2 - 3x - 9 = 0$, the quadratic formula is used as follows:

$$x = \frac{3 \pm \sqrt{9 + 36}}{2}$$

$$x = \frac{3 \pm \sqrt{45}}{2}$$

$\sqrt{45} = \sqrt{9} \cdot \sqrt{5} = 3\sqrt{5}$. The truth set is $\left\{\frac{3 + 3\sqrt{5}}{2}, \frac{3 - 3\sqrt{5}}{2}\right\}$. Solve $x^2 - 5x - 5 = 0$ and simplify the radical. ______________ .

$\left\{\frac{5 + 3\sqrt{5}}{2}, \frac{5 - 3\sqrt{5}}{2}\right\}$

339 Solve $x^2 - x - 11 = 0$ and simplify the radical.

______________ .

$\left\{\frac{1 + 3\sqrt{5}}{2}, \frac{1 - 3\sqrt{5}}{2}\right\}$

340 Solve $x^2 - 7x + 1 = 0$. ______________ .

$\left\{\frac{7 + 3\sqrt{5}}{2}, \frac{7 - 3\sqrt{5}}{2}\right\}$

341 To solve $x^2 - 6x - 4 = 0$, the quadratic formula is used as follows:

$$x = \frac{6 \pm \sqrt{36 + 16}}{2}$$

$$x = \frac{6 \pm \sqrt{52}}{2}$$

$$x = \frac{6 \pm 2\sqrt{13}}{2}$$

(To simplify $\frac{6 \pm 2\sqrt{13}}{2}$, both terms of the numerator, 6 and $2\sqrt{13}$, must be divided by the denominator, 2.)

$$x = \frac{6}{2} \pm \frac{2\sqrt{13}}{2}$$

$$x = 3 \pm \sqrt{13}$$

Solve $x^2 + 2x - 6 = 0$. ____________________ . $\{{}^{-}1 + \sqrt{7}, {}^{-}1 - \sqrt{7}\}$

342 Solve $x^2 - 6x + 4 = 0$. ____________________ . $\{3 + \sqrt{5}, 3 - \sqrt{5}\}$

343 Solve $x^2 - 4x + 2 = 0$. ____________________ . $\{2 + \sqrt{2}, 2 - \sqrt{2}\}$

344 To solve $2x^2 + 4x - 3 = 0$, the quadratic formula is used as follows:

$$x = \frac{^-4 \pm \sqrt{16 + 24}}{4}$$

$$x = \frac{^-4 \pm \sqrt{40}}{4}$$

$$x = \frac{^-4 \pm 2\sqrt{10}}{4}$$

(To simplify $\frac{^-4 \pm 2\sqrt{10}}{4}$, both terms of the numerator, $^-4$ and $2\sqrt{10}$, must be divided by the denominator, 4.)

$$x = \frac{^-4}{4} \pm \frac{2\sqrt{10}}{4}$$

$$x = {^-1} \pm \frac{\sqrt{10}}{2}$$

Solve $2x^2 + 6x - 5 = 0$. ______________ .

$\left\{\frac{^-3 + \sqrt{19}}{2}, \frac{^-3 - \sqrt{19}}{2}\right\}$ or $\left\{\frac{^-3}{2} + \frac{\sqrt{19}}{2}, \frac{^-3}{2} - \frac{\sqrt{19}}{2}\right\}$

345 Solve $3x^2 - 4x - 1 = 0$. ______________ .

$\left\{\frac{2 + \sqrt{7}}{3}, \frac{2 - \sqrt{7}}{3}\right\}$ or $\left\{\frac{2}{3} + \frac{\sqrt{7}}{3}, \frac{2}{3} - \frac{\sqrt{7}}{3}\right\}$

346 Solve $4x^2 + 2x - 5 = 0$. ______________ .

$\left\{\frac{^-1 + \sqrt{21}}{4}, \frac{^-1 - \sqrt{21}}{4}\right\}$ or $\left\{\frac{^-1}{4} + \frac{\sqrt{21}}{4}, \frac{^-1}{4} - \frac{\sqrt{21}}{4}\right\}$

347 Solve $3x^2 - 2x - 2 = 0$. ______________.

$\left\{\frac{1 + \sqrt{7}}{3}, \frac{1 - \sqrt{7}}{3}\right\}$

or

$\left\{\frac{1}{3} + \frac{\sqrt{7}}{3}, \frac{1}{3} - \frac{\sqrt{7}}{3}\right\}$

348 Solve $4x^2 - 7x + 1 = 0$. ______________.

$\left\{\frac{7 + \sqrt{33}}{8}, \frac{7 - \sqrt{33}}{8}\right\}$

349 Solve $3x^2 + 5x + 2 = 0$. ________.

$\left\{\frac{{}^-2}{3}, {}^-1\right\}$

350 To solve $x^2 - 3x + 7 = 0$,

$$x = \frac{3 \pm \sqrt{9 - 28}}{2} = \frac{3 \pm \sqrt{{}^-19}}{2}$$

There is no rational or irrational number that if multiplied by itself will give ${}^-19$. Therefore, the truth set of $x^2 - 3x + 7 = 0$ is $\{\ \}$. Solve $x^2 + 2x + 3 = 0$. ________.

$\{\ \}$

351 Solve $2x^2 - 3x + 8 = 0$. ________.

$\{\ \}$

352 Solve $x^2 - 3x - 5 = 0$. ______________.

$\left\{\frac{3 + \sqrt{29}}{2}, \frac{3 - \sqrt{29}}{2}\right\}$

353 Solve $x^2 + 5x - 3 = 0$. ______________.

$\left\{\frac{{}^-5 + \sqrt{37}}{2}, \frac{{}^-5 - \sqrt{37}}{2}\right\}$

354 Solve $x^2 - 4x + 5 = 0$. ________.

$\{\ \}$

355 Solve $x^2 - 4x - 2 = 0$. ______________.

$\{2 + \sqrt{6}, 2 - \sqrt{6}\}$

356 The quadratic formula is for equations in the form $ax^2 + bx + c = 0$. To solve $x^2 - 7x = 2$, first add $^-2$ to both sides of the equation so that the right side will be zero. Solve $x^2 - 7x = 2$ by first adding $^-2$ to both sides of the equation. ________ .

$\left\{\frac{7 + \sqrt{57}}{2}, \frac{7 - \sqrt{57}}{2}\right\}$

357 Solve $3x^2 - x = 3$ by first adding $^-3$ to both sides of the equation. ________ .

$\left\{\frac{1 + \sqrt{37}}{6}, \frac{1 - \sqrt{37}}{6}\right\}$

358 Solve $x^2 + 3x = 5$. ________ .

$\left\{\frac{^-3 + \sqrt{29}}{2}, \frac{^-3 - \sqrt{29}}{2}\right\}$

To solve $x^2 - 3x - 4 = 0$ using the quadratic formula, the following steps are taken:

$$x^2 - 3x - 4 = 0$$

$$x = \frac{3 \pm \sqrt{9 - 4 \cdot 1 \cdot {}^-4}}{2} = \frac{3 \pm \sqrt{25}}{2} = \frac{3 \pm 5}{2}$$

The truth set is $\{4, {}^-1\}$.

To solve $x^2 - 3x - 4 = 0$ using the factoring method, the following steps are taken:

$$x^2 - 3x - 4 = 0$$

$$(x - 4)(x + 1) = 0$$

$$x - 4 = 0 \quad \text{or} \quad x + 1 = 0$$

The truth set is $\{4, {}^-1\}$.

The quadratic formula can be used to find the truth set of any quadratic equation. However, it is easier to use the factoring method whenever the left side of the equation is factorable.

In the following frames, use the factor method whenever the polynomial can be factored.

---

359 Solve $x^2 - 8x + 12 = 0$. ________ .

$\{2, 6\}$

360 Solve $x^2 + 13x + 30 = 0$. ________ .

$\{{}^-3, {}^-10\}$

361 Solve $x^2 - 2x = 15$. ________.

$\{^-3, 5\}$

362 Solve $x^2 + 3x - 5 = 0$. ________________.

$\left\{\frac{^-3 + \sqrt{29}}{2}, \frac{^-3 - \sqrt{29}}{2}\right\}$

363 Solve $2x^2 + 5x + 4 = 0$. ________.

$\{\ \}$

364 Solve $5x^2 + 9x = {^-3}$. ________________.

$\left\{\frac{^-9 + \sqrt{21}}{10}, \frac{^-9 - \sqrt{21}}{10}\right\}$

365 Solve $x^2 - 7x + 10 = 0$. ________.

$\{2, 5\}$

366 Solve $3x^2 - 2x - 4 = 0$. ________________.

$\left\{\frac{1}{3} + \frac{\sqrt{13}}{3}, \frac{1}{3} - \frac{\sqrt{13}}{3}\right\}$

367 Solve $2x^2 - 7x + 3 = 0$. ________.

$\left\{\frac{1}{2}, 3\right\}$

368 Solve $x^2 + x - 5 = 0$. ________________.

$\left\{\frac{^-1 + \sqrt{21}}{2}, \frac{^-1 - \sqrt{21}}{2}\right\}$

369 Solve $3x^2 - 2x + 6 = 0$. ________.

$\{\ \}$

370 Solve $x^2 + 9 = 8x$. ________________.
[*Hint*: First write the equation in the form $ax^2 + bx + c = 0$.]

$\{4 + \sqrt{7}, 4 - \sqrt{7}\}$

371 Solve $x^2 - x - 2 = 0$. ________.

$\{^-1, 2\}$

To check $\frac{^-3 + \sqrt{17}}{2}$ as a solution of $x^2 + 3x = 2$, the following steps are used:

$$x^2 + 3x = 2$$

becomes

$$\left(\frac{^-3 + \sqrt{17}}{2}\right)^2 + 3\left(\frac{^-3 + \sqrt{17}}{2}\right) = 2$$

when $x = \frac{^-3 + \sqrt{17}}{2}$

$$\frac{9 - 6\sqrt{17} + 17}{4} + \frac{^-9 + 3\sqrt{17}}{2} = 2$$

$$\frac{26 - 6\sqrt{17}}{4} + \frac{^-18 + 6\sqrt{17}}{4} = 2$$

$$\frac{26 - 6\sqrt{17} - 18 + 6\sqrt{17}}{4} = 2$$

$$\frac{8}{4} = 2$$

$$2 = 2 \qquad \text{is true}$$

The other solution of $x^2 + 3x = 2$ is $\frac{^-3 - \sqrt{17}}{2}$ and should be checked in the same manner.

---

372 Solve and check $x^2 + 5x = 3$. ______________ .

$\left\{\frac{^-5 + \sqrt{37}}{2}, \frac{^-5 - \sqrt{37}}{2}\right\}$

373 Solve and check $2x^2 - 3x - 1 = 0$.
______________ .

$\left\{\frac{3 + \sqrt{17}}{4}, \frac{3 - \sqrt{17}}{4}\right\}$

Self-Quiz # 8

The following questions test the objectives of the preceding section. 100% mastery is desired.

Solve:

1. $x^2 - 5x + 3 = 0$. ________ .

2. $3x^2 - x - 3 = 0$. ________ .

3. $x^2 - 8x + 6 = 0$. ________ .

4. $x^2 + 2x = 15$. ________ .

5. $2x^2 - 4x + 5 = 0$. ________ .

Chapter Summary

In this chapter the solving of quadratic equations with irrational solutions was studied.

Every quadratic equation in the form $ax^2 + bx + c = 0$ can be solved by using perfect square trinomials or by the quadratic formula. These methods of solving equations provide the necessary information to determine whether or not the truth set is { }.

Although the quadratic formula is extremely useful for solving quadratic equations, whenever the solutions are rational numbers it is generally easier to solve the equation by factoring.

CHAPTER POST-TEST

The following questions test the objectives of this chapter. A score of 90% indicates sufficient mastery, and the student may proceed to the next chapter.

Solve:

1. $x^2 = 16$. ________.

2. $x^2 = 43$. ________.

3. $x^2 = {}^{-}4$. ________.

4. $(x + 6)^2 = 1$. ________.

5. $(x - 5)^2 = 13$. ________.

6. $x^2 - 8x + 7 = 0$. ________.

7. $x^2 + 5x - 1 = 0$. ________.

8. $x^2 - 3x + 6 = 0$. ________.

9. $x^2 - 3x - 2 = 0$. ________.

10. $3x^2 + x - 1 = 0$. ________.

11. $4x^2 + 9x + 6 = 0$. ________.

12. $x^2 - 8x = 9$. ________.

1
2
3
4
5
6
7
8
9

10

Applications

CHAPTER PRE-TEST

The following questions indicate the objectives of this chapter. A score of 90% indicates sufficient mastery, and the student may immediately take the Chapter Post-Test.

1. Write an open expression for "8 less than 3 times a number."

2. Write an open expression for "7 added to the product of 9 and a number."

3. Write an open expression for "the sum of 6 times a number and 17."

4. A second number is equal to 3 times a first number. The sum of the first number and twice the second number is 14. What are the numbers?

5. The sum of 3 times the first number and the second number is 19. The first number subtracted from the second number is 7. What are the numbers?

6. The sum of a first number and second number is 15. The sum of 2 times the first number and the second number is 19. What are the numbers?

7. A second number is 4 times a first number. 4 times the second number added to 3 times the first number is 38. What are the numbers?

The first section of this chapter will be devoted to writing *open expressions* that are equivalent to verbal statements.

---

1 "$x + 5$" is an open expression for "5 more than x." Write an open expression for "8 more than x" ________ .

$x + 8$

2 Write an open expression for "15 more than z." ________ .

$z + 15$

3 "$x + 13$" is an open expression for the "sum of x and 13." Write an open expression for the "sum of z and 6." ________ .

$z + 6$

4 Write an open expression for the "sum of w and 24." ________ .

$w + 24$

5 "5 added to a number" can be shown by the open expression "$x + 5$." Write an open expression for "7 added to a number." ________ .

$x + 7$

6 Write an open expression for "15 added to a number." ________ .

$x + 15$

7 "A number decreased by 7" may be shown by the open expression "$x - 7$." Write an open expression for "a number decreased by 5." ________ .

$x - 5$

8 Write an open expression for "a number decreased by 31." ________ .

$x - 31$

9 "6 less than a number" can be shown by the open expression "$x - 6$." Write an open expression for "2 less than a number." ________ .

$x - 2$

10 Write an expression for "35 less than a number." ________ .

$x - 35$

11 "4 subtracted from a number" can be shown by the open expression $x - 4$. Write an open expression for "9 subtracted from a number." ________ . $x - 9$

12 Write an open expression for "14 subtracted from a number." ________ . $x - 14$

13 "5 times a number" can be shown by the open expression $5x$. Write an expression to indicate "7 times a number." ________ . $7x$

14 Write an expression to show "18 times a number." ________ . $18x$

15 "The product of 8 and a number" can be shown by the open expression $8x$. Write an open expression to show the "product of 17 and a number." ________ . $17x$

16 Write an expression to indicate the "product of 23 and a number." ________ . $23x$

17 "5 multiplied by a number" can be shown by $5x$. Write an expression to show "12 multiplied by a number." ________ . $12x$

18 "8 multiplied by a number" can be shown by ________ . $8x$

The expression "13 plus 5 times 7" is somewhat ambiguous because it does not specifically indicate whether the addition or the multiplication should be completed first.

$$(13 + 5) \cdot 7 \text{ is not equal to } 13 + 5 \cdot 7$$

Throughout the remainder of this chapter the ambiguity of expressions such as "13 plus 5 times 7" will be eliminated by the following agreement: When an expression indicates both addition (subtraction) and multiplication, the multiplication operation always precedes the addition operation.

---

19 "13 plus 5 times a number" can be shown by the expression $13 + 5x$. Write an open expression to show "8 plus 6 times a number." ________ .

$8 + 6x$

20 "19 more than twice a number" can be shown by the open expression $2x + 19$. Write an expression to show "7 more than 4 times a number." ________________ .

$7 + 4x$ or $4x + 7$

21 "8 less than 9 times a number" can be shown by the open expression $9x - 8$. Write an open expression to show "13 less than a number multiplied by 7." ________ .

$7x - 13$

22 "6 subtracted from the product of 13 and a number" can be written as $13x - 6$. Write an expression for "11 subtracted from the product of 7 and a number." ________ .

$7x - 11$

23 $15x - 14$ can be written to show "14 subtracted from the product of 15 and a number." Write an expression to show "17 subtracted from the product of 5 and a number." ________ .

$5x - 17$

24 Write an expression to show "12 less than the product of 2 and a number." ________ .

$2x - 12$

25 Write an expression to show "15 subtracted from the product of 3 and a number." ________ .

$3x - 15$

26 Write an expression to show "19 plus 5 times a number." ________ .

$19 + 5x$

27 Write an expression to show "27 more than twice a number." ________ .

$27 + 2x$

Self-Quiz # 1

The following questions test the objectives of the preceding section. 100% mastery is desired.

Write an open expression for each of the following:

1. 72 more than w. ________ .

2. The sum of m and 71. ________ .

3. A number decreased by 14. ________ .

4. 21 less than a number. ________ .

5. The product of 4 and a number. ________ .

6. 23 plus 7 times a number. ________ .

7. 12 more than 9 times a number. ________ .

8. 23 subtracted from 6 times a number. ________ .

9. 13 more than the product of 10 and a number. ________ .

10. 37 subtracted from 9 times a number. ________ .

The following section will be devoted to writing equalities from verbally stated sentences.

---

28 The sentence "5 plus 4 equals 9" may be written as the equality $5 + 4 = 9$. Write an equality for the sentence "the sum of 4 and 7 is equal to 11."

________________ .

$4 + 7 = 11$

29 The sentence "8 less than 23 equals 15" can be written as the equality $23 - 8 = 15$. Write a sentence to show the equality "17 less than 31 equals 14."

________________ .

$31 - 17 = 14$

30 The sentence "3 times 12 is equal to 36" may be written as $3 \cdot 12 = 36$. Write an equality for the sentence "the product of 4 and 7 is equal to 28." ____________ .

$4 \cdot 7 = 28$

31 The sentence "7 less than 11 equals the sum of 1 and 3" can be written as the equality $11 - 7 = 1 + 3$. Write an equality for "15 decreased by 5 is equal to the sum of 8 and 2." ____________ .

$15 - 5 = 8 + 2$

32 Write an equality to show "the product of 3 and 6 is equal to the sum of 10 and 8." ____________ .

$3 \cdot 6 = 10 + 8$

33 The word "is" may often be translated to "equals" or "is equal to." "5 is the sum of 3 and 2" may be written as $5 = 3 + 2$. Write a mathematical statement for the sentence "9 is 4 less than 13." ____________ .

$9 = 13 - 4$

34 Write the sentence for "the sum of 9 and 6 is the product of 5 and 3." ____________ .

$9 + 6 = 5 \cdot 3$

35 The following sentence can be written as an equation: "The sum of a number and 5 is 13."

$$x + 5 \quad = 13$$

the sum of x and 5 is 13

Write an equation for the sentence "the sum of a number and 4 is 15." ____________ .

$x + 4 = 15$

36 Write an equation for the sentence "8 subtracted from x is 11." ____________ .

$x - 8 = 11$

37 "A number times 5 is 40" may be written as

$$5x \quad = 40$$

x times 5 is 40

Write an equation for "7 times a number is 42." ______ .

$7x = 42$

38 Write an equation for "the product of a number and 5 is 45." ________ .

$5x = 45$

39 "3 more than 5 times a number is 23" may be written as

$$\underset{\text{3 more than } 5x}{5x + 3} \quad \underset{\text{is}}{=} \quad \underset{}{23}$$

Write an equation for "7 more than 3 times a number is 34." ______________ .

$3x + 7 = 34$

40 Write an equation for "17 more than 5 times a number is 37." ______________ .

$5x + 17 = 37$

41 "7 less than the product of 4 and a number is 13" may be written as

$$\underset{\text{7 less than } 4x}{4x - 7} \quad \underset{\text{is}}{=} \quad 13$$

Write an equation for "6 less than the product of 8 and a number is 50." ______________ .

$8x - 6 = 50$

42 "The product of a number and 7 can be increased by 6 and have a sum of 34" can be written as

$$\underset{7x \text{ increased by 6}}{7x + 6} \quad \underset{\text{is}}{=} \quad 34$$

Write an equation for the product of 4 and a number that can be increased by 12 to obtain 32.

______________ .

$4x + 12 = 32$

43 "9 times a number decreased by 7 is equal to 11" can be written as $9x - 7 = 11$. Write an equation for "5 times a number decreased by 8 is equal to 37."

______________ .

$5x - 8 = 37$

44 "6 added to the product of 3 and a number is 39" can be written as $3x + 6 = 39$. Write an equation for "7 added to the product of 5 and a number is 42."

______________ .

$5x + 7 = 42$

45 To solve $5x + 7 = 42$, the following steps are used:

$$5x + 7 = 42$$
$$5x + 7 - 7 = 42 - 7$$
$$5x = 35$$
$$\{7\}$$

Solve $5x - 8 = 37$. ________ . $\{9\}$

46 Solve $4x + 12 = 32$. ________ . $\{5\}$

47 Write an equation for "a number increased by 5 is 17." ________________ . $x + 5 = 17$

48 Solve $x + 5 = 17$. ________ . $\{12\}$

49 Write an equation for "a number decreased by 8 is 21." ________________ . $x - 8 = 21$

50 Solve $x - 8 = 21$. ________ . $\{29\}$

51 Write an equation for "7 times a number is 28." ________ . $7x = 28$

52 Solve $7x = 28$. ________ . $\{4\}$

53 Write an equation for "3 more than 6 times a number is 45." ________________ . $6x + 3 = 45$

54 Solve $6x + 3 = 45$. ________ . $\{7\}$

55 Write an equation for "the product of 9 and a number can be decreased by 8 and the answer is 28." ________________ . $9x - 8 = 28$

56 Solve $9x - 8 = 28$. ________ . $\{4\}$

To answer the following word problems correctly the following two steps are necessary:

1. Write an equation for the verbal statement.
2. Solve the equation obtained in step 1.

57 A number increased by 13 is 21. Find the number.
_______________, ________ .

$x + 13 = 21$
8

58 A number decreased by 4 is 19. Find the number.
_______________, ________ .

$x - 4 = 19$
23

59 The sum of a number and 14 is 29. Find the number.
_______________, ________ .

$x + 14 = 29$
15

60 9 subtracted from a number is 13. Find the number.
_______________, ________ .

$x - 9 = 13$
22

61 5 less than a number is 18. Find the number.
_______________, ________ .

$x - 5 = 18$
23

62 13 added to a number gives a sum of 34. Find the number. _______________, ________ .

$x + 13 = 34$
21

Self-Quiz # 2

The following questions test the objectives of the preceding section. 100% mastery is desired.

Write an equation and solve each of the following:

1. The sum of a number and 15 is 29. What is the number? ________ .

2. A number decreased by 29 is 14. What is the number? ________ .

3. 5 less than a number is 17. What is the number? ________ .

4. The product of a number and 9 is 72. What is the number? ________ .

5. 19 subtracted from a number is 5. What is the number? ________ .

6. 15 times a number is 75. What is the number? ________ .

7. 5 more than 3 times a number is 20. What is the number? ________ .

8. 9 subtracted from the product of 4 and a number is 23. What is the number? ________ .

In the preceding sections of this chapter expressions were written which involved only one variable or letter. In this section expressions involving two variables are explained.

---

63 The expression $x + y$ represents the sum of two unknown numbers. Use the letters w and k to write an expression for the sum of two numbers. ________ .

$w + k$

64 Write an expression for y more than x. ________ .

$x + y$

65 The expression $x - y$ represents the subtraction of one number (y) from another number (x). Write an expression using the letters m and n to indicate that n is being subtracted from m. ________ .

$m - n$

66 Write an expression for y less than x. ________ . $x - y$

67 The expression $2x + y$ represents 2 times x increased by y. Write an expression for 5 times k increased by d. ________ . $5k + d$

68 Write an expression for s increased by the product of 8 and y. ________ . $s + 8y$

69 Write an expression for the sum of 3 times x and the number y. ________ . $3x + y$

70 Write an expression for y less than 4 times x. ________ . $4x - y$

71 Write an expression for x decreased by the product of 6 and y. $x - 6y$

72 Write an expression for 5 times y subtracted from x. ________ . $x - 5y$

73 Write an expression for y more than 9 times x. ________ . $9x + y$

74 Write an expression for a first number increased by a second number. ________ . $x + y$

75 Write an expression for 5 times a number increased by a second number. ________ . $5x + y$

76 Write an expression for a first number decreased by 3 times a second number. ________ . $x - 3y$

77 The word "is" may often be used as "equals" or "is equal to." The sentence "the sum of two numbers is 13" may be written as the equation $x + y = 13$.

$$\underbrace{x + y \qquad\qquad = 13}_{\text{the sum of two numbers is 13}}$$

Write an equation for "the sum of two numbers is 25."

_______________ .

$x + y = 25$

78 The sentence "one number subtracted from another number is 12" may be written as $x - y = 12$. Write an equation for "one number decreased by another number is 7." _______________ .

$x - y = 7$

79 Write an equation for "2 times a first number increased by a second number is 18."

_______________ .

$2x + y = 18$

80 Write an equation for "a first number decreased by 4 times a second number is 9." _______________ .

$x - 4y = 9$

81 Write an equation for "the sum of 3 times a first number and 7 times a second number is 31."

_______________ .

$3x + 7y = 31$

82 Write an equation for "3 times a first number decreased by twice a second number is 17."

_______________ .

$3x - 2y = 17$

83 Write an equation for "2 times a first number subtracted from a second number is 19."

_______________ .

$y - 2x = 19$

84 Write an equation for "7 times a first number decreased by a second number is 17."

_______________ .

$7x - y = 17$

85 Write an equation for "the product of a first number and 5 increased by a second number is 85."

________________ .

$5x + y = 85$

86 Write an equation for "the sum of 5 times a first number and 4 times a second number is 68."

________________ .

$5x + 4y = 68$

87 The sentence "one number is 4 more than another" is written as the equation $x = y + 4$. Write an equation for "one number is 6 less than another."

________________ .

$x = y - 6$

88 Write an equation for "the sum of a first number and 9 is equal to a second number." ________________ .

$x + 9 = y$

89 Write an equation for "6 less than a first number is equal to a second number." ________________ .

$x - 6 = y$

90 Write an equation for "the sum of 5 times a first number and 12 is 8 times a second number."

________________ .

$5x + 12 = 8y$

91 Write an equation for "5 less than 3 times a first number is the second number." ________________ .

$3x - 5 = y$

92 The sentence "one number is 3 times another number" may be written $x = 3y$. Write an equation for "one number is the product of 7 and another number."

________________ .

$x = 7y$

93 Write the equation for "8 times a first number is equal to the second number." ________________ .

$8x = y$

94 Write the equation for "5 times a number is equal to a second number." ________________ .

$5x = y$

Self-Quiz # 3

The following questions test the objectives of the preceding section. 100% mastery is desired.

Write equations for the following sentences:

1. The sum of two numbers is 17. _______ .
2. One number decreased by another number is 14. _______ .
3. Three times a first number decreased by a second number is 37. _______ .
4. Two times a first number increased by 7 times a second number is 19. _______ .
5. The sum of three times a first number and 5 times a second number is 73. _______ .
6. One number is 8 more than another number. _______ .
7. One number is 9 less than a second number. _______ .
8. Three times a first number increased by 18 is twice the second number. _______ .
9. One number is 11 more than another number. _______ .
10. Eight times a number is equal to a second number. _______ .

Problems involving two unknown numbers are written as equations and solved. Solutions to such problems are dependent upon an ability to find common solutions for pairs of equations with two variables.

---

95 To find the common solution for

$$3x - y = 7$$
$$2x + y = 3$$

the equations are added to eliminate the y's, as follows:

$$\begin{array}{r} 3x - y = 7 \\ 2x + y = 3 \\ \hline 5x \quad\;\; = 10 \\ x = 2 \end{array}$$

Use 2 as a replacement for x in either of the two original equations to find the value of y. If $x = 2$, then $y =$ ________ .

$^{-}1$

96 Find the common solution (x, y) for the following pair of equations:

$x - 5y = {}^{-}19$
${}^{-}x + 2y = 7$

________ .

(1,4)

97 For the equations

$$4x - 3y = 17$$
$$2x - \;\; y = 9$$

neither x nor y can be eliminated by addition using the equations in their present form. To eliminate the y's, each term of the second equation must be multiplied by $^{-}3$. Multiply each term of the second equation, $2x - y = 9$ by $^{-}3$. What is the resulting equation?

________________ .

${}^{-}6x + 3y = {}^{-}27$

98 To solve

$$4x - 3y = 17$$
$$2x - y = 9$$

each term of the second equation must be multiplied by $^{-}3$. This gives

$$4x - 3y = 17$$
$$^{-}6x + 3y = {}^{-}27$$

Find the common solution (x, y) for the equations. ________ .

(5, 1)

99 Find the common solution (x, y) for

$$3x - 2y = 22$$
$$x - 7y = 20$$

by first multiplying each term of the second equation by $^{-}3$. ________ .

(6, $^{-}2$)

100 Find the common solution (x, y) for

$$3x + 2y = 0$$
$$5x - 4y = 22$$

by first multiplying each term of the first equation by 2. ________ .

(2, $^{-}3$)

101 Find the common solution (x, y) for
$3x - 4y = 13$
$x + y = 9$
________ .

(7, 2)

102 To eliminate the y's from the equations below, both equations must be written in different forms.

$$3x - 2y = 22$$
$$5x + 7y = {}^{-}15$$

The coefficients of the x's are 3 and 5. If the x's were to be eliminated, one coefficient should be made a 15 and the other a ${}^{-}15$. The coefficients of the y's are ${}^{-}2$ and 7. To eliminate the y's, one coefficient should be made a 14 and the other a ________ .

${}^{-}14$

103 To find the common solution (x, y) for

$$3x - 2y = 22$$
$$5x + 7y = {}^{-}15$$

the y's can be eliminated by multiplying each term of the first equation by 7 and each term of the second equation by ________ .

2

104 To solve

$$3x - 2y = 22$$
$$5x + 7y = {}^{-}15$$

both equations may be altered by multiplying each term of the first equation by 7 and each term of the second equation by 2.

$$\begin{array}{rl} 21x - 14y & = 154 \\ 10x + 14y & = {}^{-}30 \\ \hline 31x \quad\quad & = 124 \\ x & = 4 \end{array}$$

Complete the solution (x, y) by replacing x by 4 in one of the original equations. ________ .

$(4, {}^{-}5)$

105 To solve

$$2x - 5y = 0$$
$$3x - 4y = {}^{-}7$$

either the x's or the y's must be eliminated. To eliminate the x's, the first equation can be multiplied by 3 and the second equation by ______ .

$^{-}2$

106

$$2x - 5y = 0$$
$$3x - 4y = {}^{-}7$$

Multiplying the terms of the first equation by 3 gives

$$6x - 15y = 0$$

Multiplying the terms of the second equation by $^{-}2$ gives ______________ .

$^{-}6x + 8y = 14$

107 The common solution of

$$2x - 5y = 0$$
$$3x - 4y = {}^{-}7$$

is found by first writing the equations as

$$6x - 15y = 0$$
$${}^{-}6x + 8y = 14$$

Complete the common solution (x, y) for the equations. ________ .

$({}^{-}5, {}^{-}2)$

108 Find the common solution (x, y) for
$3x - 5y = 3$
$5x + 2y = 36$
________ .

$(6, 3)$

109 Find the common solution (x, y) for
$4x - 3y = 19$
$5x - 2y = 22$
______________ .

$(4, {}^{-}1)$

110 The pair of equations

$$x = 3y$$
$$2x - 7y = {}^{-}2$$

may be solved by the substitution method. The equation $x = 3y$ is used to substitute ________ for x in the second equation.

$3y$

111 To solve

$$x = 3y$$
$$2x - 7y = {}^{-}2$$

by substitution, $3y$ is used to replace x in the second equation, $2x - 7y = {}^{-}2$.

$$2x - 7y = {}^{-}2$$
$$2(3y) - 7y = {}^{-}2$$
$$6y - 7y = {}^{-}2$$
$${}^{-}y = {}^{-}2$$
$$y = 2$$

Complete the solution (x, y) by replacing y by 2 in either of the original equations. ________________ .

(6, 2)

112 Find the common solution for

$$x = 5y$$
$$2x - 3y = 14$$

________________ .

(10, 2)

113 To solve

$$x = y + 4$$
$$2x - 5y = 11$$

by the substitution method, the equation $x = y + 4$ is used to substitute ________ for x in the second equation.

$y + 4$

114 To solve

$$x = y + 4$$
$$2x - 5y = 11$$

by substitution, $y + 4$ is used to replace x in the second equation, $2x - 5y = 11$.

$$2x - 5y = 11$$
$$2(y + 4) - 5y = 11$$
$$2y + 8 - 5y = 11$$
$$^{-}3y + 8 = 11$$
$$^{-}3y = 3$$
$$y = {}^{-}1$$

Complete the common solution (x, y) by replacing y by $^{-}1$ in either of the original equations. ________ .

(3, $^{-}$1)

115 Find the common solution (x, y) for

$$x = y - 6$$
$$2x + 7y = 6$$

________ .

($^{-}$4, 2)

Self-Quiz # 4

The following questions test the objectives of the preceding section. 100% mastery is desired.

Find the common solution (x, y) for the following pairs of equations.

1. $4x + y = 9$
$3x - y = 5$

________________ .

2. $2x - 5y = 16$
$4x + 3y = 6$

________________ .

3. $3x + 5y = 3$
$2x - 3y = {}^{-}17$

________________ .

4. $y = 3x$
$3x - 4y = {}^{-}18$

________________ .

5. $x = y - 2$
$2x - 7y = {}^{-}9$

________________ .

The remaining frames contain verbal problems involving two unknown numbers. Each frame is divided into three parts: (a), (b), and (c). Part (a) asks for an equation to be written for the first sentence of the problem. Part (b) asks for an equation to be written for the second sentence of the problem. Part (c) asks for the common solution of the two equations obtained in parts (a) and (b).

---

116 A first number increased by a second number is 15. The first number decreased by the second number is 7. Find the numbers.

(a) Write an equation for the first sentence of the problem. ________ . $x + y = 15$

(b) Write an equation for the second sentence of the problem. ________ . $x - y = 7$

(c) Find the common solution (x, y) for

$$x + y = 15$$
$$x - y = 7$$

________ . $(11, 4)$

117 Two times a first number decreased by a second number is 13. Three times the first number increased by two times the second number is 44. Find the numbers.

(a) Write an equation for the first sentence of the problem. ________ . $2x - y = 13$

(b) Write an equation for the second sentence of the problem. ________ . $3x + 2y = 44$

(c) Find the common solution (x, y) for

$$2x - y = 13$$
$$3x + 2y = 44$$

________ . $(10, 7)$

118 The sum of 5 times a first number and a second number is 28. The first number decreased by twice the second number is 10. Find the numbers.

(a) Write an equation for the first sentence of the problem. ________ . $5x + y = 28$

(b) Write an equation for the second sentence of the problem. ________ . $x - 2y = 10$

(c) Find the common solution (x, y) for

$$5x + y = 28$$
$$x - 2y = 10$$

________ . $(6, {}^{-}2)$

119 Four times a first number diminished by 3 times a second number is 11. Five times the first number increased by 2 times the second number is 31. Find the numbers.

(a) Write an equation for the first sentence of the problem. ________ . $4x - 3y = 11$

(b) Write an equation for the second sentence of the problem. ________ . $5x + 2y = 31$

(c) Find the common solution (x, y) for

$$4x - 3y = 11$$
$$5x + 2y = 31$$

________ . $(5, 3)$

120 One number is 2 times another number. The sum of the two numbers is 21. Find the numbers.

(a) Write an equation for the first sentence of the problem. ________ . $x = 2y$

(b) Write an equation for the second sentence of the problem. ________ . $x + y = 21$

(c) Find the common solution (x, y) for

$$x = 2y$$
$$x + y = 21$$

________ . $(14, 7)$

121 One number is 4 less than another number. The sum of twice the first number and three times the second number is 17. Find the numbers.

(a) Write an equation for the first sentence of the problem. ________ . $x = y - 4$

(b) Write an equation for the second sentence of the problem. ________ . $2x + 3y = 17$

(c) Find the common solution (x, y) for

$$x = y - 4$$
$$2x + 3y = 17$$

________ . $(1, 5)$

122 The difference between twice the first number and the second number is 7. The sum of the first number and 3 times the second number is $^{-}7$. Find the numbers.

(a) Write an equation for the first sentence of the problem. ________ .

$2x - y = 7$

(b) Write an equation for the second sentence of the problem. ________ .

$x + 3y = {}^{-}7$

(c) Find the common solution (x, y) for

$2x - y = 7$

$x + 3y = {}^{-}7$

________ .

$(2, {}^{-}3)$

123 Three added to the first number is equal to the second number. The sum of twice the first number and 3 times the second number is 4. What are the numbers?

(a) Write an equation for the first sentence of the problem. ________ .

$y = x + 3$

(b) Write an equation for the second sentence of the problem. ________ .

$2x + 3y = 4$

(c) Find the common solution (x, y) for

$y = x + 3$

$2x + 3y = 4$

________ .

$({}^{-}1, 2)$

124 The sum of two numbers is 18. The second number subtracted from the first number is 2. Find the numbers.

(a) Write an equation for the first sentence of the problem. ________ .

$x + y = 18$

(b) Write an equation for the second sentence of the problem. ________ .

$x - y = 2$

(c) Find the common solution (x, y) for

$x + y = 18$

$x - y = 2$

________ .

$(10, 8)$

Self-Quiz # 5

The following questions test the objectives of the preceding section. 100% mastery is desired.

Write the equations and find the common solutions (x, y) for the following:

1. The sum of twice the first number and the second number is 4. Three times the first number diminished by the second number is 1. What are the numbers? ________ .
2. The sum of twice the first number and three times the second number is 13. Three times the first number increased by the second number is 9. What are the numbers? ________ .
3. The second number is 5 times the first number. Nine times the first number diminished by the second number is 8. What are the numbers? ________ .

Chapter Summary

Verbal problems were studied in this chapter. The final goal was the solution of verbal problems involving two variables.

The chapter included writing open expressions and equations involving one and two variables. Verbal problems involving two unknown numbers were solved by writing two equations with two variables and finding their common solution.

CHAPTER POST-TEST

The following questions test the objectives of this chapter. A score of 90% indicates sufficient mastery.

1. Write an open expression for "17 less than 5 times a number."
2. Write an open expression for "5 added to the product of 8 and a number."
3. Write an expression for "9 less than the product of 3 and a number."
4. "8 more than 5 times a number is 38." What is the number?
5. A second number is equal to 5 times a first number. Three times the first number subtracted from 2 times the second number is 14. What are the numbers?
6. The sum of twice a first number and the second number is 7. The second number subtracted from 3 times the first number is 3. What are the numbers?
7. A first number increased by a second number is 12. The first number decreased by the second number is 2. What are the numbers?

Appendix A: Graphing Equations of the Form $y = mx + b$

The method taught in Chapter 1 for graphing linear equations finds the two points where the line crosses the axes and uses those two points to draw the line. Shown below is another method for graphing linear equations. It also depends upon finding two points on the line.

To graph $5x - 2y = 6$ the following steps may be used:

1. Solve the equation for y.

$$\begin{array}{rcl} 5x - 2y &=& 6 \\ {}^{-}5x \qquad && {}^{-}5x \\ \hline {}^{-}2y &=& {}^{-}5x + 6 \end{array}$$

${}^{-}5x$ is added to both sides of the equation.

$$\frac{{}^{-}1}{2} \cdot {}^{-}2y = \frac{{}^{-}1}{2}({}^{-}5x + 6)$$

$\frac{{}^{-}1}{2}$ is multiplied by both sides of the equation.

$$y = \frac{5}{2}x - 3$$

2. *Any* two numbers may now be substituted for x to find two solutions for the equation. In this case, even numbers are easier replacements to make for x because the denominator of $\frac{5}{2}$ is 2.

$y = \frac{5}{2}x - 3$ if $x = 2$, $y = \frac{5}{2} \cdot 2 - 3 = 5 - 3 = 2$

Therefore, (2, 2) is a solution.

$y = \frac{5}{2}x - 3$ if $x = 6$, $y = \frac{5}{2} \cdot 6 - 3 = 15 - 3 = 12$

Therefore, (6, 12) is a solution.

3. The points (2, 2) and (6, 12) are then graphed.
4. The line of $5x - 2y = 6$ is the line that goes through the points (2, 2) and (6, 12).

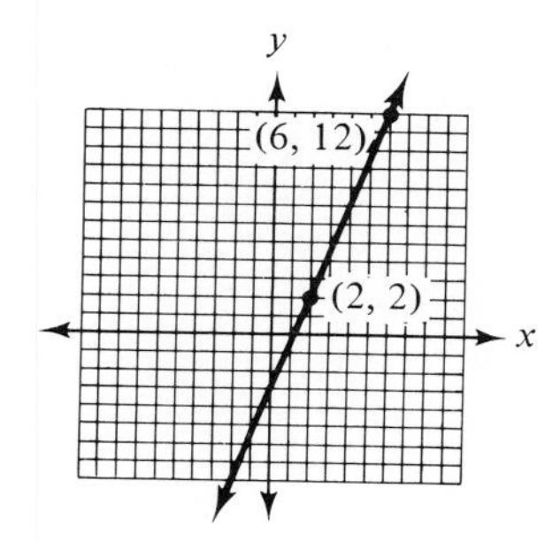

Appendix B: The Quadratic Formula

The general quadratic equation $ax^2 + bx + c = 0$ can be solved using the completing-the-square method to find solutions for x in terms of the letters a, b, and c. This solution is shown below.

$$ax^2 + bx + c = 0$$

$$ax^2 + bx = {}^{-}c$$

${}^{-}c$ is added to both sides of the equation.

$$x^2 + \frac{b}{a}x = \frac{{}^{-}c}{a}$$

$\frac{1}{a}$ is multiplied by both sides of the equation.

$$x^2 + \frac{b}{a}x + \frac{b^2}{4a^2} = \frac{b^2}{4a^2} - \frac{c}{a}$$

$\frac{b^2}{4a^2}$ is added to both sides of the equation to make a trinomial perfect square on the left.

$$\left(x + \frac{b}{2a}\right)^2 = \frac{b^2 - 4ac}{4a^2}$$

The left side of the equation is factored. The terms on the right side are given a common denominator.

Then two linear equations can be written.

$$x + \frac{b}{2a} = \frac{\sqrt{b^2 - 4ac}}{2a} \quad \text{or} \quad x + \frac{b}{2a} = \frac{-\sqrt{b^2 - 4ac}}{2a}$$

To solve the linear equation, $\frac{{}^{-}b}{2a}$ is added to both sides of each equation.

$$x = \frac{{}^{-}b + \sqrt{b^2 - 4ac}}{2a} \quad \text{or} \quad x = \frac{{}^{-}b - \sqrt{b^2 - 4ac}}{2a}$$

$$x = \frac{{}^{-}b \pm \sqrt{b^2 - 4ac}}{2a}$$

The last equality is called the quadratic formula which can be used to solve any quadratic equation in the form $ax^2 + bx + c = 0$. The numbers in the places for a, b, and c are used in their respective positions in the formula. The $\pm$ symbol means that two solutions can be found, one using the $+$ sign and the other using the $-$ sign.

Answers to Pre-Tests, Self-Quizzes, and Post-Tests

CHAPTER 1: GRAPHING EQUATIONS WITH TWO VARIABLES

PRE-TEST

1. Yes
2. Yes
3. No
4. No
5. $(6, \underline{9})$
6. $(\underline{12}, 5)$
7. $\left(7, \dfrac{1}{\underline{2}}\right)$
8. $\left(\dfrac{1}{\underline{5}}, {}^-4\right)$
9. *A* $({}^-3, 5)$
   B $(2, {}^-3)$
   C $({}^-2, 0)$
   D $({}^-5, {}^-2)$
   E $(5, 1)$

10.

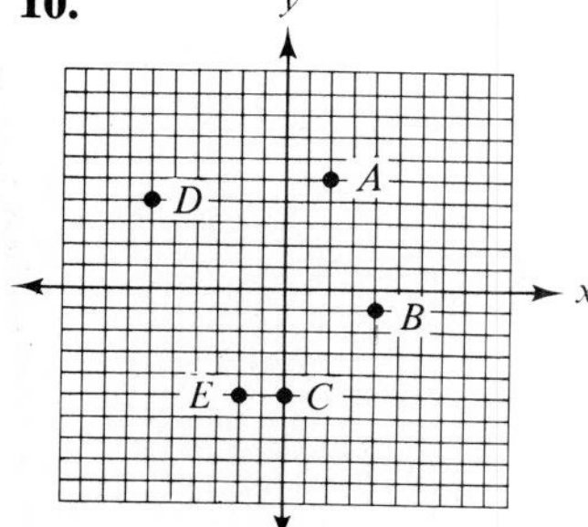

11.

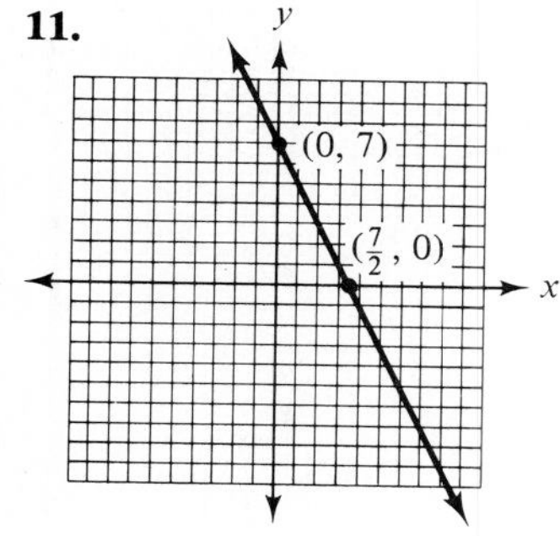

12.

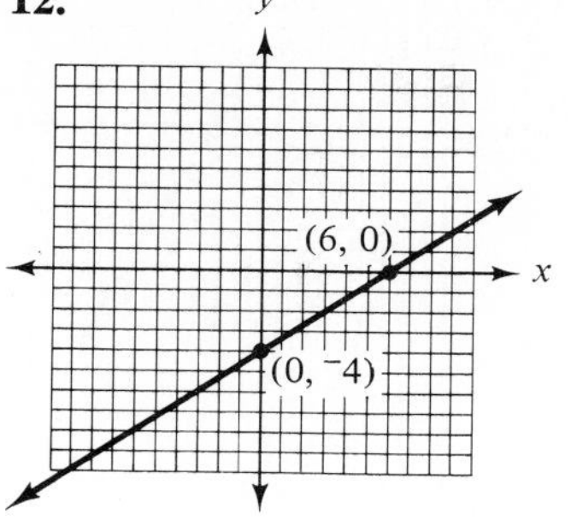

13.

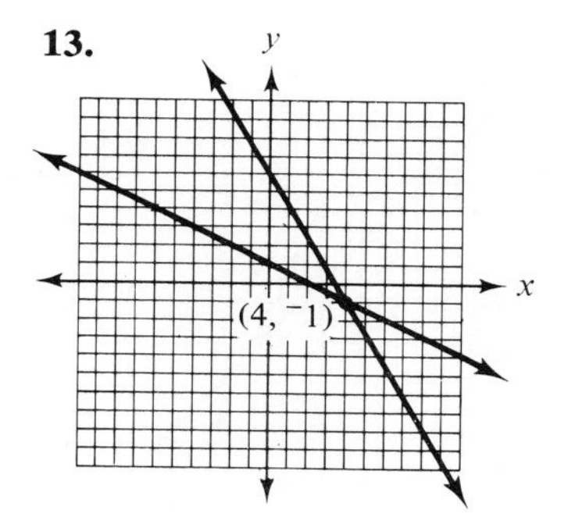

SELF-QUIZ # 1

1. $(3, 5)$
2. $(16, 1)$
3. Yes
4. Yes
5. Yes

SELF-QUIZ # 2

1. $(26, {}^-5)$
2. $(2, {}^-4)$
3. (a) $(3, \underline{9})$
4. (b) $(\underline{4}, 7)$
   (c) $({}^-1, \underline{17})$
   (d) $\left(8\dfrac{1}{\underline{2}}, {}^-2\right)$

SELF-QUIZ # 3

1. *A* $(6, 2)$
   B $(3, {}^-5)$
   C $({}^-7, 4)$
   D $({}^-2, {}^-3)$
   E $(0, 8)$
   F $({}^-5, 0)$
2.

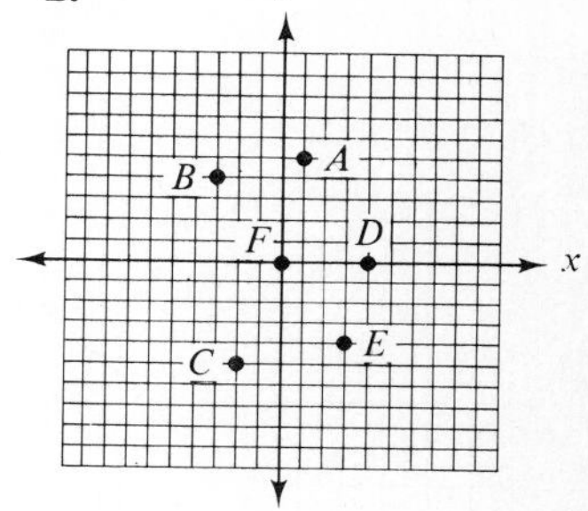

SELF-QUIZ # 4

1. *A* $(0, -1)$
   B $(3, 5)$
   C $({}^-2, {}^-5)$
2. $({}^-2, 3)$
3. No
4. Yes
5. No

SELF-QUIZ # 5

1.

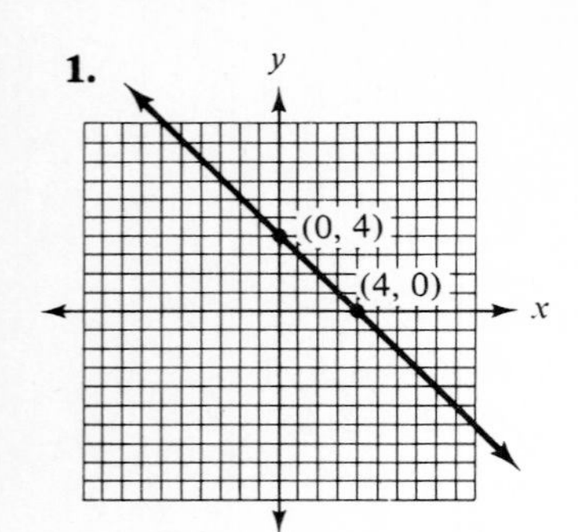

2.

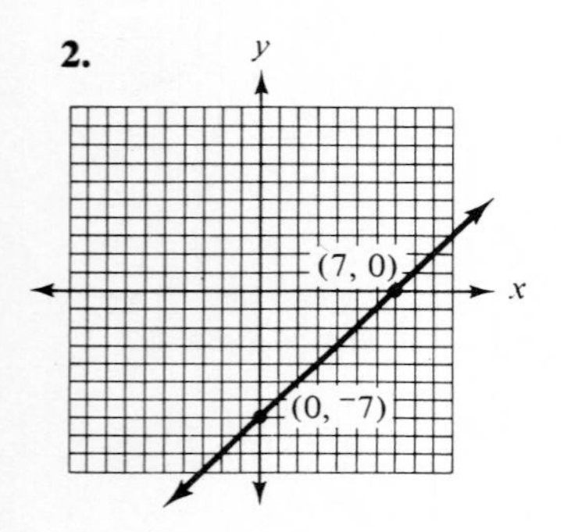

3.

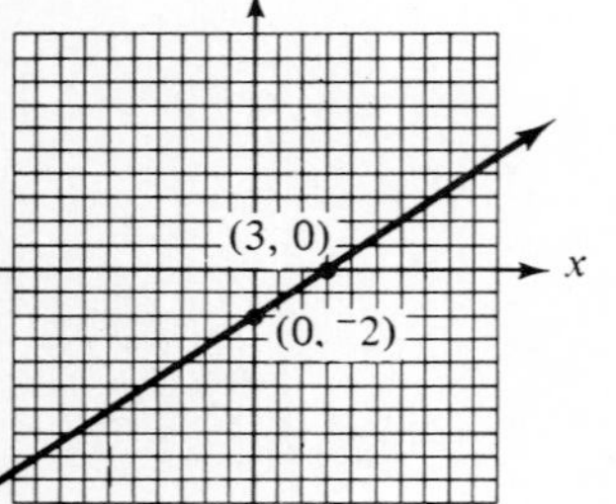

4.

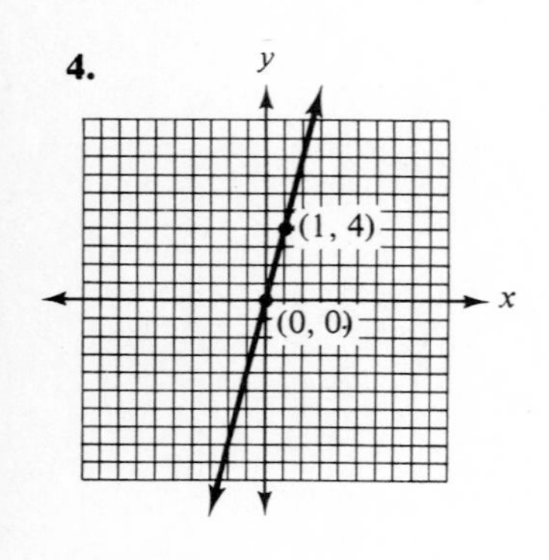

5.

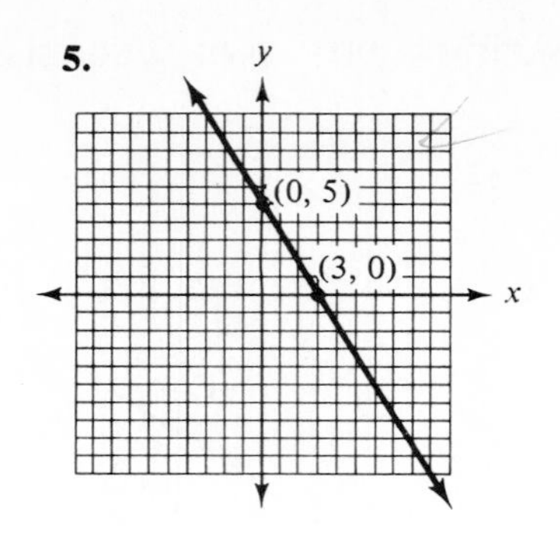

SELF-QUIZ # 6

1.

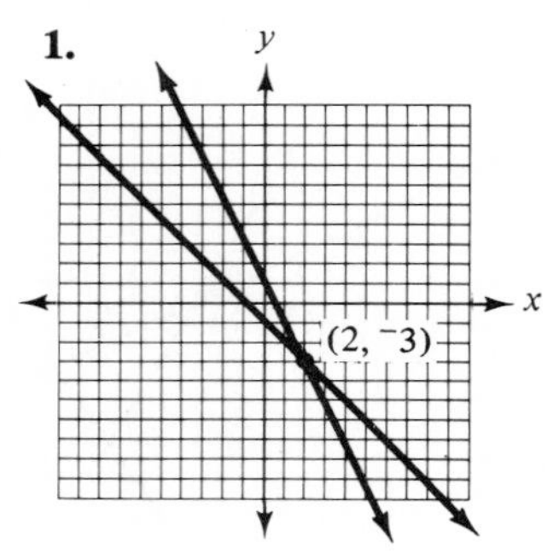

2.

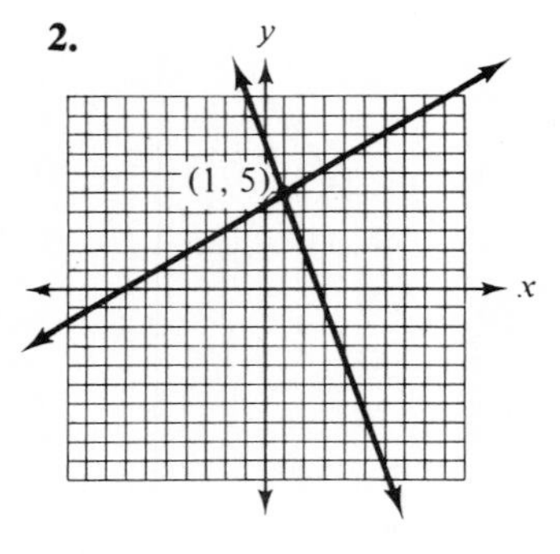

3.

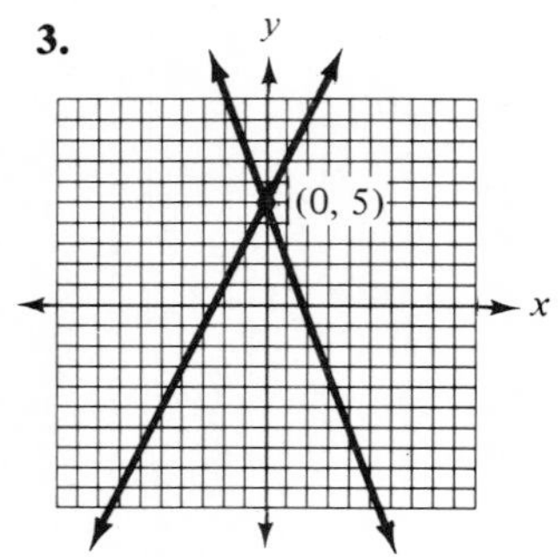

4.

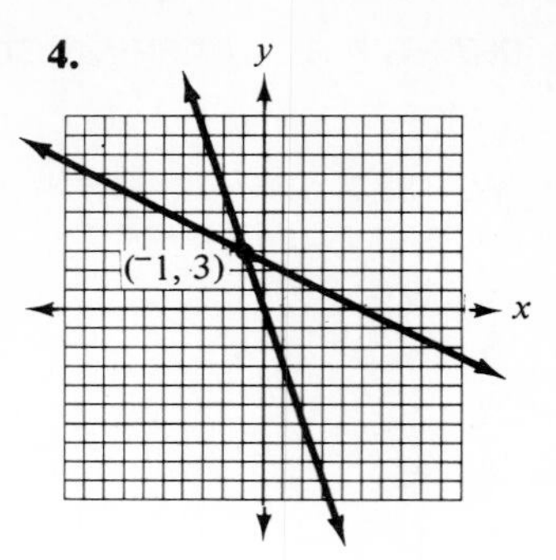

POST-TEST

1. Yes

2. Yes

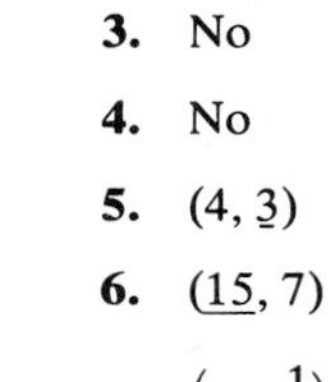

3. No

4. No

5. (4, 3)

6. (15, 7)

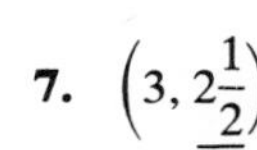

7. $\left(3, 2\frac{1}{2}\right)$

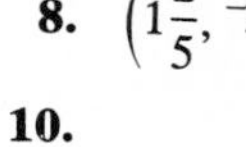

8. $\left(1\frac{2}{5}, {}^{-}2\right)$

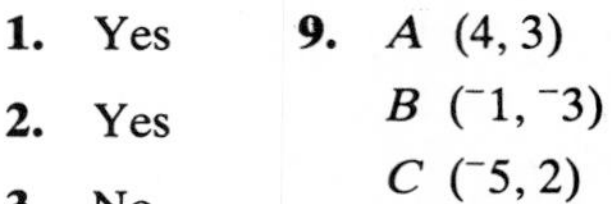

9. *A* (4, 3)
B (⁻1, ⁻3)
C (⁻5, 2)
D (0, 4)
E (4, ⁻5)

10.

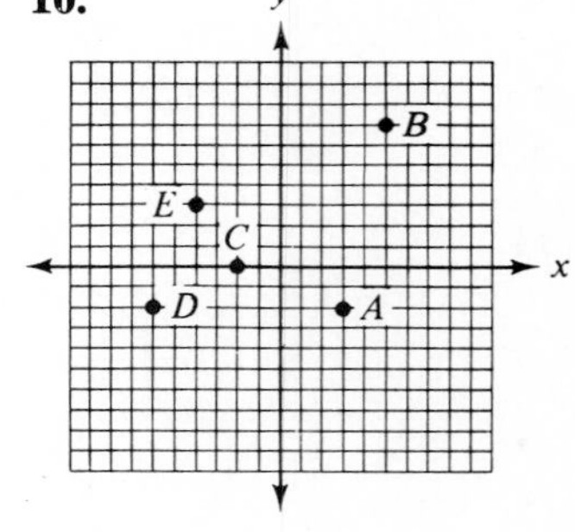

11.

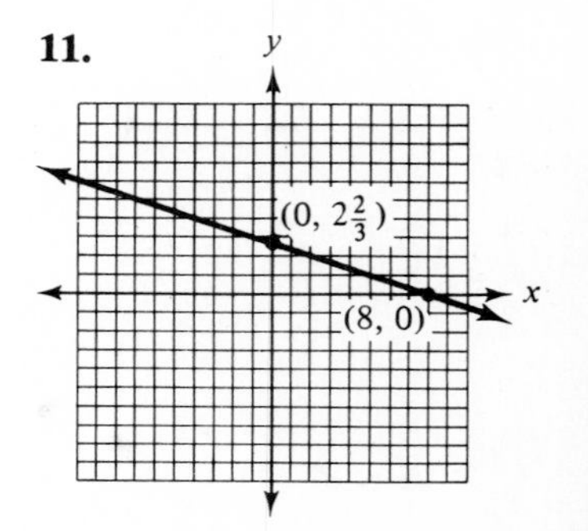

12.

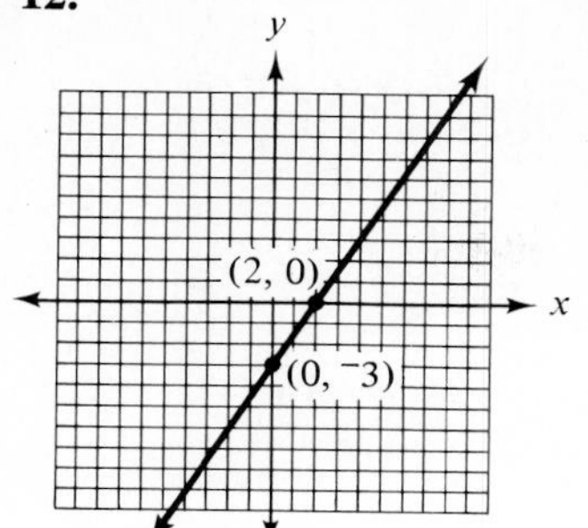

13.

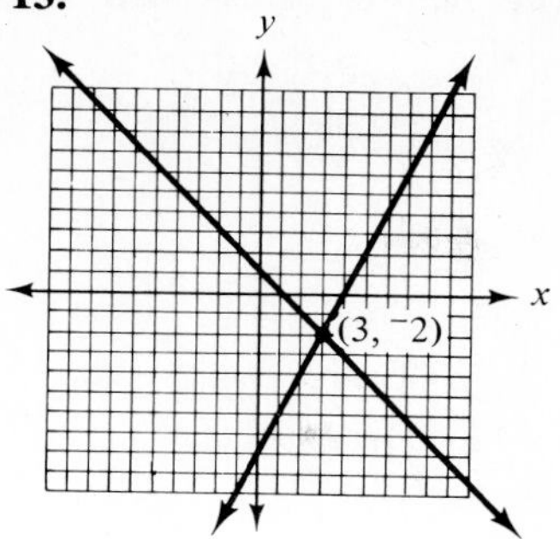

CHAPTER 2: SOLVING PAIRS OF EQUATIONS

PRE-TEST

1. $(25, {}^{-}6)$
2. $(5, 2)$
3. $({}^{-}1, {}^{-}2)$
4. No solution
5. $(8, 2)$
6. $(2, 5)$
7. $(5, {}^{-}1)$
8. $(2, 7)$
9. No solution
10. $(3, {}^{-}2)$

SELF-QUIZ # 1

1. $(2, 3)$
2. $(4, {}^{-}2)$
3. $({}^{-}1, 3)$
4. $({}^{-}3, {}^{-}1)$
5. $(5, {}^{-}2)$

SELF-QUIZ # 2

1. $(2, 1)$
2. $(3, {}^{-}2)$
3. $({}^{-}4, 2)$
4. $({}^{-}5, {}^{-}3)$
5. $(4, 0)$

SELF-QUIZ # 3

1. $(7, {}^{-}5)$
2. $(2, 3)$
3. $({}^{-}2, 5)$
4. $(6, 7)$
5. $(1, {}^{-}1)$

SELF-QUIZ # 4

1. $(6, 2)$
2. $(1, 4)$
3. $(1, {}^{-}1)$
4. $({}^{-}4, {}^{-}2)$

SELF-QUIZ # 5

1. $(2, 7)$
2. $(12, 4)$
3. No solution
4. $(4, 1)$
5. $({}^{-}2, 3)$

POST-TEST

1. $(19, {}^{-}2)$
2. $(4, 3)$
3. $(5, 2)$
4. No solution
5. $({}^{-}9, {}^{-}3)$
6. $(5, 3)$
7. $(7, 5)$
8. $({}^{-}3, 5)$
9. $\left(\frac{3}{2}, \frac{5}{2}\right)$
10. $\left(\frac{13}{4}, \frac{{}^{-}3}{2}\right)$

CHAPTER 3: AN INTRODUCTION TO POLYNOMIALS

PRE-TEST

1. $^{-}18x^5$
2. $4x^6y^7$
3. $^{-}18x + 12$
4. $7x^3 - 21x$
5. $^{-}2x^2 + 10x + 6$
6. $x^2 - 12x + 27$
7. $3x^2 + 19x - 14$
8. $6x^2 + x - 40$
9. $9x^2 - 6x + 1$
10. $x^3 - 8x^2 + 23x - 24$
11. $\dfrac{15 - xy}{5x}$
12. $\dfrac{^{-}x^2 + 6x + 9}{7x - 35}$
13. $\dfrac{4x^2 - 4x + 11}{x^2 - x - 6}$
14. $x - 7$
15. $2x + 1$ R $^{-}4$

SELF-QUIZ # 1

1. $^{-}8x^6$
2. $^{-}5x^6y^8$
3. $8x - 32$
4. $^{-}3x + 5$
5. $2x^3 + 14x$
6. $15x - 12$
7. $^{-}5x^2 - 15x + 10$
8. $x^4y - 2x^2y^3 + x^3y^2$

SELF-QUIZ # 2

1. True
2. True
3. True
4. False
5. True
6. Yes
7. No
8. Yes
9. No

SELF-QUIZ # 3

1. $6x - 9$
2. $15x - 10y + 35$
3. $x^2 + 8x + 15$
4. $x^2 - 11x + 28$
5. $x^2 + 4x - 12$
6. $x^2 - 5x - 24$
7. $2x^2 + 9x - 18$
8. $10x^2 + 29x - 21$
9. $x^2 - 25$
10. $9x^2 - 4$
11. $x^2 + 6x + 9$
12. $4x^2 - 12x + 9$
13. $x^3 + x^2 - 4x + 6$
14. $x^3 + 2x^2 - 11x + 6$

SELF-QUIZ # 4

1. $\dfrac{23x}{6}$
2. $\dfrac{6x - 20y}{15}$
3. $\dfrac{7x + 7}{10}$
4. $\dfrac{2x^2 - 4x + 26}{x^2 - 2x - 8}$
5. $\dfrac{^{-}3x - 47}{30}$
6. $\dfrac{ad + bc}{bd}$
7. $\dfrac{9x^2 + 15x + 7}{3x^2 + 11x - 4}$
8. $\dfrac{21x - 8y}{28}$

SELF-QUIZ # 5

1. $x - 4$
2. $x - 6$
3. $x + 7$ R 2
4. $5x - 1$
5. $3x - 2$ R 5
6. $x + 6$

POST-TEST

1. $^{-}5x^6$
2. $6x^3y^4$
3. $^{-}10x + 35$
4. $3x^3 + 12x$
5. $6x^2 + 42x - 12$
6. $x^2 - x - 42$
7. $2x^2 + x - 21$
8. $6x^2 + 13x - 5$
9. $4x^2 + 12x + 9$
10. $x^3 + x^2 - 7x + 20$
11. $\frac{42 - xy}{7x}$
12. $\frac{^{-}x^2 + 3x + 43}{5x - 35}$
13. $\frac{3x^2 + 9x - 42}{x^2 + 6x - 27}$
14. $x - 6$
15. $3x - 5$ R 14

CHAPTER 4: FACTORING POLYNOMIALS

PRE-TEST

I
1. $6(x - 4)$
2. $x(8x - 9)$
3. $^{-}4(2x + 3)$
4. $5(x + 1)$

II
1. $(x + 5)(x + 3)$
2. $(x - 7)(x + 3)$
3. $(x - 10)(x - 7)$
4. $(x + 7)(x - 1)$

III
1. $(2x + 5)(x + 1)$
2. $(3x + 2)(x - 2)$
3. $(3x - 1)(2x - 1)$
4. $(2x - 9)(x + 4)$

IV
1. $(x + 10)(x - 3)$
2. $x(4x^2 - 9x - 1)$
3. $(x + 5)(x - 5)$
4. $(3x - 2)(x - 3)$
5. $(4x + 1)(4x - 1)$
6. $5(x + 3)(x + 1)$
7. $3(2x + 3)(x - 2)$
8. $(x + 8)(x + 3)$

SELF-QUIZ # 1

1. $3(x + 4)$
2. $5(x - 7)$
3. $2x(x - 3)$
4. $x(x + 6)$
5. $^{-}5x(2x + 3)$
6. $4(x^2 + 2x - 4)$
7. $7x(x - 3)$
8. $x(5x^2 - 6x + 7)$
9. $4x(x + 4)$
10. $2x(3x^2 - 4x - 5)$
11. $(x + 2)(x - 3)$
12. $(x - 5)(3x - 2)$
13. $(x - 7)(x + 1)$
14. $(x + 3)(5x - 1)$
15. $(2x - 3)(7x + 4)$

SELF-QUIZ # 2

1. $(x + 5)(x + 3)$
2. $(x + 4)(x + 2)$
3. $(x - 7)(x + 3)$
4. $(x + 7)(x - 5)$
5. $(x - 5)(x + 2)$
6. $(x + 6)(x + 3)$
7. $(x + 5)(x - 2)$
8. Prime
9. $(x - 3)^2$
10. $(x + 11)(x - 2)$

SELF-QUIZ # 3

1. $x(x - 3)(x - 2)$
2. $3(x + 4)(x - 2)$
3. $5(x + 2)(x - 2)$
4. $2(x + 5)(x - 3)$
5. $x(x^2 + 3x - 21)$
6. Prime

SELF-QUIZ # 4

1. $(2x - 1)(x + 3)$
2. $(3x + 2)(x + 3)$
3. $(4x + 1)(x - 3)$
4. $(5x + 3)(x + 2)$
5. $(3x - 1)(2x - 3)$
6. $(4x + 3)(2x - 3)$
7. $(3x + 5)(3x - 5)$

SELF-QUIZ # 5

1. $3(x - 8)$
2. $(x + 5)(x - 2)$
3. $3x(x - 3)$
4. $(x + 5)(x - 5)$
5. $(3x - 1)(2x + 3)$
6. $(x - 8)(x - 5)$
7. $3(2x + 5)(x - 3)$
8. $(3x - 1)(x + 4)$
9. Prime
10. $3x(x + 4)(x - 4)$
11. $(2x - 5)(x + 3)$
12. $(3x + 4)(2x - 3)$

POST-TEST

I **1.** $7(x - 5)$
2. $x(6x^2 - 5)$
3. $^{-}3(3x - 4)$
4. $4(x + 1)$

II **1.** $(x + 3)(x + 1)$
2. $(x + 9)(x - 9)$
3. $(x - 7)(x + 4)$
4. $(x - 8)(x - 3)$

III **1.** $(3x - 4)(2x - 1)$
2. $(5x + 2)(x - 2)$
3. $(3x + 1)(x - 4)$
4. $(2x + 1)^2$

IV **1.** $(x - 9)(x + 4)$
2. $x(2x^2 - 9x - 1)$
3. $(5x + 1)(5x - 1)$
4. $(3x - 1)(x - 7)$
5. $(x + 12)(x + 2)$
6. $3(x - 4)(x - 1)$
7. $(x + 1)(x - 1)$
8. $2(3x^2 - 8x - 15)$

CHAPTER 5: POLYNOMIAL FRACTIONS

PRE-TEST

I **1.** $\frac{2x^4}{3}$
2. $\frac{1}{x - 9}$
3. $x + 7$

II **1.** $\frac{2}{7x^3}$
2. $\frac{x + 9}{3}$
3. $\frac{^{-}7}{x + 7}$

III **1.** $\frac{^{-}7x + 49}{15}$
2. $\frac{23}{24x}$
3. $\frac{2x^2 + 9x + 43}{(x - 5)(x - 1)(x + 5)}$

SELF-QUIZ # 1

1. $5x^2$
2. $\frac{2}{x + 5}$
3. $\frac{x - 4}{x + 3}$
4. $\frac{x + 6}{x - 4}$
5. $\frac{x + 2}{x - 3}$
6. $\frac{x - 8}{x}$

SELF-QUIZ # 2

1. $\frac{x - 1}{x + 7}$
2. $\frac{x}{2}$
3. $\frac{x + 5}{x + 7}$
4. $\frac{1}{x + 2}$
5. $^{-}(x + 3)$

SELF-QUIZ # 3

1. $\frac{15}{x}$
2. $\frac{^-3}{2y}$
3. $\frac{7}{x - 3}$
4. $\frac{2x + 2}{x + 14}$
5. $\frac{x - 4}{5x}$
6. $\frac{6x - 3}{x^2 + 3x - 7}$
7. $\frac{3x - 10}{x^2 - 19}$
8. $\frac{^-33x + 4}{x + 2}$

SELF-QUIZ # 4

1. 72
2. 60
3. $40x$
4. $12x^2$
5. $5(x + 3)(x - 3)$
6. $(x - 8)(x + 3)(x + 8)$
7. $5(x + 2)(x - 2)$
8. $(x + 6)(x + 4)(x - 3)$

SELF-QUIZ # 5

1. $\frac{4x + 3}{6x}$
2. $\frac{^-5x + 11}{12}$
3. $\frac{2x^2 + 4x - 2}{(x + 4)(x - 3)(x + 2)}$
4. $\frac{^-x^2 - 24x - 7}{(x + 5)(x - 4)(x - 3)}$
5. $\frac{2x^2 + 2x - 4}{(x + 1)(x - 1)(x + 2)}$

POST-TEST

I 1. $\frac{9x^3}{7}$

2. $\frac{1}{x - 3}$

3. $x - 5$

II 1. $\frac{3x^2}{8}$

2. $\frac{4}{x + 2}$

III 1. $\frac{9x + 74}{30}$

2. $\frac{41}{36x}$

3. $\frac{2x^2 + 10x + 11}{(x + 7)(x - 4)(x + 4)}$

CHAPTER 6: SOLVING LINEAR EQUATIONS

PRE-TEST

1. $\{^-7\}$
2. $\left\{\frac{^-16}{7}\right\}$
3. $\left\{\frac{15}{4}\right\}$
4. $\left\{\frac{1}{4}\right\}$
5. $\{^-15\}$
6. $\left\{\frac{23}{8}\right\}$
7. $\left\{\frac{19}{2}\right\}$
8. $\left\{\frac{4}{9}\right\}$
9. $\left\{\frac{23}{12}\right\}$
10. $\left\{\frac{^-27}{68}\right\}$
11. $\left\{\frac{^-52}{105}\right\}$
12. $\left\{\frac{155}{36}\right\}$
13. $\left\{\frac{14}{15}\right\}$
14. $\left\{\frac{24}{5}\right\}$

SELF-QUIZ # 1

1. (a) {12}
   (b) { }
   (c) {11}
   (d) { }
2. (a) $\{^-4\}$
   (b) $\{^-3\}$
   (c) { }
   (d) {4}
   (e) $\{^-6\}$
   (f) {2}

SELF-QUIZ # 2

1. $\left\{\frac{^-1}{10}\right\}$
2. $\left\{\frac{17}{21}\right\}$
3. $\left\{\frac{15}{16}\right\}$
4. $\{^-6\}$
5. $\left\{\frac{^-14}{15}\right\}$
6. $\left\{\frac{^-3}{4}\right\}$
7. {3}
8. $\left\{\frac{15}{4}\right\}$
9. $\left\{\frac{22}{5}\right\}$
10. {2}

SELF-QUIZ # 3

1. $\left\{\frac{^-9}{8}\right\}$
2. $\left\{\frac{2}{5}\right\}$
3. {15}
4. {10}
5. $\left\{\frac{4}{3}\right\}$

POST-TEST

1. $\{^-13\}$
2. $\left\{\frac{^-13}{9}\right\}$
3. $\left\{\frac{^-11}{3}\right\}$
4. $\left\{\frac{22}{7}\right\}$
5. $\left\{\frac{^-4}{3}\right\}$
6. $\left\{\frac{3}{10}\right\}$
7. $\left\{\frac{^-6}{7}\right\}$
8. $\left\{\frac{3}{5}\right\}$
9. $\left\{\frac{^-7}{10}\right\}$
10. $\left\{\frac{10}{39}\right\}$
11. $\left\{\frac{17}{8}\right\}$
12. $\left\{\frac{^-4}{21}\right\}$
13. $\left\{\frac{112}{25}\right\}$
14. $\left\{\frac{^-13}{5}\right\}$

CHAPTER 7: FRACTIONAL AND QUADRATIC EQUATIONS

PRE-TEST

1. {6}
2. {29}
3. $\{^-12\}$
4. $\{^-3\}$
5. $\left\{\frac{23}{6}\right\}$
6. $\{^-8, 2\}$
7. {0, 13}
8. $\{8, ^-8\}$
9. $\{^-4, ^-7\}$
10. {4, 8}
11. $\{2, ^-2\}$
12. $\{7, ^-2\}$
13. {11, 2}
14. {6, 7}
15. $\left\{\frac{1}{3}, ^-4\right\}$

SELF-QUIZ # 1

1. $\{^-3\}$
2. {4}
3. {4}
4. { }
5. {0}

SELF-QUIZ # 2

1. {2}
2. {7}
3. {3}
4. { }
5. {2}
6. {20}

SELF-QUIZ # 3

1. $\{5, {}^-2\}$
2. {3, 4}
3. $\{0, {}^-7\}$
4. $\{{}^-6, {}^-1\}$
5. {4}
6. {0, 8}
7. $\left\{\frac{3}{2}, {}^-9\right\}$
8. $\{{}^-5, 6\}$
9. $\left\{\frac{{}^-5}{2}\right\}$
10. $\left\{0, \frac{{}^-2}{3}\right\}$

SELF-QUIZ # 4

1. $\{{}^-3, {}^-10\}$
2. $\{{}^-6, 6\}$
3. $\{8, {}^-3\}$
4. $\{0, {}^-7\}$
5. {3, 4}
6. $\{0, {}^-6\}$
7. $\{5, {}^-5\}$
8. $\left\{\frac{1}{2}, 5\right\}$

SELF-QUIZ # 5

1. $\{{}^-5, 3\}$
2. {6, 1}
3. $\{{}^-9, 9\}$
4. $\{{}^-3, 3\}$
5. {2}
6. $\left\{\frac{1}{3}, 2\right\}$
7. $\left\{0, \frac{3}{5}\right\}$
8. $\{4, {}^-3\}$
9. $\{{}^-5, 1\}$
10. $\{{}^-5, 4\}$

POST-TEST

1. $\left\{\frac{7}{5}\right\}$
2. $\left\{\frac{{}^-15}{8}\right\}$
3. $\left\{\frac{{}^-1}{4}\right\}$
4. $\{{}^-20\}$
5. $\left\{\frac{30}{7}\right\}$
6. $\{{}^-7, 2\}$
7. {0, 17}
8. $\{{}^-10, 10\}$
9. $\{{}^-7, {}^-8\}$
10. {3, 7}
11. $\{2, {}^-2\}$
12. $\{{}^-5, 2\}$
13. {2, 7}
14. {5, 7}
15. $\left\{\frac{4}{3}, 2\right\}$

CHAPTER 8: RATIONAL AND IRRATIONAL NUMBERS

PRE-TEST

I a, d, e
II a, c, d
III a, c, e
IV **1.** $15\sqrt{2}$
2. ${}^-21\sqrt{3}$
3. ${}^-8$
4. $\sqrt{5}$
5. $\frac{7\sqrt{11}}{11}$
6. $\sqrt{5}$
7. $\frac{\sqrt{3}}{5}$
8. $\frac{\sqrt{10}}{5}$
9. $16 - 2\sqrt{13}$
10. ${}^-11\sqrt{5}$
11. ${}^-2\sqrt{7}$
12. $18\sqrt{3}$
13. $2\sqrt{2}$
14. $11\sqrt{7}$
15. ${}^-4\sqrt{2}$

SELF-QUIZ # 1

1. a, d
2. b, d
3. a, e
4. {49, 1, 196, 25}
5. {9, 324, 144, 0}

SELF-QUIZ # 2

1. b, e
2. a, b, d
3. a, d, e
4. 6
5. $^-10$
6. $^-8$
7. 2
8. 3
9. 0
10. $^-1$

SELF-QUIZ # 3

1. a, d, e
2. a, b, d
3. b, d
4. a, b

SELF-QUIZ # 4

1. 4 and 5
2. $^-7$ and $^-8$
3. $^-9$ and $^-10$
4. 7 and 8
5. 9 and 10

SELF-QUIZ # 5

1. 4
2. $^-12$
3. $12\sqrt{2}$
4. $^-5\sqrt{2}$
5. $10\sqrt{2}$
6. $^-8\sqrt{2}$
7. $6\sqrt{6}$
8. $20\sqrt{5}$
9. $^-9$
10. $15\sqrt{3}$

SELF-QUIZ # 6

1. $3\sqrt{5}$
2. $\frac{\sqrt{5}}{\sqrt{2}}$
3. $\frac{1}{\sqrt{3}}$
4. $4\sqrt{3}$
5. $^-9$
6. $\frac{1}{\sqrt{7}}$
7. $\frac{\sqrt{2}}{\sqrt{3}}$
8. $\sqrt{5}$
9. c, f, j

SELF-QUIZ # 7

1. $\frac{5\sqrt{14}}{7}$
2. $\sqrt{6}$
3. $^-9$
4. $\frac{2\sqrt{2}}{3}$
5. $\frac{2\sqrt{7}}{7}$
6. $\frac{3\sqrt{5}}{10}$
7. $\frac{\sqrt{6}}{3}$
8. $\sqrt{7}$
9. $\frac{\sqrt{15}}{4}$
10. $2\sqrt{2}$

SELF-QUIZ # 8

1. $11\sqrt{2}$
2. $2\sqrt{17}$
3. $2\sqrt{5} + 3\sqrt{7}$
4. $11\sqrt{13} - 5$
5. $13\sqrt{3}$
6. $4\sqrt{2}$
7. $9\sqrt{5}$
8. $8\sqrt{7}$

POST-TEST

I a, b, e

II a, c, d

III a, c, e

IV

1. $24\sqrt{2}$

2. $^{-}15\sqrt{6}$

3. $^{-}7$

4. $\sqrt{7}$

5. $\frac{3\sqrt{13}}{13}$

6. $\sqrt{3}$

7. $\frac{4\sqrt{15}}{25}$

8. $\frac{\sqrt{6}}{3}$

9. $19 - 5\sqrt{11}$

10. $^{-}4\sqrt{5}$

11. $^{-}\sqrt{7}$

12. $11\sqrt{3}$

13. $32\sqrt{2}$

14. $19\sqrt{7}$

15. $^{-}7\sqrt{2}$

CHAPTER 9: SOLVING QUADRATIC EQUATIONS WITH IRRATIONAL SOLUTIONS

PRE-TEST

1. $\{3, ^{-}3\}$

2. $\{\sqrt{31}, ^{-}\sqrt{31}\}$

3. $\{\ \}$

4. $\{9, ^{-}5\}$

5. $\{^{-}8 + \sqrt{19}, ^{-}8 - \sqrt{19}\}$

6. $\left\{\frac{^{-}7 + \sqrt{33}}{2}, \frac{^{-}7 - \sqrt{33}}{2}\right\}$

7. $\left\{\frac{^{-}1 + \sqrt{21}}{2}, \frac{^{-}1 - \sqrt{21}}{2}\right\}$

8. $\{1, 9\}$

9. $\left\{\frac{5 + \sqrt{37}}{6}, \frac{5 - \sqrt{37}}{6}\right\}$

10. $\left\{\frac{4 + \sqrt{2}}{2}, \frac{4 - \sqrt{2}}{2}\right\}$

11. $\left\{\frac{1 + \sqrt{21}}{5}, \frac{1 - \sqrt{21}}{5}\right\}$

12. $\{^{-}10, 3\}$

SELF-QUIZ # 1

1. $\{4, ^{-}4\}$

2. $\{10, ^{-}10\}$

3. $\{\ \}$

4. $\{\sqrt{11}, ^{-}\sqrt{11}\}$

5. $\{2\sqrt{2}, ^{-}2\sqrt{2}\}$

6. $7, ^{-}7$

7. $\sqrt{17}, ^{-}\sqrt{17}$

SELF-QUIZ # 2

1. $\{3, ^{-}7\}$

2. $\{10, ^{-}4\}$

3. $\{^{-}5 + \sqrt{17}, ^{-}5 - \sqrt{17}\}$

4. $\{7 + \sqrt{29}, 7 - \sqrt{29}\}$

5. $\left\{\frac{5}{2} + \sqrt{17}, \frac{5}{2} - \sqrt{17}\right\}$

6. $\{^{-}8, ^{-}10\}$

7. $\{^{-}16 + \sqrt{13}, ^{-}16 - \sqrt{13}\}$

8. $\{^{-}5, 13\}$

9. $\{^{-}5, ^{-}25\}$

10. $\left\{\frac{^{-}11}{2} + \sqrt{19}, \frac{^{-}11}{2} - \sqrt{19}\right\}$

SELF-QUIZ # 3

1. 9
2. 25
3. $\frac{49}{4}$
4. 16
5. $\frac{81}{4}$
6. $\frac{225}{4}$

SELF-QUIZ # 4

1. $4; (x + 2)^2$
2. $81; (x - 9)^2$
3. $\frac{1}{4}; \left(x + \frac{1}{2}\right)^2$
4. $49; (x - 7)^2$
5. $\frac{25}{4}; \left(x - \frac{5}{2}\right)^2$
6. $\frac{9}{4}; \left(x + \frac{3}{2}\right)^2$

SELF-QUIZ # 5

1. $\{9, {}^{-}1\}$
2. $\{{}^{-}3 + \sqrt{13}, {}^{-}3 - \sqrt{13}\}$
3. $\{\ \}$
4. $\left\{\frac{5}{2} + \sqrt{3}, \frac{5}{2} - \sqrt{3}\right\}$
5. $\{{}^{-}2, {}^{-}5\}$

SELF-QUIZ # 6

1. $\{{}^{-}2 + \sqrt{3}, {}^{-}2 - \sqrt{3}\}$
2. $\{3 + \sqrt{10}, 3 - \sqrt{10}\}$
3. $\{2 + \sqrt{5}, 2 - \sqrt{5}\}$
4. $\{2 + \sqrt{6}, 2 - \sqrt{6}\}$
5. $\{{}^{-}2 + \sqrt{7}, {}^{-}2 - \sqrt{7}\}$

SELF-QUIZ # 7

1. $3, 4, \sqrt{65}$
2. ${}^{-}5, 14, \sqrt{53}$
3. ${}^{-}3, 2, \sqrt{45} = 3\sqrt{5}$
4. $1, 4, \sqrt{41}$
5. ${}^{-}1, 2, \sqrt{25} = 5$
6. $10, 4, \sqrt{60} = 2\sqrt{15}$
7. ${}^{-}7, 2, \sqrt{37}$

SELF-QUIZ # 8

1. $\left\{\frac{5 + \sqrt{13}}{2}, \frac{5 - \sqrt{13}}{2}\right\}$
2. $\left\{\frac{1 + \sqrt{37}}{6}, \frac{1 - \sqrt{37}}{6}\right\}$
3. $\{4 + \sqrt{10}, 4 - \sqrt{10}\}$
4. $\{{}^{-}5, 3\}$
5. $\{\ \}$

POST-TEST

1. $\{4, {}^{-}4\}$
2. $\{\sqrt{43}, {}^{-}\sqrt{43}\}$
3. $\{\ \}$
4. $\{{}^{-}5, {}^{-}7\}$
5. $\{5 + \sqrt{13}, 5 - \sqrt{13}\}$
6. $\{7, 1\}$
7. $\left\{\frac{{}^{-}5 + \sqrt{29}}{2}, \frac{{}^{-}5 - \sqrt{29}}{2}\right\}$
8. $\{\ \}$
9. $\left\{\frac{3 + \sqrt{17}}{2}, \frac{3 - \sqrt{17}}{2}\right\}$
10. $\left\{\frac{{}^{-}1 + \sqrt{13}}{6}, \frac{{}^{-}1 - \sqrt{13}}{6}\right\}$
11. $\{\ \}$
12. $\{9, {}^{-}1\}$

CHAPTER 10: APPLICATIONS

PRE-TEST

1. $3x - 8$
2. $9x + 7$
3. $6x + 17$
4. 2 and 6
5. 3 and 10
6. 4 and 11
7. 2 and 8

SELF-QUIZ # 1

1. $w + 72$
2. $m + 71$
3. $x - 14$
4. $x - 21$
5. $4x$
6. $23 + 7x$
7. $9x + 12$
8. $6x - 23$
9. $10x + 13$
10. $9x - 37$

SELF-QUIZ # 2

1. $x + 15 = 29$; 14
2. $x - 29 = 14$; 43
3. $x - 5 = 17$; 22
4. $9x = 72$; 8
5. $x - 19 = 5$; 24
6. $15x = 75$; 5
7. $3x + 5 = 20$; 5
8. $4x - 9 = 23$; 8

SELF-QUIZ # 3

1. $x + y = 17$
2. $x - y = 14$
3. $3x - y = 37$
4. $2x + 7y = 19$
5. $3x + 5y = 73$
6. $x = y + 8$
7. $x = y - 9$
8. $3x + 18 = 2y$
9. $x = y + 11$
10. $8x = y$

SELF-QUIZ # 4

1. (2, 1)
2. $(3, {}^{-}2)$
3. $({}^{-}4, 3)$
4. (2, 6)
5. $({}^{-}1, 1)$

SELF-QUIZ # 5

1. $$\begin{array}{r} 2x + y = 4 \\ \underline{3x - y = 1} \\ \text{1 and 2} \end{array}$$
2. $$\begin{array}{r} 2x + 3y = 13 \\ \underline{3x + y = 9} \\ \text{2 and 3} \end{array}$$
3. $$\begin{array}{r} y = 5x \\ \underline{9x - y = 8} \\ \text{2 and 10} \end{array}$$

POST-TEST

1. $5x - 17$
2. $8x + 5$
3. $3x - 9$
4. 6
5. 2 and 10
6. 2 and 3
7. 7 and 5

Index